Electronic Project
Design and Fabrication

Electronic Project Design and Fabrication

Sixth Edition

RONALD A. REIS
Los Angeles Valley College

PEARSON

Prentice Hall

Upper Saddle River, New Jersey
Columbus, Ohio

Library of Congress Cataloging-in-Publication Data

Reis, Ronald A.
  Electronic project design and fabrication / Ronald A. Reis.—6th ed.
    p. cm.
  Includes index.
  ISBN 0-13-113054-4
    1. Electronic apparatus and appliances—Design and construction. 2. Prototypes,
  Engineering. 3. Science projects—Design and construction. I. Title.

TK7870.R425 2005
621.381'078—dc22

                                                            2004044643

Editor in Chief: Stephen Helba
Assistant Vice President and Publisher: Charles E. Stewart, Jr.
Assistant Editor: Mayda Bosco
Production Editor: Kevin Happell
Production Coordination: *The GTS Companies*/York, PA Campus
Design Coordinator: Diane Ernsberger
Cover Designer: Mark Shumaker
Cover Art: Digital Vision
Production Manager: Matthew Ottenweller
Marketing Manager: Ben Leonard

This book was set in Century Schoolbook by *The GTS Companies*/York, PA Campus. It
was printed and bound by Courier Kendallville, Inc. The cover was printed by Coral
Graphic Services, Inc.

Electronics Workbench™ and Multisim™ are trademarks of Electronics Workbench.

10 9 8 7 6 5 4 3 2 1

ISBN: 0-13-113054-4

To Austin, my grandson—keep those electrons flowing.

Preface

Electronic Project Design and Fabrication, Sixth Edition, offers a complete experience in the design and fabrication of electronic prototype projects. Whether you are enrolled in a traditional lecture and laboratory class or a fabrication course devoted entirely to the subject, this book provides the hands-on experience in project building and documentation every prospective electronics technician requires.

A project, however, is not all you will be developing as you progress through the text. You will also be building confidence in your ability to construct electronic devices. The know-how to take a project from an initial concept through to final construction with support documentation is sure to give you pride and self-assurance as you enter the electronics industry. *Electronic Project Design and Fabrication* will prepare you for that exciting moment.

Think of yourself as a technician on the job. Your supervisor presents you with a project idea designed to fill a need or to solve a problem. You must determine if the idea is feasible and, if so, build a working prototype project and develop an accompanying Project Report.

The project should resemble the final, mass-produced product. You must design the cicuit, simulate its operation, breadboard it, design and fabricate a printed circuit board, and then enclose the project in a functional and attractive case. The Project Report is to include appropriate written and graphic documentation. It delineates various troubleshooting and test procedures, summarizes project developments, and contains the 10-drawing set used to construct the prototype project.

In project construction, you have a choice of three approaches. In the first approach, you choose to build the Sample Project—the Variable Power Supply. This project progresses from chapter to chapter, starting with the two-step design process and ending with a completely packaged and tested prototype project, along with necessary documentation.

The Exercise Project, the 3-Channel Color Organ, is at a slightly higher level of difficulty. Because less help is provided in building this project, you'll have to do a bit of digging and developing on your own. The Exercise Project is also carried from chapter to chapter.

Finally, a dozen Elective Projects are presented in Part III. Only Concepts and Requirements Documents and circuit design sketches are provided. Here you will need to develop specific project design and fabrication details on your own.

Although the project construction and documentation motivates you to progress from one chapter to the next, *Electronic Design and Fabrication* is more than a project book. The fundamentals of prototype project design, fabrication, and documentation are a part of every chapter. Even if you elect not to construct a project (a most unlikely occurrence), you will gain a firmer understanding of electronic device fabrication after studying the book.

The sixth edition of *Electronic Project Design and Fabrication* represents an up-to-date revision of a well-received text. New material has been added, including the following:

- [ ] All project drawings have been revised using computer-aided design (CAD) software.
- [ ] Circuit simulation, using Multisim 7.0, has been added for the Sample Project.
- [ ] A section on "product versus project design" has been added to Chapter 3.
- [ ] The section on printed circuit board artwork, in Chapter 7, has been completely revised.
- [ ] The Press-n-Peel™ PC Board Transfer Film method had been added to Chapter 8.
- [ ] Materials on CAD, in Chapter 11, have been revised.
- [ ] Two new elective projects have been added in Chapter 17.
- [ ] All appendices have been revised, where appropriate.
- [ ] New terms have been added to the glossary.

I hope you will find the sixth edition of *Electronic Project Design and Fabrication* a great tool as you design and create new electronic products.

ACKNOWLEDGMENTS

Many people at Prentice Hall have provided me with invaluable assistance in the preparation of this new edition. In particular, I want to thank Charles Stewart, Assistant Vice President and Publisher, for his technical assistance, patient encouragement, and, most importantly, enthusiastic support. I would also like to thank the following reviewers for their valuable feedback: Richard L. Windley, ECPI College of Technology, Virginia Beach, Virginia; and James Howen, Modesto Junior College, Modesto, California.

Ronald A. Reis
Los Angeles Valley College

Contents

PART I
Project Design and Fabrication

1 An Overview— Making It in Electronics

OBJECTIVES

In this chapter you will learn

- [ ] How an electronics design and fabrication experience can enhance skill level in a traditional "lecture and laboratory" classroom.
- [ ] Just what it is that electronics technicians do.
- [ ] About a career path for technicians with fabrication experience.
- [ ] What the five-stage design and fabrication process is all about.
- [ ] How the three-tier project approach provides maximum flexibility for project design and fabrication.
- [ ] How the drawings and written documentation come together in the Project Report.
- [ ] What traditional drafting tools are required for making the 10-drawing set contained in the Project Report.
- [ ] How CAD has replaced traditional drafting equipment.
- [ ] What tools are necessary for the mechanical and electrical fabrication of the prototype projects.

"You built that? I didn't know you could do such things, that you were so into electronics. How smart and clever!" Has anyone ever said something like that to you? And have you ever experienced the glow, the sense of pride, accomplishment, and satisfaction that comes when others compliment you on something you have built? Actually, you don't need to hear praise from someone else—although that's always nice. As you lean back from your workbench, surveying your new project, an inner voice tells you: "I did it; I made it myself"—and that sense of accomplishment feels great.

If you haven't quite reached that nirvana yet, don't despair—you will be there soon enough. Why? Because that's what electronics design and fabrication is all about: building—taking a project from the concept and design stage all the way to a finished, working prototype. And, in addition to getting the practical exposure actually needed to work as an electronics technician, the electronics design and fabrication experience will give you that all-important sense of accomplishment. After all, the completed project is the culmination of all you have been doing and learning. For the prospective technician, theory and laboratory exercises are a means to an end. And that end is the fabricated project: the device you will learn to develop and build.

In this chapter we first look at the traditional electronics course and see how the electronics design and fabrication experience can be made a part of it. We then explore this process, from the design and drawing stages on through experimenting, prototyping, testing, troubleshooting, and documenting the working project. Finally, we examine the "tools of the trade," on the board, at the computer, and on the bench. We will see what tools and equipment you will need to "make it" in electronics.

1.1 ELECTRONICS DESIGN AND FABRICATION IN THE SCHOOL LABORATORY

This section examines the traditional electronics course with its "lecture and lab" approach to learning. You will then see how your knowledge can be expanded by supplementing such a method with an electronics design and fabrication experience.

The Traditional Electronics Course

Let's review what is done in the standard lecture and lab class and, where it exists, the separate electronics design and fabrication course.

Lecture and Lab Class

Most electronics courses taught at the college or technical school level are organized around the lecture and lab structure. During the lecture portion, students listen to the professor speak, take notes, enter into discussions, and read the textbook. This aspect of the course is known as the "theory" part.

Lectures on the subject matter are supplemented by laboratory experience. In addition to familiarizing the student with the type of equipment he or she will use in industry, the lab experience is designed to verify, or "prove out," the theoretical concepts discussed in the classroom. For example, you learn in the lecture that the output frequency of a 555 IC oscillator is determined by its R/C (resistance/capacitance) time constant. To "prove" it, you go to the

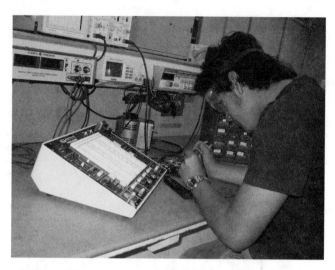

FIGURE 1–1
Experimenting in the laboratory

lab, run a simulation, using a simulation software program, and construct, that is, breadboard, the actual circuit (Figure 1–1). Building such circuits, as every student knows, is an extremely important part of any electronics learning experience.

Yet this experimenting in the lab is often the only hands-on experience the electronics student is likely to get. In the traditional course, students do not ordinarily build permanent, working *prototype projects,* that is, projects that are hand-assembled to resemble as much as possible the final, mass-produced equipment (Figure 1–2). Nonetheless,

FIGURE 1–2
Finished prototype project and Project Report

building such projects is an experience every electronics student should have.

Separate Electronics Design and Fabrication Course

In a separate electronics design and fabrication course, students often build a complete project, taking it from the design stage to a finished prototype. Such a course usually involves printed circuit board design and manufacture, chassis and enclosure construction, project packaging, and the necessary documentation. It is likely to be a popular course, for all of the obvious reasons. Students are building the types of electronic devices they will assemble on the job. They are doing what technicians do.

We believe that students who are unable to take a separate electronics design and fabrication course should be building prototype projects in their lecture and lab classes. This book is written for use in *both* types of courses. Before we examine how it can be used in such courses, let's see why electronics fabrication is such an important part of what students should be doing.

The Need for an Electronics Design and Fabrication Experience

In order to appreciate why, as future electronics technicians, you need to have an electronics design and fabrication background, let's examine briefly just what it is that technicians do. Let's see how the familiar "learning by doing" concept fits nicely into the activities that lie ahead and examine a most interesting career path, that of fabrication technologist.

What Electronics Technicians Actually Do

Electronics technicians build, repair, test, modify, and install electronic devices and equipment. That is, they do electronics fabrication. They are a hands-on group. If electronic theory is your overriding interest, perhaps you should strive to be an engineer rather than a technician. But if you want to "do" as well as "know," produce with your hands as well as your head, then working with the actual hardware is where you need to be. A soldering iron, along with a hand calculator, will be an important tool.

As a working technician, you may not do all the things covered in this book. For example, you may never actually design a printed circuit board, but you will undoubtedly work (or "interface") with those who do. You need to appreciate their concerns and problems and be able to provide technical input when required. In addition, although you may rarely make a printed circuit board yourself, you will certainly treat the boards with greater respect if you

are familiar with the process. The point is, as an electronics technician you will be working with a variety of people in the long chain of the design and fabrication process. Thus, even as a student, you will want to know about and experience that process.

Learning by Doing

If your background in electronics is limited, or if the field is new to you, you may be reluctant to plunge right into electronics design and fabrication. Don't be. Learning by doing is not hard; in fact, it can be downright enjoyable. Think a moment. You probably got interested in electronics because it seemed like a field in which you could get paid for doing the fun and interesting things you always wanted to do. If you have been an electronics hobbyist or enthusiast for some time, you are already familiar with the basics of electronics fabrication and are that much farther ahead. If signing up for an electronics course is your first experience, don't worry; this book will show you all you need to know to get started.

Fabrication Technologist

Ordinarily, the electronics technician is expected to do it all: installation, maintenance, testing, troubleshooting, repair, and prototyping. Whether in medical, communications, industrial, computer, military, or consumer electronics (to name but a few specialties), the technician deals with all aspects of the equipment, except perhaps its initial design. This is particularly true, for instance, with regard to troubleshooting and repair, two areas that are often seen as merging into one. Until recently, if you could find the problem (diagnosis), repair was just a simple follow-up: grab some solder, a soldering iron, a couple of additional hand tools, and the repair was quick and easy. With just a little training almost anyone could do it!

Increasingly, this is no longer the case. In many situations, prototyping and repair have become so complex, sophisticated, and demanding of special skills that the typical generalist electronics technician cannot be expected to do them well, if at all. Rising to meet the challenge, fortunately, is the *fabrication technologist,* a skilled technician specializing in the prototyping, rework, and repair of electronic devices.

It is true in most cases, especially in small- and medium-sized firms, that an electronics technician is still required to handle routine repairs and do an occasional prototype. However, today's microminiaturized components and multilayered circuit boards generate a host of specialized problems in repair, rework, and prototyping. The fabrication technologist deals with this complex reality.

As a fabrication technologist (F.T.), you need to become highly proficient in three areas: rework, repair, and prototyping.

Rework involves reestablishing a device's functional and physical characteristics without deviating from the original manufacturer's specifications.

Repair means reestablishing the functionality, quality, and reliability of an assembly that has failed or has been damaged. It may require deviations from the original manufacturer's specifications, however.

Prototyping, of course, involves the creation of a hand-assembled project that will resemble as much as possible the final, mass-produced equipment.

Although the skills necessary to produce a quality prototype differ somewhat from those required for circuit repair and rework, there is, as you would expect, considerable commonality among the three areas. Circuit board fabrication, component assembly, sheet metal cutting and bending, the application of external finishes and nomenclature—all these techniques are necessary for project repair, rework, and prototyping. The knowledge and skills acquired in your fabrication course should go a long way in helping you to become a unique kind of electronics technician—a fabrication technologist.

Something to Think About

With virtual electronics a reality (if you'll excuse the pun), do we need to prototype anything? What does the term hands-on *now mean?*

1.2   THE ELECTRONICS DESIGN AND FABRICATION PROCESS

Let's examine the five stages of electronics design and fabrication, from thought to finished project, and see how these stages are covered in the book. Keep in mind that when you have completed the process involved, you will have both a finished **prototype project** and a completed **Project Report** (Figure 1–2).

From Thought to Finished Prototype: A Five-Stage Process

This book explores the five stages required to design and fabricate a working prototype project: design; drawing; experimenting; prototyping; and testing, troubleshooting, and final documentation. What follows is a brief look at each stage, defining what the stage is and explaining why it is needed.

Doing It All: From Design to Documentation

Figure 1–3 illustrates the five-stage design and fabrication process and how it relates to the written and graphics documentation of the Project Report. At each stage, the project progresses, as does the documentation for it. The figure also gives the individual chapters in which these stages are discussed. You will probably refer to Figure 1–3 often. Now let's see what each stage comprises.

Design Stage. The project design is usually, but not always, done by the electronics engineer. The technician, however, is often intimately involved. At the very least, the technician is expected to spot errors and inconsistencies in design, particularly when they concern product packaging, testing, installation, and maintenance.

At this stage, a two-step design process takes place. It involves producing a Concepts and Requirements Document and a set of three design drawings. One of these drawings, the circuit design sketch (schematic), even if drawn with CAD, is to be considered preliminary. This sketch must be finished and cleaned up, at a later time, before any serious work on the project can begin.

If, in the design stage, you are preparing computer-generated circuit simulations, using Circuit Maker or Multisim, for example, you will, of course, want to include your graphic results as part of your design drawing set. Such a set will thus contain at least four drawings instead of three.

Drawing Stage. A group of 10 drawings (including the design drawing set) is required to explain graphically how the project will be built. A working schematic, breadboard drawing, printed circuit board design layout and artwork, and fabrication and assembly drawings must be produced. Furthermore, a sheet metal layout, wiring diagram, and packaging illustration will be necessary. Although all these drawings are eventually completed and become part of the Project Report, only the design drawing set and working schematic drawing will actually be done before proceeding to the experimenting stage.

Experimenting Stage. This stage, often referred to as the *breadboarding* stage, is essential in proving out the design; that is, to see if the circuit actually works. Various quick and easy circuit assembly

FIGURE 1–3
Five-stage electronics design and fabrication process to produce a working prototype and Project Report

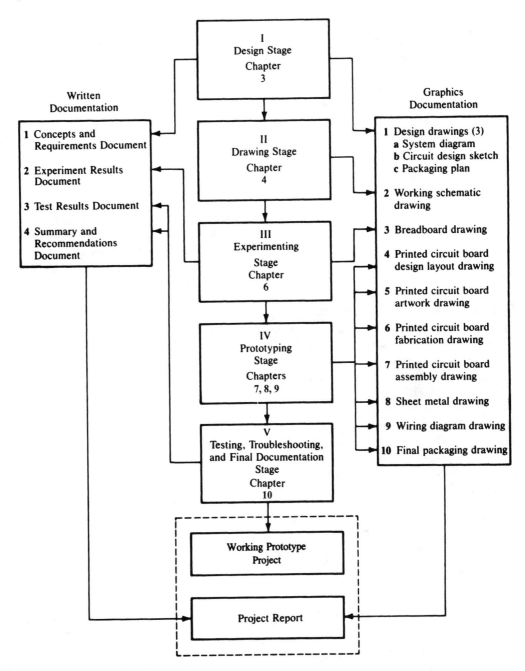

methods, the most popular of which involves the use of engineering solderless circuit board, have been developed specifically for use in the experimenting stage. With solderless circuit board, circuits can be assembled and altered in very little time, in many cases, without requiring any soldering at all. More permanent methods of circuit assembly involving cut-slash-and-hook and wire wrapping are also extensively used. A breadboard drawing and Experiment Results Document are produced at this time.

Prototyping Stage. It is here that a permanent, working project is actually built. Prototype devices

are almost always made with **printed circuit boards,** circuits in which the interconnecting wires have been replaced by conductive strips etched onto an insulating board. The design and manufacture of printed circuit boards are a major part of the entire fabrication process. All the remaining drawings are completed at this stage.

Testing, Troubleshooting, and Final Documentation Stage. Once the prototype is completed, it must be tested. Does it meet the design requirements? Does it work under adverse environmental conditions? Often, extensive laboratory and field testing is needed to determine the answers. If the device does not work correctly, it must be fixed. Troubleshooting becomes a very necessary part of this stage. The Test Results and Summary and Recommendations Documents are now produced. Finally, all the documentation must come together. The Project Report, with its written and graphics documentation, is assembled at this time.

These, then, are the five stages required to produce a working prototype project and Project Report. Of course, there is, strictly speaking (when overall product development and manufacturing is considered), a sixth stage. It is the production stage, in which the project is mass-produced. Because assemblers, rather than technicians, are usually involved here, this book does not cover production in any great detail.

But I'm an Electronics Technician!

Why, you ask, does an electronics technician need to know about drafting, sheet metal fabrication, and nomenclature application techniques? The answer is simple: the problems a technician encounters on the job are not exclusively electronic. The technician works with mechanical assemblies, enclosures, and systems. Often, the most sought-after technicians are those with an electromechanical background. At the very least, the technician is required to read mechanical drawings, assemble and disassemble enclosures, and do minor mechanical adjustments and calibrations. This situation is especially true at small- and medium-sized companies, where the technician is expected to do it all. As a student preparing to be such a technician, you will want to gain as broad a design and fabrication experience as possible—now.

The Electronics Design and Fabrication Process and This Book

Let's look at the five-stage design and fabrication process as it is discussed in individual chapters, see how student projects form an integral part of the book, and investigate the role of the Project Report.

Five Stages—Seven Chapters

Except for this overview chapter, Chapter 2 (on safety), Chapter 5 (on technical writing), and the concluding chapter (on computer-aided design, or CAD), Part I of this text is devoted to exploring the five stages of electronics design and fabrication.

1. **Design Stage**—Chapter 3. This chapter, in addition to reviewing basic electronic principles, introduces concepts of electronics design and looks at how we think up problems to be solved.

2. **Drawing Stage**—Chapter 4. Chapter 4 introduces an important language of engineering—the drawing—and examines the 10-drawing set that you will develop throughout the book. The working schematic is also produced at this stage.

3. **Experimenting Stage**—Chapter 6. In this chapter the different methods used to assemble projects in order to test out their design are studied. Various experimenting systems, including cut-slash-and-hook, are reviewed.

4. **Prototyping Stage**—Chapters 7, 8, and 9. Chapters 7 and 8 cover the design and manufacture of printed circuit boards. Chapter 9 deals with printed circuit board assembly, project packaging, and final assembly.

5. **Testing, Troubleshooting, and Final Documentation Stage**—Chapter 10. Here, the testing and troubleshooting of the completed prototype project are investigated. Preliminary testing and troubleshooting, at the project experimentation level, will already have taken place. Also, here is where all the necessary documentation that has been produced in the appropriate chapters is pulled together in the final Project Report.

As you can see, the design and fabrication of an electronics prototype project means producing two items: a Project Report, consisting of written and graphics documentation, and the finished working prototype project. Both items are developed concurrently as you move back and forth between the drawing board, or computer screen, and the workbench.

Student Construction Projects

Although the seven chapters just reviewed contain miniexercises designed to help you become familiar with the various processes discussed, it is through the building of either a Sample or Exercise Project, one of the dozen projects in the Elective Projects group, or one of the three Surface Mount Technology Projects in Appendix E that you will derive the greatest benefit from this book (Figure 1–4). Let's see how project building fits into the learning scheme.

A Sample Project, a regulated Variable Power Supply, is carried from chapter to chapter, starting with the two-step design process and ending with a completely packaged and tested prototype. All documentation is brought together in the Project Report, presented in Part III. With this Sample Project, you see a real project develop before your eyes as you proceed through the chapters. You may wish to recreate this project, along with its documentation, as your class project.

Sample Project

Variable Power Supply

Step-by-step construction details are provided, along with a complete project report.

or

Exercise Project

3-Channel Color Organ

Step-by-step construction hints and suggestions are given, along with a complete project report.

or

Elective Projects (12)

- ☐ Adjustable Dual Power Supply
- ☐ Burglar Alarm
- ☐ Carport Night-Light Controller
- ☐ Clap-On/Clap-Off Switch
- ☐ Events Timer
- ☐ LED Attention-Getter
- ☐ Light-Controlled Switch
- ☐ Mosquito Repeller
- ☐ Push-Button Combination Lock
- ☐ Sound-Activated Switch
- ☐ 10-Note Tunable Electronic Organ
- ☐ 2-W Stereo Amplifier
- ☐ Variable Strobe Light

No special construction details are provided, and only the Concepts and Requirements Document is included for each project. (See Appendix E for Surface Mount Technology Projects.)

FIGURE 1–4
Student construction projects

An Exercise Project, a 3-Channel Color Organ, is also developed in the appropriate chapters; however, the information presented for this project is not as complete as that for the Sample Project. This situation, of course, is what is intended. If you choose to build the Exercise Project, you will be challenged at a slightly higher level because less assistance is provided. Nonetheless, since the 3-Channel Color Organ parallels, in design and construction difficulty, the Variable Power Supply, you will find the exercise well suited for a first construction project. Furthermore, if you wish to refer to it, a completed Project Report for the 3-Channel Color Organ is provided in Part III.

Also in Part III are an additional dozen Elective Projects to choose from. Only the Concepts and Requirements Document and the three design drawings are provided for each project. You must take it from there.

As you can see, then, you are given a choice, at three different levels, of appropriate projects to build, or "fabricate."

Project Report

Along with actually fabricating a project, you must produce the proper documentation. It is true that for the Variable Power Supply and 3-Channel Color Organ Projects, complete documentation is already provided (Part III); but to gain the full benefit and experience in the design and fabrication process, you will need to complete the appropriate graphics and written documentation, regardless of which project you construct.

With respect to drawings, you can simply redo the 10-drawing set for the Sample Project. With the Exercise Project, you can copy the drawings from the Project Report in Part III or, preferably, draw your own set. If you choose a project from the Elective Projects group, however, you will have to produce all but the design drawings yourself.

In addition to the graphics documentation, the Project Report requires written certification in the form of Concepts and Requirements, Experiment Results, Test Results, and Summary and Recommendations Documents. Such paperwork is not extensive; in some cases just a page or two is all that is needed; however, it must be done, and it must be concise and complete.

If all this sounds a bit like work, well, it is; but when you bring that completed project, along with its Project Report, to your first job interview, you will know it has all been worthwhile.

Beginning at the Beginning

Although electronics design and fabrication involves effort, this book is not going to make you run before you walk—or crawl. It is going to take you through the process step by step and keep it simple. You will learn the basics—not everything there is to know about electronics design and fabrication. There is nothing to be gained by cluttering up your mind with a description of a dozen types of soldering lugs, for example. Although consistency with industry standards is needed, the main concern is to provide the instruction that will enable electronics design and fabrication to take place in the typical school setting. The purpose of this book is to get you started in the right direction; the "bells and whistles" can come later.

To design and fabricate electronic projects requires four things. You will need *tools* for building and repairing; *plans* in the form of drawings and written documentation; *materials* such as hardware and electronic components; and the *knowledge* and *skills* necessary to complete the task. Because we promised to take first things first, let's now turn our attention to the tools of the trade on the board, at the computer, and on the bench.

Something to Think About

Why not add to the list of elective projects? Searching magazines, books, and, of course, the Internet, see if you can come up with additional, interesting, yet relatively simple, projects to include.

1.3   TOOLS OF THE TRADE: ON THE BOARD

The tools needed to design an electronic project are of two types: traditional and computer-based. With the former, often referred to as *drafting tools,* or *board tools,* any of the drawings in the 10-drawing set can be made.

In this section we list each traditional drafting tool needed and explain what it does. Discussion of tool use, however, comes later, in chapters where the applicable drawing procedure is described. In the next section, we look at the CAD drawing process.

The Drafting Table

There are two types of drawing tables: the permanent table mounted on "all fours" (or at least twos) and the portable drawing table, one that literally can go anywhere.

Sturdy and Smooth

When it comes to a "stay-in-one-place" drafting table, four legs are usually better than two. Simply put, such tables are more stable, and less likely to wobble while you are trying to draw a straight line. A smooth surface to draw on is a must. This surface is ordinarily obtained in one of two ways: either by coating the table with a roll of Borco, a vinyl-like material especially formulated to provide a smooth drawing surface, or by using a *drawing board.* A drawing board is made of seasoned, straight-grained soft woods and is designed to be used with a T-square and triangles.

Although some drafters prefer to stand at their table, most usually sit. Above all, the chair you choose should be comfortable. Chairs with an adjustable seat height and back support—the ideal choice—are available from many commercial sources.

You should choose lighting with great care. The key factor here is to provide direct, as well as indirect, illumination. Drafting lamps with articulated support arms are available. They can accommodate a 100-watt (W) bulb, have an adjustable arm, and include a clamp so that they can be mounted on the edge of the drawing table.

Portable Drafting Tables

Various portable drafting tables are available that make drawing convenient and inexpensive for the part-time or on-the-go drafter. Many of these units consist of far more than a drafting table. A number of them contain a small drafting machine, an instrument that allows you to quickly draw straight lines at designated lengths and angles. If you are primarily interested in electronics fabrication, the economy of the portable drafting table is its main attraction; but before you make a decision, read on. An even more preferable solution, the mini-drafting machine, may be your answer.

Drafting Equipment

Here, both new and traditional drafting equipment are discussed. (Additional drafting materials, as they pertain specifically to surface mount technology design, are discussed in Chapter 13.)

Equipment for Drawing Straight Lines

The standard drafting machine is large and expensive, and its cost is rarely justified for the part-time electronics drafter. A much better way to go is to buy a traditional package of drafting tools consisting of a T-square, two triangles (30°/60°, and 45°), protractor, scale, and drawing board—all that is necessary to draw straight lines at any particular length or angle.

You can, however, combine all these features in a fascinating product known as a mini-drafting machine that sells for less than $12. These machines, one of which is illustrated in Figure 1–5, are made of durable plastic and contain two pairs of knurled rollers that contact the paper surface. No special table or paper is required. With just a few minutes' practice, you will be making penciled or inked lines just like a professional.

Pencils, Paper, and Things

Whether you use basic drafting equipment or a drafting machine, you will still need additional "bits and pieces."

☐   **Pencils.** Because they allow for the interchanging of leads, mechanical refill pencils are preferred over wood pencils. The lead itself comes

FIGURE 1–5
Mini-drafting machine

in a variety of grades: from very hard (7H and 6H), which draws a light line, to extremely soft (HB down to 7B), which lays a dark line on the paper. Select a 4H for drawing dimension and hidden lines; use a 2H or H for object lines.

- **Paper.** For the 10-drawing set you will be producing, graph paper with a 0.100" × 0.100" cross-sectional grid is required. It is available in $8\frac{1}{2}$" × 11" (A size) sheets of vellum or layout bond paper. The grid should be printed in non-reproducible blue. This means that the grid pattern will not reproduce when the drawing is copied.
- **Compass and Circle Guides.** Compass and circle guides are used to draw circles and arcs. For circles greater than $1\frac{1}{2}$" in diameter, the compass should be used. For small circles, rounds, or fillets, the circle guide is ideal.
- **Drafting Templates.** Templates save time and effort. You will need two types: one to aid in making schematic symbols, the other for creating printed circuit board design layouts. The latter should be chosen in the full, or 1× scale.
- **Scale.** You are going to need a scale. Avoid the ones available at the local drugstore; quality is of the utmost importance here. A mechanical engineer's scale, graduated in 0.02" divisions, is required. It doesn't really matter whether it's triangular or flat, as long as it is of good quality.

There are a few additional items: pencil sharpener, eraser, erasing shield, drafting tape, lettering guides, and so forth. Your local stationery or drafting supply center will help you with these supplemental supplies.

A Note on Quality

Traditional drafting instruments and tools should last you a lifetime. In fact, much of the equipment available from top manufacturers is guaranteed for the life of the owner. Look for tools that are sturdy and fit together well. To find such equipment, your best bet is to seek out only quality, name-brand manufacturers. Avoid cheap goods; they will wind up costing you much more in the end.

> ### Something to Think About
>
> *Traditional drafting equipment rarely wears out. A quality triangle is good for a lifetime. With a whole generation of mechanical engineers now in retirement, you may be able to purchase a complete set of drafting tools at a garage sale down the block. It's worth looking into.*

1.4 TOOLS OF THE TRADE: AT THE COMPUTER

In creating a prototype project there is a design stage and a drafting stage. In design, new ideas are formulated; in drafting, these ideas are documented on paper in the form of drawings. Computers can help with both phases. With a sophisticated CAD (known both as computer-aided design *and* computer-aided drafting) system, every drawing in the Project Report can be entirely produced with the aid of a computer. Let's see what CAD is and then examine the computer hardware necessary to produce computer-generated drawings.

The What and Why of CAD

CAD does for graphics documentation what word processing does for written documentation. A written document is much easier to create, store, alter, produce, and transmit with a word processor than on an electric typewriter. With CAD, we can do all these same things for graphics documentation as well.

CAD: What Is It?

In the beginning, in the early 1960s, there was just computer-aided drafting. It was run on a large mainframe computer, it was slow, and it was very, very costly. At the latter end of the decade, computer-aided design came along. It, too, was available only on the large machines. But with the advent of fast, memory-filled microcomputers in the late 1980s and early 1990s, both computer-aided design

FIGURE 1–6
A computer used to produce CAD drawings

and drafting were put within the reach of practically everyone who needed them.

CAD uses a computer as a design and drafting tool (Figure 1–6). With it, you can design new or modify old circuits and create finished working drawings. The result is a product that is less expensive, more accurate, of higher quality, of greater consistency, and easier to store, alter, and transmit than would be possible without a computer. The discussion in this chapter is confined to electronic schematic and PC board drawings, but keep in mind that any type of drawing in the mechanical, architectural, civil engineering, and aerospace fields also can be produced with a powerful enough CAD system.

Why Use CAD? Beyond the T-Square

CAD is not just a drawing aid but a design assistant. With a powerful computer-aided design program, you can save yourself a great deal of bench time and breadboarding. Many software packages allow you to do waveform analysis and troubleshooting without ever handling an electronic component. At the touch of a keyboard or stroke of a light pen, design modifications and improvements can easily be made. In addition, circuit simulations allow real-world circuit operations to be performed. You can also play "what if" games. Change one circuit parameter—for example, a transistor value—and the program will tell you what effect that change has on the entire circuit operation.

CAD really comes into its own as a drafting aid. Actually, creating a drawing for the first time with a CAD system is probably no faster than with the paper-and-pencil method. Where CAD shines is

when changes, modifications, and updates are required. It's similar to what happens in word processing. I could probably write this text just as fast on an IBM Selectric typewriter as I can on an IBM-PC. But when using the Selectric, if I make a few mistakes, or want to insert a paragraph here, delete one there, the whole chapter will probably need to be retyped. Not so with the computer. With the PC, I just correct mistakes and insert or delete text right on the video display terminal. When I am satisfied with the results, I call for a print operation, lean back, sip a cup of coffee, and watch my printer turn out the document at roughly four pages a minute.

The concept behind computer-aided drafting is the same as the one behind word processing. You create a drawing on the screen and store it in the computer's memory. When you need modifications, you just make the necessary changes on the video display terminal and then ask the printer or plotter to make a new drawing. The work is done neatly, quickly, and accurately. The old-fashioned way requires laborious redrawing (often in ink) by hand— a wasteful, messy, inaccurate process at best.

There are other advantages in using CAD to create drawings. Because all the information about the drawing (what components are located where, for example) is stored in the computer in digital bits, the drawing can be transmitted across town, or around the world, more or less instantaneously. Furthermore, drawings produced with CAD are much more consistent in lettering, line weight, and symbol configuration (Figure 1–7). Finally, think of the savings in storage space that result from keeping a drawing on a disk rather than on a large sheet of paper.

Computer Systems/CAD Systems

A CAD system is, naturally, a computer system. So what's a computer system? That's the logical first question that we need to answer before we can go on to see how computers are configured specifically for CAD.

A Basic Computer System

A computer is an electronic device that performs arithmetic and logic functions. In so doing, it manipulates numbers, letters, and practically any other graphics symbols. All these symbols are converted into binary 1s and 0s (high and low voltages) that, under software control, are processed inside the computer. This processing takes place in what is called the *CPU*, the *central processing unit*. The CPU, made up of the ALU (arithmetic logic unit), a

FIGURE 1–7
High-quality CAD
schematic

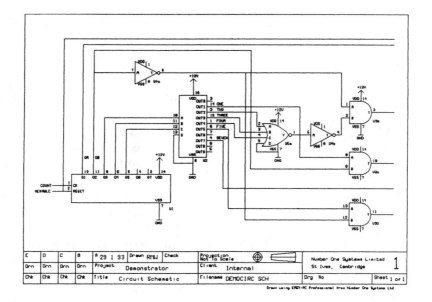

few registers, and control circuitry, forms a computer (Figure 1–8a). The CPU is not, however, a computer system.

To create a **computer system,** we must, as with any electronic system (see Chapter 3), in addition to providing a circuit (CPU), include at least one input and one output device, or transducer. A computer system (Figure 1–8b) is made up of a CPU (and associated circuitry), where electrical signals are manipulated digitally, with memory, and various input and output transducers, called *peripherals.*

A functional diagram of a typical computer system is shown in Figure 1–9. The keyboard is the input peripheral, used to enter data into the computer. The video display terminal (VDT) displays the data as they are being entered or after they have been manipulated (processed). The printer produces a hard copy of what appears on the screen.

But the computer won't accept or manipulate anything unless it is told specifically what to do. The instructions come in the form of a *program,* known as *software.* The program, with its lines of code, or instructions, is stored in the computer's memory, as are the results of any data processing.

(a)

(b)

FIGURE 1–8
Computer and computer system

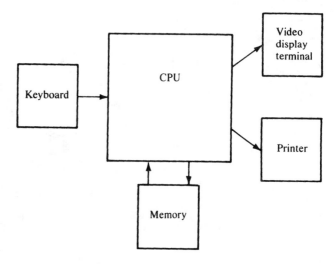

FIGURE 1–9
Computer system as word processor

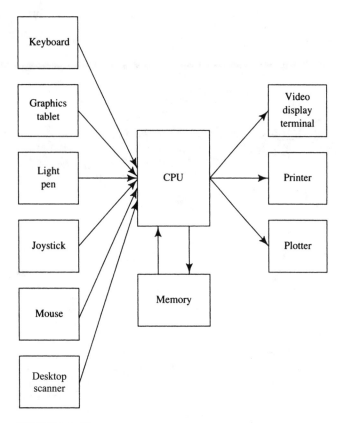

FIGURE 1–10
CAD system

A Basic CAD System

Let's examine a basic computer system configured to do computer-aided design.

In Figure 1–10 we see a functional diagram of a CAD computer system, in Figure 1–11, an illus-

FIGURE 1–11
CAD hardware

tration of representative hardware. Like any computer system, it contains a CPU, along with memory to store a program and the results of data manipulation. What makes the CAD system different, at least from a hardware standpoint, are the input and output peripherals. These devices are used primarily to enter and display graphics symbols—lines, geometric shapes, schematic symbols, and so forth. Let's examine the more common types of peripherals used for CAD generation.

Included among the input peripherals are such devices as the keyboard, graphics tablet (digitizer), light pen, joystick, mouse, and desktop scanner.

Keyboard. The keyboard was mentioned earlier; for entering numbers, letters, and basic computer commands, it can't be beat.

Graphics Tablet. The graphics tablet is used for inputting graphics with a stylus. Diagrams or freehand drawings can be created and appear instantly on a video display terminal. The typical tablet consists of a grid of wires partially embedded in a thin substrate. When the tablet is touched by a stylus, the wires are brought into electrical contact. The result is a pulse that enables the computer to register the position of the stylus. With a sketch placed atop the tablet, practically any shape can be digitized, stored, and displayed in the computer.

Light Pen. The light pen is an electronic stylus. It contains a light sensor that, when aimed at the video screen, specifies a position on the screen.

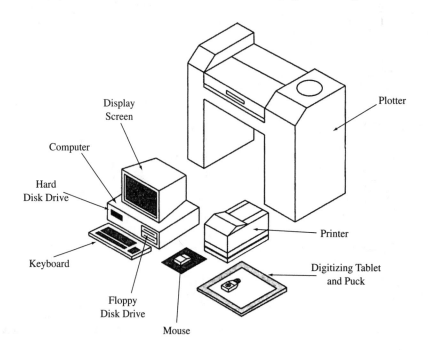

Joystick. The joystick, an upgraded version of the type used with video games, is a lever whose motions control the movement of a cursor (small square of light) on the VDT. It can also be used to draw diagrams and geometric shapes.

Mouse. Probably the most popular input peripheral for graphics generation, the mouse is moved by an operator over the surface of a mouse pad, graphics tablet, or table. Its position is recorded by the computer. The mouse is used to move text and illustrations about on the screen. It is also used to enter various commands into the computer.

Desktop Scanner. The first modern scanners, known as *drum scanners,* were made for the publishing industry. They are still around, and do a great job, but are terribly expensive. In today's desktop scanners, some selling for less than $100, a document is placed facedown on a glass bed, and a scanning array consisting of a lamp, a mirror, a lens, and an image pickup moves back and forth underneath the glass (Figure 1–12). The image sensor may be a charge-coupled device (CCD) or a compact image sensor (CIS). Either way, CCD or CIS, digitization, via an analog-to-digital converter, or ADC, takes place, and the resulting image is sent to the scanner's own hardware. The results are quite impressive.

Video Display Terminal. For CAD work, a video display terminal should be a video color monitor with at least a 15" screen. Its VGA (video graphics array) card should be capable of presenting a video resolution of at least 1024 by 768 pixels—with up to 256 colors displayed simultaneously.

FIGURE 1–12
Inexpensive scanner

Printer. Although an ink-jet printer can produce acceptable schematic drawings and PC board artwork, for truly professional results, a laser printer is a must. Such a printer should have a minimum resolution of 300 dpi (dots per inch), with 400 dpi being better, and 600 dpi being even better.

Plotter. The plotter is a true wonder of printing technology. Most use an ink pen, often more than one (each of a different color), that is driven on X-Y axes. The plotter rapidly and tirelessly draws the most intricate illustrations with perfect precision and consistency. With what seems like a mind of its own, the plotter produces drawings with a quality no human being can match.

> **Something to Think About**
>
> *New, improved CAD systems are constantly coming on the market. Now is the time to gather vendor literature on what's available. Why not contact product manufacturers and distributors to start building a file?*

1.5 TOOLS OF THE TRADE: ON THE BENCH

Like tools for the drawing board, tools for the workbench must be chosen for their quality. As a working technician, you will be judged by the tools you keep. Your reputation is at stake; select tools that will give you pride of ownership.

In this section the workbench and the tools for mechanical and electrical assembly that go with it are examined. In Chapter 10, the basic instruments needed to test and troubleshoot electronic projects are briefly introduced. (Additional tools and equipment, necessary to work with surface mount devices, are discussed in Chapter 13.)

The Bench: More Than the Kitchen Table

Whether selecting a workbench for the home, school, or workplace, the basic criteria for its selection are the same. You must be concerned with its location within a given setting, its structure and storage features, and lighting and power requirements (Figure 1–13).

Location and Structure

Ideally, the workbench should be located in an area that will provide a minimum of distractions. Such a location will increase the likelihood that the bench

FIGURE 1–13
Workbench

will be used only for its intended purpose, electronics fabrication. If you must use a common area, such as the kitchen table, assemble your project on a large plywood board that can be moved. In that way hardware, components, and small tools will remain undisturbed.

The workbench area should remain relatively clean. The dirt and mess associated with normal mechanical assembly (such as is found in the typical auto shop) is incompatible with modern electronics fabrication. Electronic assembly is considered a "light industry"—light and clean.

In terms of structure, the workbench must provide comfort to lessen fatigue. It should have a solid, nonmetallic surface and be of such height that you can rest your elbows comfortably on the surface.

Adequate storage must also be provided. Shelves should be placed immediately in front of the work area to house test equipment. For storing electronic components and small hardware items, modular plastic drawers are ideal. An alternative approach is to arrange a baker's dozen of boxes or drawers. Twelve of the boxes contain the dozen basic electronic components used in electronics: resistors, capacitors, transistors, and the like. The thirteenth box is labeled "miscellaneous." Nothing should remain in this particular junk box too long; if you can't file something within a few days, it probably isn't worth keeping.

Lighting and Power Requirements

The workbench should be bathed in direct and indirect light. You might not think this lighting is necessary, but when you are working with tiny circuits

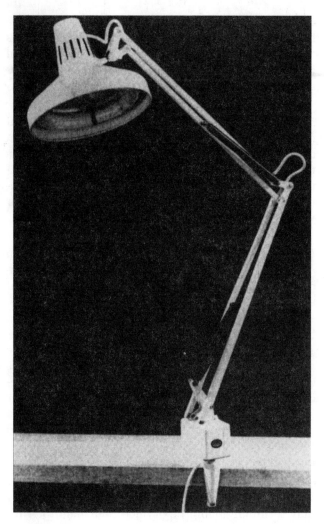

FIGURE 1–14
Drafting lamp

and minute letters and numbers printed on small components, you will need all the brightness you can muster. A drafting lamp (Figure 1–14) can provide a concentrated direct beam of light. An even better choice is a swivel lamp with an illuminated magnifier. Such lamps are particularly useful when you are inspecting or assembling printed circuit boards (Figure 1–15).

Obviously, power outlets are a must. Your workbench should have at least six, all fused (or provided with a circuit breaker) and rated at 20–25 amperes (A). Of course, they should be of the three-conductor, or grounded socket, type. Bench-mounted power distribution strips are inexpensive (from $12 to $20). Finally, an isolation transformer will provide the optimum in safety, since it totally isolates the incoming line voltage from the project in use. No complete test bench should be without one.

FIGURE 1–15
Magnifier lamp

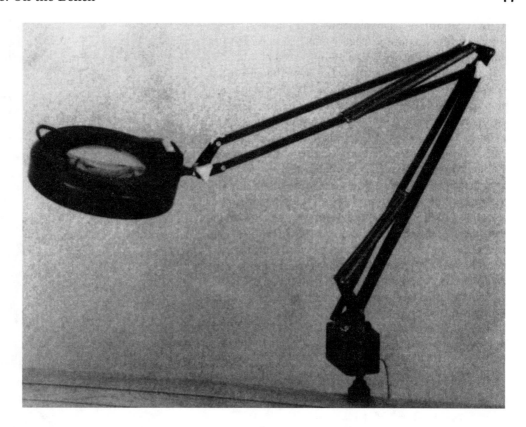

Tools for the Workbench

Now it's time to take a look at the tools that go on the workbench. The discussion divides them into two broad categories: those used for *mechanical work* and those required for *electrical work;* however, keep in mind that the distinction between the two is not always clear-cut. A needle nose pliers is useful both for holding small pieces of hardware and for bending wire.

Tools for Mechanical Work

What follows is a list of must-have tools, those that are essential to the mechanical fabrication associated with project building.

1. Drivers. There are two types of *drivers:* screwdrivers and nutdrivers. The former are used to turn screws; the latter, to twist nuts. You will need a set of both (Figure 1–16).

Screwdrivers are either *blade* or *Phillips head.* The blade type has a chisel-shaped head that mates with straight-slotted screws. Phillips screwdrivers are designed to fit screws with star-shaped holes in their heads. Sets of each type are available in two packaging styles: one containing individual screwdrivers, the other having a handle into which any of several driver shafts can be inserted.

A set of inexpensive *jeweler's screwdrivers* also is useful in electronic assembly. They are necessary to turn the tiny setscrews holding shaft couplers and control knobs.

Nutdrivers are almost as indispensable as screwdrivers. They, too, come individually or as a set with driver shafts that plug into a common handle. Choose those with a hollow shaft. This type allows you to keep a grip on the nut even though the screw on which the nut is mounted is protruding.

2. Pliers, Cutters, and Wire Strippers. *Pliers* are small pincers with long, jagged jaws. They are used for holding small objects or for bending and cutting wire (Figure 1–17). In electronics fabrication, two types are required: *slip-joint* pliers handle the bulk of heavy mechanical assembly; for holding wires in place during soldering and for acting as a heat sink, *needle nose* (or long-nose) pliers are the logical choice. At the start, all you will need is one of each.

Diagonal cutters, also called *dykes,* are used to cut wires or component leads. A diagonal cutter should be made of high-quality tool steel, and its jaws should be well aligned. With a little practice, you can also use diagonal cutters to strip wire. A better approach, however, is to purchase an inexpensive *wire stripper,* either the simple plier type or a compound unit that automatically adjusts to wire

FIGURE 1–16
Drivers: both screw and nut

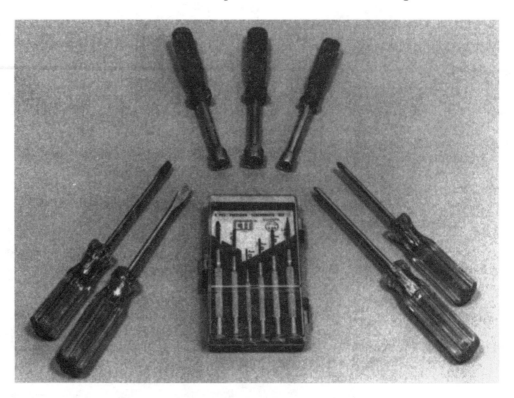

FIGURE 1–17
Wire stripper, pliers, and
diagonal cutter

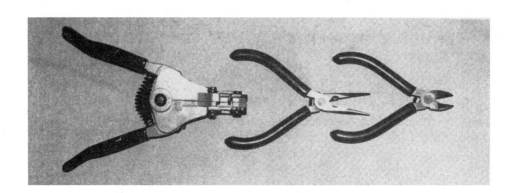

FIGURE 1–18
An Allen wrench has many
uses.

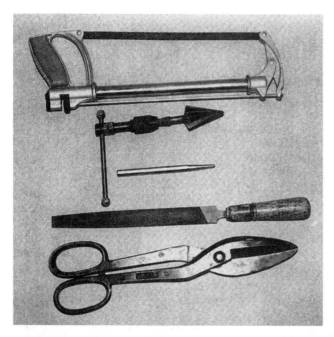

FIGURE 1–19
Metalworking tools

used to install or remove control knobs with small Allen setscrews. A complete set, with a dozen or more wrenches, sells for just a few dollars. No technician should be without one.

4. Metalworking Tools. The more common metalworking tools needed by the electronics technician are shown in Figure 1–19. A *hacksaw* for cutting metal, a *center punch* for marking the center of holes to be drilled, a *reamer* to enlarge or shape a hole, a *tin snip* to cut sheet metal, and a set of *files* for forming or smoothing metal surfaces are all tools likely to be familiar to the electronics enthusiast. Two metalworking tools designed to punch and cut holes in sheet metal, however, may require an introduction.

A set of *chassis punches* (Figure 1–20) makes cutting large holes in chassis for sockets, meters, panel lights, and so forth, a breeze. They make neat, clean holes. Choose a set that provides the hole sizes you are likely to use. Keep in mind, though, that you can add individual punches as the need arises.

Another handy chassis cutter is the *nibbling tool,* also shown in Figure 1–20. With it, bite-size chunks of metal are taken out of the chassis in the process of forming practically any shape hole of any required size. A nibbler can be purchased for less than $10. It is an excellent investment.

5. Drills and Drill Bits. Two types of drills need to be considered. The *power hand drill* is indispensable for general drilling in wood, plastic, and metal (Figure 1–21). Select one that has a variable speed control, a chuck capacity of $\frac{1}{4}$" or $\frac{3}{8}$", and either a grounded three-conductor cord or a two-conductor cord with double-insulated plastic body. Drill bits, which actually do the cutting, should be of high-speed

size, specifically designed to do the job. With such a tool, the possibility of nicking or gouging a wire is greatly reduced.

3. Wrenches. A *wrench*—a tool with adjustable jaws, lugs, or sockets at one or both ends—is used for holding, twisting, or turning a bolt or nut. One type, of particular interest to us, is the *Allen wrench,* consisting of an L-shaped hexagonal bar of hardened steel either end of which fits the socket of a screw or bolt (Figure 1–18). The Allen wrench is

FIGURE 1–20
Sheet metal hole cutters.
Left: chassis punch; right:
nibbler

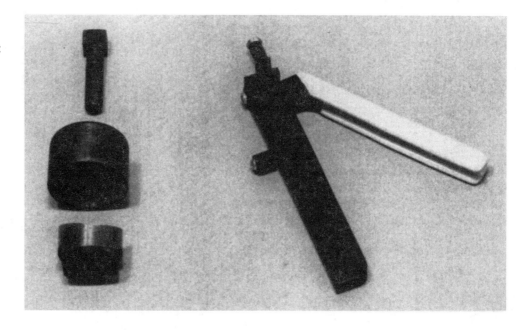

FIGURE 1–21
Drills and drill bits

steel, not carbon steel, as high-speed steel bits have the superior cutting and wearing characteristics required in electronics fabrication. Buy a set that has bits ranging in size from $\frac{3}{32}$" to $\frac{3}{8}$".

A drill particularly adaptable for printed circuit board work is the *Dremel Moto Tool,* a high-speed drill and grinder (Figure 1–21). This drill, which can be held comfortably in one hand, accepts drill bits ranging in size from #80 to #30 (the larger number being the smaller bit size). The Dremel Moto Tool can be converted into an excellent printed circuit board drill press by mounting it in one of the many *drill stands* designed specifically to accommodate it (Figure 1–21). A Dremel Moto Tool is essential for those wanting to manufacture their own printed circuit boards.

Tools for Electrical Work

Tools for use in electrical work are needed to secure connections between conductors. Because most connections are still made with solder, soldering and desoldering equipment is emphasized in this book. Nonetheless, wire-wrapping materials and solderless circuit boards (breadboards) are also explored.

1. Soldering Equipment. Soldering is the process of joining metal leads, wires, and terminals by melting an alloy of tin and lead (solder) over them. To heat the connection, a *soldering iron*

is used (Figure 1–22). Soldering irons come in two basic configurations: guns and pencils. *Guns* are basically used to supply heat to large electrical connections and metal chassis. For general electronics fabrication, only a soldering *pencil* should be used.

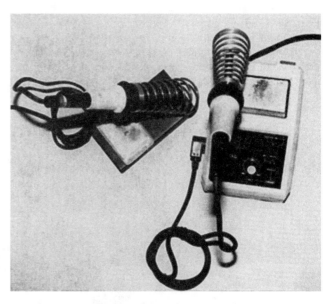

FIGURE 1–22
Soldering equipment

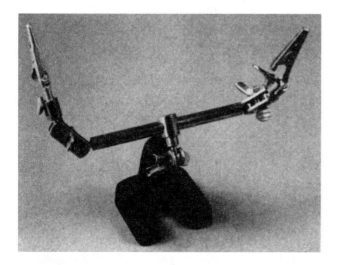

FIGURE 1-23
Third hand

Soldering irons have a wattage rating that measures how quickly they transfer heat. A 25- to 35-W iron is a good choice, particularly for printed circuit board fabrication. It should have a well-insulated handle (your hand should be kept cool) and, preferably, a grounded tip to protect static-sensitive components (such as CMOS ICs and JFET transistors). Although controlled temperature and cordless soldering irons are also available, the former (Figure 1–22) are quite expensive (often more than $50) and the latter are used only rarely. As a beginner, you should perhaps stick to the basics, a good-quality solder pencil.

When purchasing solder, choose a 60/40 or 63/37 rosin core solder with a #21 gauge (0.031").

You may also wish to buy a "third hand" soldering aid to hold the circuit board during project assembly (Figure 1–23). Such a unit is quite helpful and inexpensive (often less than $5).

The *soldering tip,* the element that actually transfers heat to the connection, is available in two materials and three basic configurations. You can purchase raw copper or nickel-plated copper tips. Choose the plated type; they last up to 10 times as long. Tip shape is, to a large extent, a matter of personal preference. Blunt chisel, pyramid, and needlelike tips are the most popular. For printed circuit board soldering, the needlelike tip is the preferred choice.

2. Desoldering Equipment. Unfortunately, what is often assembled sometimes must be disassembled. When it is necessary to desolder connections, special (though inexpensive) implements are required (Figure 1–24). The rubber suction *bulb* with a Teflon tip draws up molten solder that has been reheated with a soldering iron. A combination bulb and soldering iron is also available. The *desoldering pump* (also known as a solder sucker) uses the action of a spring-loaded plunger to suck up, in one quick action, melted solder from a connection. Finally, there is the *desoldering wick,* a braided copper material that uses capillary action to pull up the hot solder. Any of these tools will do the job of removing solder from an electrical connection. Their selection is based both on the particular job to be done and on personal preference.

3. Wire-Wrapping Equipment. In wire wrapping the stripped ends of a thin wire are tightly wrapped around a square or rectangular terminal post. No soldering is involved in making the connection.

FIGURE 1-24
Desoldering equipment

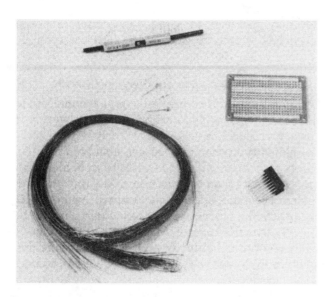

FIGURE 1–25
Wire-wrapping materials

Wire wrapping as a breadboarding method is discussed in Chapter 6. Here, we just want to alert you to the tools and materials necessary to do the job, should you choose to use this construction approach at the breadboarding stage.

You are going to need a *wire-wrapping/unwrapping/stripping tool* (often referred to simply as a wire-wrap tool). The type shown in Figure 1–25 is available for less than $10. The *wire* you will use is #30 gauge and comes in spools or individually prestripped lengths in five popular colors: red, blue, white, black, and yellow. Special *wire-wrap sockets* are needed when connecting integrated circuits. The pins must be longer than those on the solder-type socket and they must be square, not round or flat. Various *terminal posts,* designed specifically for wire wrapping, are also necessary. Component leads will be attached to the non-wire-wrapped side of the terminal post, usually by soldering. Finally, you will need a special *perforated board* on which to mount the sockets and pins. The board is a phenolic material, $\frac{1}{16}$" thick, with holes spaced 0.10" apart.

4. Solderless Circuit Board. In the early stage of breadboarding, when the circuit design may be undergoing frequent changes, a quick and easy system of circuit assembly is required. Solderless circuit board provides that method (Figure 1–26). The board consists of a plastic block with arrays of holes spaced 0.10" apart. Electrical conductors interconnect adjacent rows or columns of holes. The board readily accepts IC pins and the leads of other

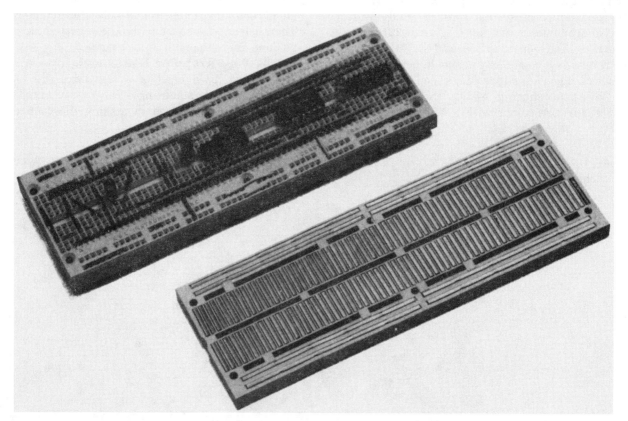

FIGURE 1–26
Solderless circuit board

electronic components. Thus the technician can build up a circuit with reliable connections without having to resort to soldering or wire wrapping.

Some solderless circuit boards are designed with interlocking ridges so that any number can be snapped together to expand the total breadboarding area. You can start out with a 2" × 5" piece and as your needs and budget permit, add additional units. In no time you will have a very professional setup for serious experimenting and breadboarding.

One final note: There are a number of tool kits, or caddies, on the market that provide an essential assortment of hand tools in a convenient carrying case. One such kit, from Graymark International, Inc., is shown in Figure 1–27. Prices range from $25 to $50 for the less expensive units.

Additional Tools for Modern Electronic Circuits

Before the advent of printed circuit technology, the tools required for electronics fabrication were basic and few. Today, a screwdriver, needle nose pliers, and soldering iron just aren't enough. The basic tools you will need have been discussed, along with

a few of the more specialized ones; but there are other tools and instruments you will want to keep an eye out for. You don't have to rush out and buy them all now; as we move through the fabrication process we will identify and discuss specific ones. Then you can make the major "buy/borrow/forget it" decision on acquiring them.

Something to Think About

Tool catalogs are available for the asking, from manufactures and distributors alike. Why not begin a collection now? When you receive a new catalog, look through it slowly and carefully, studying the description of each tool. Such catalogs can be an education in themselves.

SUMMARY

In this chapter we discussed the traditional electronics course and how the lecture and lab approach to learning can be supplemented with an electronics fabrication experience. By examining just what it is electronics technicians do, we gained an appreciation of the need for learning how to design, build, troubleshoot, and document a complete prototype project.

Next, we investigated the five-stage design and fabrication process by probing the features of the (1) design; (2) drawing; (3) experimenting; (4) prototyping; and (5) testing, troubleshooting, and final documentation stages. We then saw how this book is structured around those five stages. We discussed the three-tier project building experience and looked into the Project Report, a summary containing all the necessary written and graphics documentation for the prototype project.

We explored the tools of the trade, on the board, at the computer, and on the bench. We examined the drafting table, along with the materials that go with it. We described the basic equipment necessary to make schematic and assembly drawings, as well as what is needed to produce the design layout and artwork for printed circuit board fabrication. We probed the outlines of a computer-aided design (CAD) system. Finally, we considered the workbench and its requirements in terms of space, size, construction, and lighting and power needs. We concluded with an inventory of the tools required for mechanical and electronics fabrication.

Before you go on to that actual fabrication, however, you need to be sure that you can protect both your project and yourself. To do that, we examine lab safety, the subject of Chapter 2.

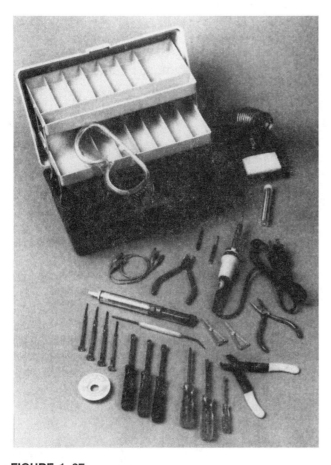

FIGURE 1–27
Tool kit

QUESTIONS

1. Most of the electronics courses taught at the college or trade school level are organized around the _____ and _____ structure.

2. Electronics _____ build, repair, test, modify, and install electronic devices and equipment.

3. A project that is hand-assembled to resemble as much as possible the final, mass-produced equipment is known as a _____.

4. Name the five stages in producing a working prototype project. _____, _____, _____, _____, _____.

5. The _____ _____ contains all the written and graphics documentation that will accompany the prototype project.

6. Many portable drafting tables have an instrument called a drafting _____ that allows you to quickly draw straight lines at designated lengths and angles.

7. List three input and three output peripheral devices used with a CAD system.

   Input   Output

   a. _____   a. _____

   b. _____   b. _____

   c. _____   c. _____

8. Both the drawing table (board) and the workbench must be bathed in direct and indirect _____.

9. Although diagonal cutters can be used for stripping wire, a better choice is a _____ stripper, specifically designed for the job.

10. For general soldering work, a soldering _____ should never be used, only a soldering _____.

2 Safety—It's Your Life

OBJECTIVES

In this chapter you will learn

- [ ] Why employers are so safety conscious and why you should be, too.
- [ ] How your eyes, ears, nose, throat, and skin are susceptible to harm from laboratory accidents.
- [ ] What electrical shock is, and how it can affect your heart.
- [ ] How today's solid-state circuits are safer than those of the vacuum tube era.
- [ ] How to avoid danger while working on electrical circuits.
- [ ] How to handle toxic and hazardous chemicals safely.
- [ ] How hand tools can be more dangerous than power tools.
- [ ] How to avoid safety problems when working with power tools.
- [ ] How to protect your voltage-sensitive projects from the "static electricity monster."
- [ ] What first-aid steps to take if an accident occurs in the laboratory.
- [ ] About environmental concerns with regard to product design and development.

Congratulations!

What for? For reading, or at least starting to read, this chapter. There are those who would just as soon skip this chapter and get right to the hands-on material. But hold on a moment; have a little patience—"safety first," as the saying goes.

To protect both yourself and the project you build, you need to become thoroughly familiar with basic electronics laboratory safety procedures. If you want to become an accomplished electronics technician, then now, while you are still in school, is the time to obtain the knowledge and develop the effective safety habits necessary to succeed on the job. We want you to have that success and, along with it, a safe and satisfying experience with electronics.

In this chapter you will see why safety is so essential, so important to you and your colleagues, whether they be students in school or coworkers at the plant. We'll examine why practicing good safety is considered an essential job requirement. We will investigate how exposure to hazardous conditions affects various parts of the human body, and how to minimize the risks arising from electrical shock, contact with toxic chemicals, and the misuse of hand and power tools. We will also analyze how you, by storing static electricity, can be the source of danger, destroying the projects around you. In Appendix A is a comprehensive *electronics safety test*. Be sure you take, and pass, that safety test immediately after reading this chapter.

2.1 SAFETY: WHO CARES?

Who cares about safety? Well, your employer for sure and, we hope, you too. For your employer, safety is primarily an economic and legal issue; for you, it is a personal and, when your colleagues are involved, a moral concern.

Safety as a Part of Your Job Description

Companies take safety very seriously. On the firm's entrance examination, you are just as likely to be asked a question on the safe handling of etching solutions as one on Ohm's law. The reason is the bottom line: Accidents cost money. Let's look at the economic and legal issues from the employer's standpoint and then consider why you should also be sharing this concern for a safe working environment.

Economic and Legal Issues

As an electronics technician, you will be required to perform such tasks as soldering, operating complex test equipment, building and repairing electronic devices, using hand and power tools, and possibly handling toxic and hazardous materials. You will be expected not only to do these assignments well but to do them in such a way that minimizes the danger to equipment, to yourself, and to others around you. In today's electronics environment, it is not uncommon to find printed circuit boards populated with dozens of sensitive integrated circuits that together cost hundreds of dollars. A bolt of static electricity, discharged from you or your clothing, could render the entire circuit mute. In some sophisticated laboratories, oscilloscopes and related pieces of test equipment can cost as much as a new automobile.

But much more serious than damage to equipment is injury to people. The lost hours due to accidents are costing U.S. industry billions of dollars a year. Although accidents will happen, try not to be "accident prone." You can be sure that any company you work for is keenly aware of problems arising from accident litigation. No matter how good you are at troubleshooting complex digital circuits or building sophisticated prototype projects, your value to the company will be seriously diminished if you are costing them money by being a safety risk. Simply stated: Practicing good safety habits makes good economic sense.

Personal and Moral Issues

Being a safety risk to yourself is one thing, but endangering fellow students and coworkers is quite another. No one wants to be around a person with a reputation for accidents; they might be injured, too. And if no one will work with you, how can you keep your job?

You have a responsibility to watch out for your safety and that of others around you. It's like driving a car. Every time you get behind the wheel, the lives of other drivers and pedestrians are in your hands. They are counting on you to stay to the right of a center line and stop at a red light. You and your colleagues in electronics are depending on each other, too. Everyone needs to be part of the safety solution, not the safety problem.

Body Parts Susceptible to Damage

The human body is a fragile thing. It is easily cut, bruised, burned, irritated, shocked, and poisoned. External organs associated with our five senses and internal organs vital to our very lives are susceptible to injury in the school or workplace.

Covering Your Five Senses

Although bionics is the current rage, and electronics has contributed mightily to its development, most would agree that the real thing is still the best. To keep our eyes, ears, nose, throat, and skin safe from harm, we must first see how they can be hurt when exposed to hazardous conditions.

Injury to the *eyes* can cause permanent blindness. Small objects, such as clipped component leads and flying solder balls, can stab and burn sensitive parts of the eye. Furthermore, a splash of acid or the spray from a cleaning solvent can, at the very least, irritate; at the worst, it can destroy the entire eye. Even a bright flash of light from a xenon strobe tube, for example, can cause a serious accident. Even if you are only momentarily blinded, it may be enough to make you susceptible to other forms of injury.

Loud sounds and high-pitched tones can be very irritating to the *ears*. In extreme cases they may cause temporary or even permanent hearing loss. It isn't just heavy metal rock bands that can produce such noises. They can be found right in the electronics laboratory.

You may not think your *nose* is vulnerable to dangers found in the laboratory, yet toxic fumes can be irritating to the membranes lining the nasal passages. More importantly, the nose is one entry leading to the lungs. Damage there can be very serious, as it may affect your ability to breathe properly.

Your mouth, or *throat,* can also be an opening for foreign and hazardous particles. Of course, you wouldn't knowingly put any acid solution or solder wire (which is 40% lead) in your mouth. But if acid gets under your fingernails, and you bite your nails, you could be in trouble. Also, inadvertent chewing on a strip of solder is not unheard of. The greatest danger, however, lies in using your mouth as a temporary vise to hold small pieces of hardware, tiny transistors, and short strips of wire. Accidental swallowing, or a friendly slap on the back from your lab partner, and presto—it's time to have your stomach pumped.

FIGURE 2–1
Current flow

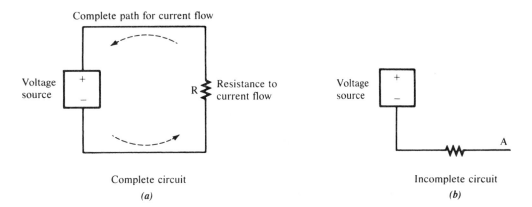

Complete path for current flow

Voltage source
+
−
R
Resistance to current flow

Complete circuit
(a)

Voltage source
+
−
A
Incomplete circuit
(b)

Then there is the *skin,* the outer layer designed to protect the entire body. Destruction of nerve endings as a result of electrical shock, burns from soldering irons and hot components, acid irritations, and, of course, common cuts, bruises, and abrasions can occur in an environment where safety is lax.

Clearly, your entire body is subject to harm; however, there is no cause for alarm, only a healthy concern. Shortly, you will see how you can eliminate most of these dangers by taking simple precautions and following commonly accepted safety practices.

At the Heart of the Matter

Of all the internal organs, the *heart* is certainly the most vital. If it stops, even for only 4 to 6 minutes, death can result. Many factors can cause the heart to stop beating. Here, however, only the most relevant one—**electrocution**, or **electrical shock**—is considered. Let's see just what electrical shock is and how it can destroy the heart.

Electrical shock is the passage of current through the body. Current flows through the human body, as it does anywhere else, when a complete circuit exists. Such a circuit is shown in Figure 2–1a. A source of voltage and a conducting path (shown in the figure as having a given amount of resistance) are all that is required. Note, however, that the path for current flow must be complete. In Figure 2–1b no current flows to point A. The reason is that although it has a place to come from (in this case, the battery's negative terminal), it has no place to go (in this case, the battery's positive terminal).

The same is true when the human body is the conducting path. As illustrated in Figure 2–2a, even though the person is touching the positive terminal of the voltage source, he is in no danger. As long as he is wearing insulated boots, current can find no

path through him to the negative, or ground, terminal of the battery; however, if he completes the circuit by, for instance, touching the negative battery terminal with the other hand (possibly through a ground connection), current will indeed flow, as in Figure 2–2b. Just how much current depends on a number of factors.

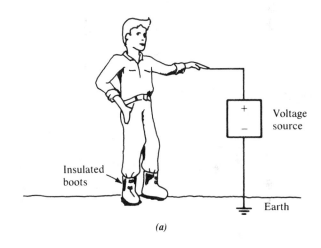

Insulated boots

Voltage source
+
−

Earth
(a)

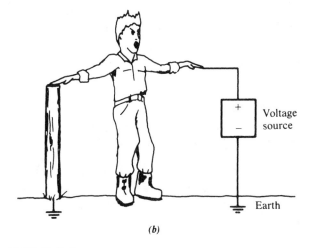

Voltage source
+
−

Earth
(b)

FIGURE 2–2
The human body as a conducting path

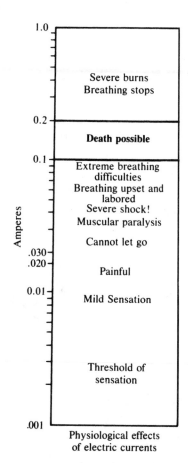

FIGURE 2–3
How much current is enough to kill?

One way to increase current flow is to reduce circuit resistance. Body resistance may be quite high if skin moisture is low and there are no cuts or abrasions at the point of electrical contact. In such cases, little current will flow and only a mild shock may result. Nonetheless, if any of these factors is reversed, resistance will be lowered and large amounts of current can result. If the path for current flow is through the chest, the heart can receive a lethal dose of current, an electrical shock. The heart will then most likely go into fibrillation (rapid, irregular muscle contraction) and stop beating.

How much current is enough to kill? Although the amount varies widely and especially depends on the organs that current passes through, it is less than you probably think. Figure 2–3 gives the effects of even mild doses of current. A mere 1 to 20 milliamperes (mA) can cause a painful sensation; at 30 mA, breathing may stop; and 100 to 300 mA is enough to cause electrocution. Clearly, even small amounts of unwanted current through the body can be very dangerous to your health.

> **Something to Think About**
>
> *The amount of current flow is equal to the voltage divided by the resistance—Ohm's law. Using an analog or digital multimeter, why not measure the current flowing between your hands by grabbing the meter's probes? Are you surprised at the results?*

2.2   WHERE THE DANGER LIES

It is time to look at where in the school laboratory or workplace the danger to your safety lies. Problem areas and practical solutions will be identified. Although specific advice in the form of "do's and don't's" will be given, keep in mind that maintaining a safe working environment isn't complicated. It involves just four simple rules:

1. Keep your work area neat and orderly. Being able to see what you have and what you are doing will eliminate many problems.
2. Be alert and attentive at all times. Distractions of any kind can only leave you open to risk.
3. Know the correct safety procedures. If you are not sure how to operate safely a piece of test equipment, a tool, or any machinery, ask your instructor. There is a safe way to do everything. Learn what it is—and do it.
4. If an accident does occur, know what action to take. Be familiar with the basic first-aid measures that could mean the difference between a quick recovery and a serious medical complication.

Electricity: The Tingle That Hurts

Let's examine how you can get into trouble when working with electricity, how to avoid major problems, and what few simple first-aid steps you should take if an accident does occur.

The Problem—The Solution

The problem is that there's both good news and bad news. The good news is that today's electronic devices, many of which are solid state and battery powered, are much safer than their predecessors of the vacuum tube era. Instead of circuits with voltages in the hundreds and currents in amperes, we now have equipment operating on 5 to 12 volts (V) dc and drawing currents of less than 50 to 100 mA. The bad news is that many systems, even though employing solid-state circuitry, do not get their power directly from a battery but rather from the ac line. Any electronic project that is plugged into the wall outlet is a potential death trap.

The problem with ac is twofold. An ordinary wall outlet supplies up to 25 A of current, certainly enough under many conditions to kill you. In addition, ac tends to "hang on." The victim can be prevented from releasing the source of voltage, thus increasing the damage to the body.

Circuits powered from an ac line are not the only problem, however. A serious shock can be had from many circuits, dc or ac, containing charged capacitors, faulty wiring, or shorted components. Treat every electronic circuit as potentially hazardous. Electricity, though not to be unduly feared, must certainly be respected.

The solution to being safe while working with electricity is to follow some basic "do's and don't's." Here are the more important ones:

The Do's

- ☐ *Do* work with one hand behind your back while testing live circuits. In that way if you complete the path for current flow, at least it won't be through your heart.
- ☐ *Do* use an isolation transformer while working on ac-powered equipment. This device isolates the powered equipment from the power source, adding a strong measure of safety.
- ☐ *Do* make sure that all capacitors (components that store an electrical charge) are discharged before beginning troubleshooting. Use an insulated screwdriver to short out capacitor leads.
- ☐ *Do* use three-conductor grounded line cords and polarized plugs with ac-operated equipment. Both items reduce the danger from short-circuited chassis.
- ☐ *Do* keep your fingers out of live chassis. Test all circuits with a voltmeter or specially designed test lamps.

The Don't's

- ☐ *Don't* install or remove any electronic components while the circuit is connected to a power source. Following this procedure will protect you as well as the component.
- ☐ *Don't* overfuse. Using a fuse with a higher rating than is recommended only defeats the fuse's purpose.
- ☐ *Don't* work with wet hands. As noted earlier, if body resistance goes down, current flow goes up. Even sweat can moisten the hands enough to cause excessive current flow.
- ☐ *Don't* cut wires carrying electricity. Again, it isn't just the ac line cord that is lethal. Assume all wires carry enough current to harm you.

- ☐ *Don't* disconnect electrical devices from the wall outlet by pulling on the line cord; pull the plug handle.

First Aid

With severe electrical shock, you must think not only of the victim but of yourself as well. You don't want to become part of the problem. After sending for help, immediately turn off the source of power. If that can't be done, do not attempt to pull the victim away; you will only get shocked yourself. Use a nonconducting material, such as a two-by-four piece of lumber, to remove the victim from the circuit. Once the victim is free and lying down, check for signs of breathing. If there are none, begin artificial respiration. If you cannot detect a heartbeat, and you are trained to do so, apply continuous chest compression cardiopulmonary resuscitation (CCC-CPR).

In less traumatic cases, where electrical shock is not life-threatening, keep the victim calm and send for expert medical help.

Exposure to Toxic and Hazardous Chemicals

Let's examine the problems associated with potentially dangerous chemicals and how best to protect yourselves when working around various sprays and liquids, and what first-aid procedures to take if something goes wrong.

The Problem—The Solution

With toxic and hazardous chemicals, you should be concerned about inhaling vapors, swallowing liquids, acid burns on the skin, contact with the eyes, and the overall danger of fire and explosion. Chemicals such as etching solutions, spray paints, component cleaners, glues, photographic developing solutions, and general household solvents, all of which are likely to be found in the electronics laboratory, can cause health hazards if improperly used or stored. To minimize risks, here are some do's and don't's:

The Do's

- ☐ *Do* read the labels on all the chemicals you use. Pay particular attention to printed warnings.
- ☐ *Do* work in well-ventilated areas. This is particularly important when using paint and chemical sprays.
- ☐ *Do* wear eye protection when working around hazardous chemicals.
- ☐ *Do* wear rubber gloves when working with acid solutions. Be sure they are washed or thrown away when work is done.

☐ *Do* hold printed circuit boards securely with tongs when placing them in or removing them from an acid solution.

☐ *Do* immediately clean all tools that have come in contact with hazardous chemicals. This will prevent someone from inadvertently touching the toxic substances.

The Don't's

☐ *Don't* pour solvents, paints, or acids down the drain. In addition to being a health hazard, in many localities it is illegal.

☐ *Don't* transfer chemicals to unlabeled containers.

☐ *Don't* inhale fumes from soldering paste. Such fumes can be harmful to your lungs.

☐ *Don't* leave any chemicals in open containers where they can be splashed or spilled. Put all unused chemicals safely away, preferably in a locked metal cabinet.

☐ *Don't* place any chemicals near sparks or open flames. Assume that all chemicals are highly combustible.

First Aid

In an emergency, follow all first-aid directions printed on the label of the chemical you have been using. Read the label before use so that you won't be confused if trouble develops. For example, some instructions tell you to induce vomiting if the substance is swallowed; others tell you not to. Do what the label says and call a physician immediately.

More specifically, in the case of acid burns and eye irritations, wash and flush thoroughly with cold running water. If your lab is equipped with an *eyewash station,* be sure you understand its proper use. When exposure to vapors has occurred, get the victim outside in the open air.

Safety Hazards Caused by the Misuse of Hand Tools

In this section hand tool safety is discussed. For convenience hand tools are defined as those that are held in the hand and are not powered by electricity.

The Problem—The Solution

Hand tools can be more dangerous than power tools. Why? Because they are used more often and we tend to dismiss their risk. Do not be deceived. A slip of a hacksaw blade or the launching of a loose hammerhead can maim you—or someone else—for life. Whether it's a puncture from a file tang, a gash from a saw blade, or a slit from a knife, hand tools can do plenty of damage. To avoid the obvious dangers, consider this list of important do's and don't's:

The Do's

☐ *Do* keep hands free of dirt, grease, or oil when using hand tools.

☐ *Do* keep hand tools sharp. A dull tool is more dangerous than a sharp one because with a dull tool you tend to apply more pressure, increasing the likelihood of slippage.

☐ *Do* cut away from your body, not toward it.

☐ *Do* secure all small pieces of work in a vise or with appropriate clamps.

☐ *Do* put sharp tools away when not in use. Leaving knives, awls, and punches lying around for someone to lean on is extremely dangerous.

The Don't's

☐ *Don't* carry hand tools in your pockets; they may injure you or someone else.

☐ *Don't* use a file without a handle. An exposed file tang can easily puncture your hand or wrist.

☐ *Don't* toss any hand tools to a neighbor. Flying objects of any kind are never to be seen in the electronics laboratory.

☐ *Don't* run your hands over the edges of sheet metal. These edges are sharp and can cause severe cuts.

First Aid

Injury from hand tools usually results in a cut or burn. What specific action to take when a laceration occurs depends on its severity. In all cases, clean the wound, apply direct pressure, and elevate the injured body part. With regard to burns, flush the affected area with cold water (do not apply ice directly) and cover with a clean cloth, then seek medical aid.

Safety Hazards Caused by the Misuse of Power Tools

Power tools are defined as those that use electricity, usually in the form of ac from the wall outlet. Let's look at the problems and solutions for this area, the do's and don't's, and the appropriate first-aid procedures.

The Problem—The Solution

Power tools can burn, cut, scrape, and even hit you with flying objects. Because most of them plug directly into the wall outlet, they can also give you a nasty electrical shock. Injuries from power tools occur when long hair or unsuitable clothing gets caught in a revolving machine, when small objects fly out of a drill press vise, when guards or safety devices are removed, or when a hot soldering iron is left where someone can lean against it. To prevent such injuries, practice the correct safety procedures as outlined by your instructor, and follow these do's and don't's:

The Do's

☐ *Do* turn on and off all power tools yourself. Do not allow others to do it for you.

☐ *Do* make sure all objects being drilled or cut are securely fastened. Small objects, and especially sheet metal, should be held in a vise, clamp, or suitable gripping tool such as pliers or vise grips.

☐ *Do* see to it that all implements and adjusting tools are removed from the work area.

☐ *Do* remove the chuck key from a drill press before turning on the machine. A flying chuck key is a frequent cause of injury in the laboratory.

☐ *Do* ease up on the drill pressure as you break through the work. This will reduce the possibility of the material's "grabbing."

☐ *Do* grasp a soldering iron only by the handle. Don't reach for a falling soldering iron—let it fall.

☐ *Do* return a soldering iron to its holder. It should be in one of two places: your hand or the holder.

The Don't's

☐ *Don't* ever leave power tools unattended. If the tool is on, stay with it until it is off—and until all parts have stopped moving.

☐ *Don't* wear clothing or accessories (jewelry) that can get tangled in revolving machine parts.

☐ *Don't* distract others while they are using power tools. If distractions cause problems with hand tools, the same is doubly true for power equipment.

☐ *Don't* stand in the direct "throw" of any machine. Don't line yourself up with a revolving saw blade or spinning grinding wheel.

☐ *Don't* remove guards or safety devices from power tools. They have been installed for a good reason; keep them in place.

☐ *Don't* put your face too close to moving machinery. Metal chips and oil spray can cause eye and skin injury.

☐ *Don't* test a soldering iron with your hand. Assume that all soldering irons are hot. Test them by attempting to melt a piece of solder.

☐ *Don't* hand a soldering iron directly to another person. Instead, return it to its holder. The other person can take it from there.

First Aid

Injuries resulting from misuse of power tools are similar to those that occur with hand tools, so the first-aid procedures are essentially the same. The accident itself, however, may be more severe. Send for help immediately, then follow the appropriate first-aid measures.

Project Safety

Although electronic circuits can do you harm, you can also damage them; you can zap as well as be zapped. In this section we will see how **static electricity** can destroy the projects we build. We will examine how this monster is created, what its effects are, and how we can cure its most troublesome characteristics by attempting to keep it in its place. Practicing good safety isn't just for us; it's also for our projects.

The Static Electricity Monster

Static electricity is a stationary charge. As such, it is a potential (voltage) waiting to go somewhere (Figure 2–4). If that charge gets to a sensitive circuit, it can destroy the circuit. In the normal course of your work, it is possible to build up a static charge of more than 50,000 V. But today's solid-state circuits don't need anything like that kind of voltage to render them useless. In some cases, less than a few volts of static electricity is all that it takes.

You have felt the effects of a static charge many times. When you walk across a shaggy carpet on a dry, winter day, electrons are literally torn off the rug and deposited all over your body. You become a walking reservoir of negative (static) charges. When you reach for a metal doorknob, which is neutral with regard to potential, the electrons on your

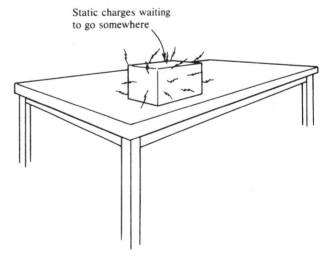

Static charges waiting to go somewhere

FIGURE 2–4
Static charge

FIGURE 2–5
Static charge

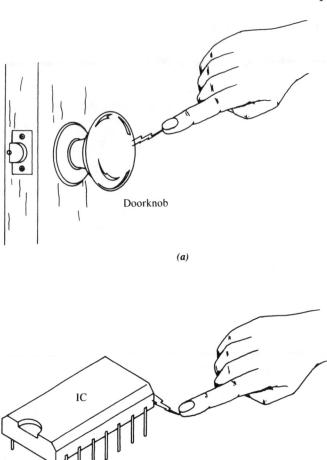

Doorknob

(a)

IC

(b)

hand jump to the knob (Figure 2–5a). An instant before the jump, the air is ionized and a spark is created. You don't have to see the spark to know that there has been a transfer of charged particles—the pain in your fingertips lets you know. That jump of electrical charges, from a more negative source (you) to a relatively neutral source, can damage sensitive circuits. If you reach for a CMOS (complementary metal-oxide semiconductor) integrated circuit instead of a doorknob, both you and the IC may feel the pain (Figure 2–5b)—in the circuit's case, perhaps enough to destroy it.

How the Static Charge Is Created

The most common way of creating a static charge is through friction, in which electrons are torn off one surface and accumulate on another. Anytime objects are rubbed together, friction is created. If the substances are dissimilar materials, the accumulated charge can be considerable. If one or both of the elements are nonconducting, the charge remains for a long time. It has no place to go. On the other hand, if

both materials are good conductors, the charges will leak off quickly, and little danger exists. The solution, then, is either to keep charged insulating materials away from sensitive circuits or to make sure the charge is bled away before it can do any damage.

Materials Susceptible to Damage

Any integrated circuit in the MOS (metal-oxide semiconductor) family can easily be destroyed by static discharge. In addition, many transistors (FETs, for example) are subject to the same effect. If it's a solid-state circuit (and what isn't today?), assume that the circuit is susceptible to damage from static electricity. This will not be true in all cases, but it is best to proceed as though it were. Following basic electrostatic discharge (ESD) safety procedures for all circuits is not difficult. Get in the habit of doing it.

Protecting Your Circuits

One way to reduce the risk from static electricity is to keep the room's humidity high. With high humidity there is a greater conducting path, and thus an electrostatic buildup will leak away much more easily.

The key to protecting circuits, however, is to try to bring everything around you to a common potential. If there is no difference in charge, there is no voltage, and therefore no static buildup. For example, if you're grounded and the integrated circuits you are working with are grounded, there is no difference in charge between you and them. There is no problem (Figure 2–6). Here is a list of what you can do to keep you and the circuits you're working on at the same potential:

☐ Be sure all MOS-type integrated circuits remain packaged in antistatic materials until used. These materials consist of specially coated plastic carriers, conductive foam, or aluminum foil. The idea is to keep all pins of an IC shorted together and thus at the same potential. In that way no charge can build up between them.

☐ When handling MOS components, avoid touching the pins. That will keep any charge you might have accumulated off the component leads, which are, of course, connected to the solid-state material within the IC package.

☐ Never install or remove integrated circuits while power is still applied to the circuit. The sudden voltage jolt that could result may damage them.

☐ If possible, use an antistatic wrist strap. The strap brings you to ground potential. Assuming your components are there too, no charge can build up.

☐ Use a soldering iron with a grounded tip. In that way no static charge will transfer to sensitive components while they are being soldered in the circuit.

The idea is to prevent a static buildup in the first place. If that is not possible, at least make sure that any such charge has a path to leak away. Doing that will ensure safe operation for your projects.

Environmental Concerns

A moment ago we discussed exposure to toxic and hazardous chemicals. We saw how handling such materials can, if one is not vigilant, have a detrimental effect on those working in a school laboratory or industrial setting. Broader issues of environmental concern have begun to surface, however, and will assume greater significance for the electronics industry and those employed in it, in the decades to come. In this concluding section you need to see why you, the prototype electronics technician, should be interested in such matters and what your role might be in workplace environmental protection.

Why You Should Be Concerned

There are at least two significant reasons why you must become familiar with and concerned about your company's role in environmental protection. *First,* the entire electronics industry is "going green," and for good reasons. Companies such as

FIGURE 2–6
Common potential

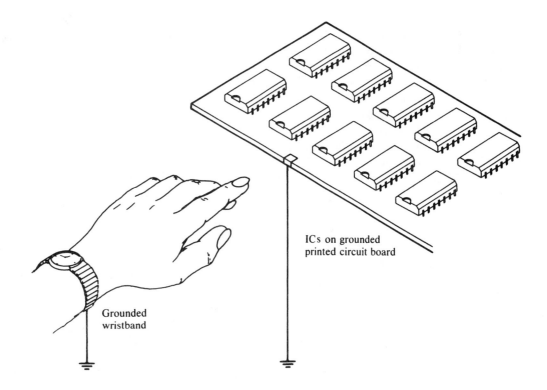

Grounded wristband

ICs on grounded printed circuit board

Intel, Tektronix, and 3M, to name but a few, are spending millions of dollars a year disposing of everyday toxic waste—the chemicals and other substances used to manufacture semiconductors, printed circuit boards, and other products. In some cases, with certain solvents and metals, it costs more to dispose of them than to buy them. Ways have to be found to either reduce the demand for such materials or recycle them more efficiently. Every employee at an electronics firm, whether it be a large or small enterprise, has to be part of that solution. As an electronics design and fabrication technician involved in prototyping, you're at the front end of the chain of production, and you will be expected to do your part in dealing with the company's environmental problems.

There is a *second,* and perhaps more relevant, reason why you in particular will be involved with environmental issues, most notably those of waste generation. Emphasis is now shifting from waste disposal on the back end to toxin use up front. In other words, environmental concerns are becoming firmly entrenched in product design and production. For example, Texas Instruments was planning to make a device with a mercury switch. Even though safeguards had been incorporated to deal with the mercury, the firm decided to redo the design using steel balls instead. The point is that you, working with the design and fabrication of prototype projects, will need to involve yourself with "green design," electronic products that are environmentally friendly from the start.

Your Role in the "Greening" of Your Company

As we have seen, your role here is primarily up front, at the design end. It is simply no longer good enough to clean up the mess—it shouldn't be generated in the first place. As the mercury switch example indicates, you want to be aware of components, materials, and fabrication processes that are environmentally sensitive—and help design these materials and processes into the products your company creates.

How do you learn about such matters? First, you adopt a positive attitude about environmental concerns—that is, don't fight the revolution, join it.

Second, read all you can on the subject, in newspapers, magazines, trade journals, and books. Third, talk to coworkers and others who share your interest and concern. Network and exchange information.

If all this seems a bit overwhelming for you and your industry, understand that those in electronics are used to rapid change and development. Fortunately, our industry is better positioned than most, in temperament, outlook, and experience, to deal with the new environmental concerns.

Something to Think About

If your school laboratory does not already have a posted listing of safety do's and don'ts, why not create one? Distill the lists of do's and don'ts presented in Chapter 2 to create the poster.

SUMMARY

Congratulations again—assuming, of course, that you did indeed read this chapter on safety. We dealt in general terms, discussing first just why it is so important to care about safety. Then, after looking at what body parts are susceptible to harm, we focused on where the dangers lie. We examined the risks related to electrical shock, toxic and hazardous chemicals, the misuse of hand and power tools, the dangers of static electricity to projects, and industrywide environmental concerns. In the chapters to come you will be reminded of what is presented here. You will also read about additional safety precautions when specific pieces of equipment and construction procedures are introduced. Now, before you rush on to Chapter 3, don't forget to take the safety test in Appendix A.

In Chapter 3, after first looking at design and designers, we examine the fundamentals of an electronic system. We explore the two-step design process while delving into the workings of the Sample and Exercise Projects. We conclude with a look at computer-based circuit design, simulation, and analysis.

QUESTIONS

1. As an electronics technician you will be required to perform your tasks in such a way as to minimize the _____ to equipment, to yourself, and to others around you.

2. Even a bright flash of light from a xenon strobe tube can cause _____ blindness.

3. Toxic _____ can irritate the membranes lining the nasal passages of your nose.

4. Electrical shock is the passage of _____ through the body.

5. Any electronic project that is plugged into the wall outlet is a potential _____ _____.

6. When testing "live" circuits, you should work with one _____ behind your back.

7. With toxic and hazardous _____, we are concerned about inhaling vapors, swallowing liquids, acid burns on the skin, contact with the eyes, and the overall danger of fire and explosion.

8. Hand tools can be more dangerous than power tools because they are used more often and we tend to _____ their risk.

9. Don't ever reach for a falling _____ _____—let it fall.

10. An integrated circuit in the MOS family can easily be destroyed by a strong _____ _____.

3 The Design Process— An Idea Is Born

OBJECTIVES

In this chapter you will learn

- [ ] How electronic products and services are created to solve a particular need.
- [ ] What personality characteristics and technical background make for a good electronics designer.
- [ ] How design problems and solutions are classified for easy identification and better understanding.
- [ ] How electronic circuits combine with input and output transducers to form electronic systems.
- [ ] How the four active components—vacuum tubes, transistors, integrated circuits, and microprocessors—switch or regulate current flow.
- [ ] How passive components, such as resistors, capacitors, diodes, coils, and transformers, help the active components do their job.
- [ ] How to select and acquire electronic components.
- [ ] About Total Quality Management (TQM).
- [ ] What a Concepts and Requirements Document does to explain the project goals, objectives, and responsibilities.
- [ ] What design drawings are required to illustrate system operation, circuit function, and packaging concept.
- [ ] How the Concepts and Requirements Document and design drawings are produced for the Exercise and Sample Projects.
- [ ] The theory of operation for the Variable Power Supply and 3-Channel Color Organ.

- [ ] How computer-based design tools allow us to design, simulate, and analyze electronic circuits.
- [ ] How **Electronics Workbench, Multisim 7.0**, works as a virtual laboratory to aid in the design process.

What will they think of next?" We are all a bit awed by the explosion of new electronic products and services. Just how did microwave ovens, CDs, ATMs, handheld wireless devices, personal GPS devices, and digital this, digital that, come about? Did the idea just pop into someone's head, or was there a perceived need for the product or service? How, in other words, is electronics design done?

In this chapter, we find out. After all, we can't build something if we don't know what to build. We first have to have an idea for a new product or service, then we have to refine that idea by specifying exactly what we want and how we intend to develop it. This chapter looks at how this is done by examining electronics design: what it is, the type of people who do it, and the basic methodology involved. We will analyze the **systems approach** to electronics design by studying the fundamentals of electronics design and by outlining the two-step design process. We will look at the important issue of Total Quality Management (TQM). Then, we will introduce the Sample Project, the Variable Power Supply, and the Exercise Project, the 3-Channel Color Organ. These projects, which are carried from chapter to chapter as the subject of electronics design and fabrication unfolds, will provide practical examples of the electronics design process.

Finally, since it is possible to both design a circuit and simulate its operation on a computer, we look at a software package that does just that—**Electronics Workbench, Multisim 7.0**. Using this program as an example, we see what computer-based circuit design, simulation, and analysis are all about.

3.1 ELECTRONICS DESIGN: PROBLEMS IN SEARCH OF SOLUTIONS

In this section we give some examples of electronics design and the kind of people who are involved in it. We will also explore a design methodology that is based on the identification of certain problem and solution categories.

Design and Designers

First, let's see how electronic products and services are created to solve a particular need and then look at the creators themselves.

Creating Products and Services

Why are our homes, offices, and factories, along with our vast communications and transportation systems, filled with electronic devices? Could it be because many of today's problems demand solutions in the form of electronic products and services? Clearly, this must be so. Let's look at some examples:

☐ Burglaries have increased dramatically in your neighborhood. What can you do? You can begin a neighborhood watch, hire a patrol service, or buy a kennel of guard dogs. But consider electronics. Electronic burglar alarm products, from the simple latch-buzzer type to sophisticated intrusion-detection systems, can be installed to discourage would-be robbers. Electronics clearly has a role to play in solving this problem.

☐ You want to know how fast you're traveling while out on your daily bike ride. You need a bicycle attachment that will monitor the bike's speed. Mechanical speedometers have been doing that reasonably well for years. Today, however, you can buy electronic versions that will not only give speed in miles per hour but, at the push of a button, tell you the elapsed time of travel, projected time of arrival, distance traveled, distance to go, and, of course, time of day.

☐ Your local bank wants to provide its customers with 24-hour basic banking services. What are the options? The bank could remain open 24 hours a day, or it could set up minibanks at gas stations that are open around the clock; however, banks all over the world discovered a better so-

lution when they developed the computerized electronic automated teller machine (ATM). After all, computers can work unattended any time, night or day.

☐ You own a medium-sized stationery store and you want instant inventory information. You could hire more personnel to monitor every item that comes in and goes out of the store. At the end of the day they would tally up what items were still on hand. Even if you were willing to hire the necessary extra workers, though, it is doubtful that they could provide the instant information you demand. An electronic cash register with computerized inventory control is the product and service you are looking for.

Electronics can't solve all our problems, and, as we know, it can create a few difficulties of its own; but in a great many cases, the electronic solution is the best solution. For the foreseeable future, countless new electronic products and services will need to be designed and produced. Let's take a look at the people who will be doing the creating.

The Designer

Who designs or creates electronic products and services? Are they engineers, designers, technicians, or just plain tinkerers? Actually, designers can be any of the above. You don't have to be an electronics engineer with a degree in electrical engineering to do design. Although it helps to have formal course work, anyone who creates something may qualify as a designer. Technicians, for example, are constantly modifying, improving on, and creating new projects. Because they, of all the electronics personnel, are closest to the actual working device, technicians are often in the best position to see what changes and improvements need to be made, and given their electronics training and experience, they can usually come up with an actual design to solve the problem.

The key to electronics design is taking the **systems approach**. This means knowing about electronic circuitry, the role of input and output transducers, and the fundamentals of good product packaging. It also means understanding and respecting the users' needs. We'll look at these and other design aspects more closely later in the chapter.

For now, it is enough to understand that being an electronics designer does not mean crossing some invisible barrier, some magic line that separates the hobbyist and technician from the college-trained engineer. Given the right attitude (always questioning, wanting to improve things, constantly fiddling and dabbling) and technical experience, you, or anyone else, can do electronics design work.

The Design Methodology

In addition to having the appropriate attitude and technical expertise, you must familiarize yourself with electronics design methodologies. One such design method involves identifying *problem* and *solution categories*. You must also be aware of the difference between a *project* and a *product*. As we will see, "A project does not a product make."

Identifying the Problem Category

Usually, design problems can be put into one of three categories: (1) a need in search of a product or service; (2) a product or service in search of a need; or (3) improvement on an existing product. Obviously, the differences among these three categories are not that well defined. You can't always be sure if you have a problem in search of a solution, or the reverse. Nonetheless, an attempt at classifying each design project into one of these three categories is bound to help you conceptualize the design problem.

In the first instance we often find ourselves saying, "We have this problem. Can electronics help us with the solution?" Illustrations of this type of thinking are endless. Suppose you are constantly "bugged" by mosquitoes at the company's annual picnic. You have tried everything, from gallons of insecticide to gas-powered flyswatters. But what about an ultrasonic electronic oscillator that drives mosquitoes batty? Small battery-powered units that actually claim to do just that are available at your local sporting goods store. Or do you constantly misplace the car keys, searching frantically for them on the mantle, between the couch pillows, or in your jacket pocket? Well, someone has come up with an electronic solution: a small circuit attached to the key chain that lets out a high-pitched audio tone when you clap your hands. Now, retrieving your lost keys is as easy as snapping your fingers.

These, and the other examples mentioned earlier, have one thing in common: first there was a need or problem, then came the solution.

It doesn't always work that way. Often we think of the product or service first and then look to see if there is a real need for it. If there is, does a product or service already exist to satisfy that need? What about a string of small dc lights that "dance" to the music? Such a color organ can be designed and built easily, but has someone already done it? If not, and you create such a product, would anyone buy it? How about an emergency strobe light that campers could carry into the outback? If lost, they could use it to signal search crews. Again, designing and building such a product isn't difficult, but is there a market for it? In these examples, first the product came quickly to mind; the need had yet to be determined.

Sometimes we just want to improve on what we already have. This could be as simple as adding an on-off indicator light to a power supply, or as complex as upgrading an analog (needle-type) tachometer to a digital version. The latter, of course, could be considered a brand-new product, not just an upgrade.

The idea behind identifying a problem category is simply to help formulate the design effort. Once the problem is clear, you can move on to singling out the best type of solution.

Identifying the Solution Category

Regardless of how the problem is categorized, one of three solution approaches is usually taken: (1) we see if the product or service already exists; (2) we see if we can modify an existing item that is close to what we want; or (3) if the first two methods don't work, we design a new product from scratch. Let's briefly examine each procedure.

Why waste time, money, and tremendous design effort if what you want already exists? Why reinvent the wheel? It is, in part, the designer's responsibility to know what's out there and whether a particular design is "common property," to be used freely by all. True, your company may still find it more advantageous to design and make its own version of a particular item, but, at the very least, the option to choose what is already available should be there.

Perhaps what you are looking for already exists, and only a slight modification is all that is needed. Designers spend a great deal of time poring over data sheets of circuit schematics that are very close to what they require. An extra stage of amplification, the addition of a variable speed control, slightly more filtering at the output—make these kinds of small modifications and you could have a custom-designed project or product that exactly meets your needs.

However, if you can't find it or modify it when you do, then starting from scratch is the only solution. Naturally, this approach involves the greatest effort; but at least this way you should get very close to exactly what you want—and, of course, it will be all yours.

A Project Does Not a Product Make

This is the theme, to paraphrase William Blake, of a well-thought-out article, written by Gerard Fonte, that appeared in the January 2003 issue of *Nuts & Volts Magazine*. It is an idea worth exploring.

Though designing and building a prototype is often the first step in developing a mass-produced product, the two are not synonymous. A product, according to Fonte, is something that is reliable, manufacturable, profitable, and salable. A project, in contrast, is something that is self-made for enjoyment or education, or to prove a concept.

When thinking "product," as opposed to project, you must consider such questions as, Is surface mount technology more appropriate than through hole technology? Is reliability more important than cost? Is size a factor? How easy will it be to manufacture? and, Is the production quantity going to be large or small?

Here, according to Fonte, are a few other differences between product and project to keep in mind:

☐ A product must be reliable. It's okay if you have to make continual adjustments on your homemade receiver, but a customer expects a product to work the first time and every time.

☐ A product must have a market. A project has no market.

☐ A project is rarely cost-effective. Most often, the labor cost involved makes the "product" impossible to manufacture.

Thus, to repeat, although with prototyping you may be on your way to a final product, your project is just the first step. As a prototype technician, always keep that in mind.

Something to Think About

Why not talk to a few designers and engineers? Identify a company, contact the engineering department, and request to speak with an electronics engineer or designer, then ask if you can make an appointment to conduct a half-hour "information interview." Believe me, they will be flattered you asked. Who knows where this could lead?

3.2 ELECTRONICS DESIGN: THE SYSTEMS APPROACH

To do electronics design, you must be able to recognize the complete electronic system and understand how it functions. We'll look at the systems approach first. In today's industrial environment you also need to be familiar with the concept of Total Quality Management (TQM). We'll examine TQM, and your role in it, next. It's then on to an exploration of the important two-step design process. We will also find out about some basic sketching techniques useful in the design stage.

The Fundamentals of Electronics Design

Here we explore the fundamentals of electronics design by defining the term *electronic system*. Then we go on to discuss electronic components: their function, description, packaging styles, and selection and acquisition. In addition to providing insight into the design process, this procedure also allows us to review some important principles of electronics.

An Electronic System Defined

Your friend enthusiastically invites you over to see his brand-new 200-W stereo amplifier. You arrive to find it sitting on the dining room table, plugged in, and producing absolutely nothing: no sound, no hum, no nothing! It's a great amplifier, assembled using the latest solid-state circuitry, but by itself it is useless. An electronic circuit is only part of the picture. Again, to paraphrase Blake: "An electronic circuit does not a system make."

The amplifier needs a signal to amplify. It gets that signal from what is called an electrical **input transducer**, a component that changes some form of energy into electricity. In a stereo system that component could be a microphone, tape head, CD player, or tuner. The amplified signal needs somewhere to go—to a component known as an electrical **output transducer**. This component—an earphone or speaker—changes electricity into another form of energy, the reverse of what the input transducer does. Together, the input transducer, electronic circuit, and output transducer make up an electronic system—in this case, a stereo system (Figure 3–1).

All electronic systems first take a form of energy, such as heat, light, or sound (movement of a microphone's diaphragm), and convert it into electricity. The circuit then manipulates the electrical current, by switching or regulating it, and sends it to an output transducer. Here it is converted into another form of energy, one we find useful, such as mechanical motion (the movement of a speaker cone), light, or heat (Figure 3–2).

To understand better this systems approach, let's look at electronic components, which combine to form electronic circuits, which must then be coupled to input and output transducers to make up the electronic system (Figure 3–3).

Electronic Component Function

Electronic components can be classified as either active or passive. **Active** components control electrons by switching or regulating them; **passive** components help the active components carry out these functions.

A giant leap from the electrical to the electronic age occurred in 1906, with the invention of the first active component, the **vacuum tube** (Figure 3–4). (We are talking here about the triode, not the diode, vacuum tube. The diode, although an active

FIGURE 3–1
Electronic stereo system

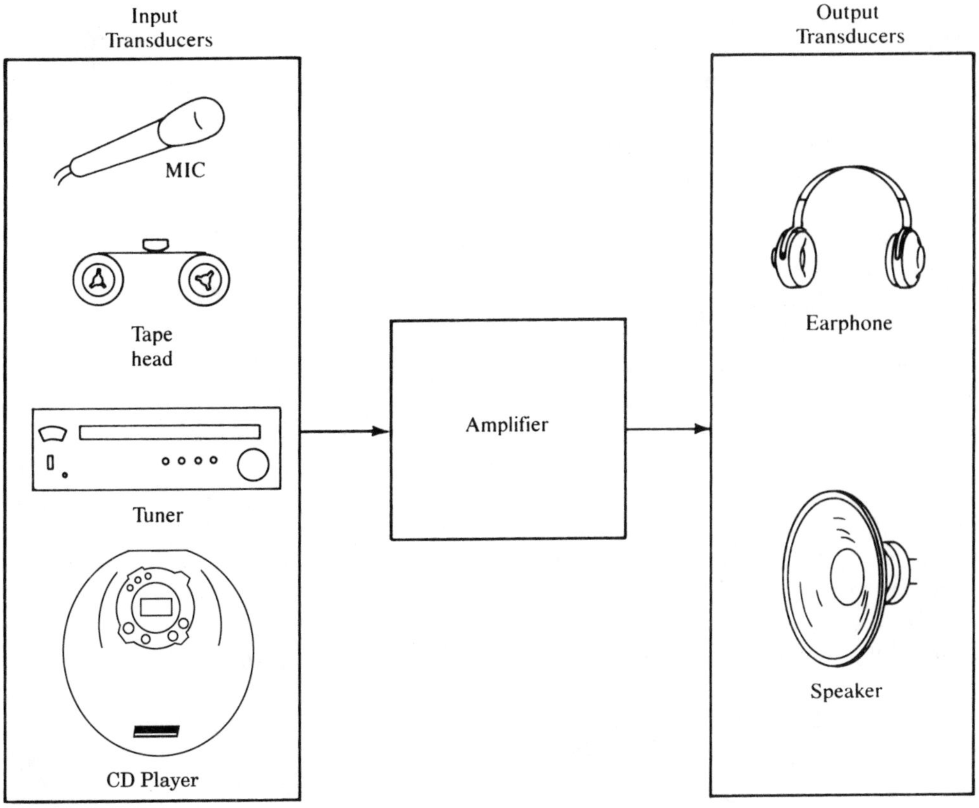

Input
Transducers

MIC

Tape
head

Tuner

CD Player

Amplifier

Output
Transducers

Earphone

Speaker

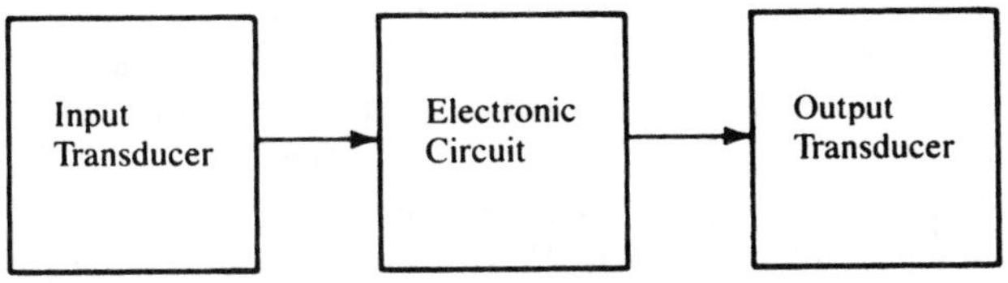

Input
Transducer → Electronic
Circuit → Output
Transducer

FIGURE 3–2
A complete electronic system

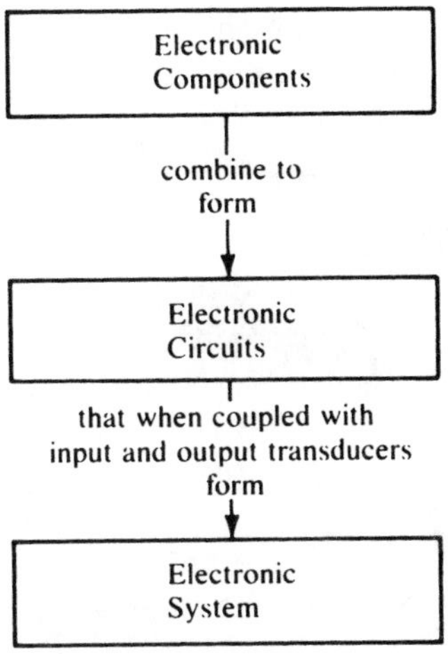

Electronic
Components

combine to
form

Electronic
Circuits

that when coupled with
input and output transducers
form

Electronic
System

FIGURE 3–3
Electronic components, circuits, system

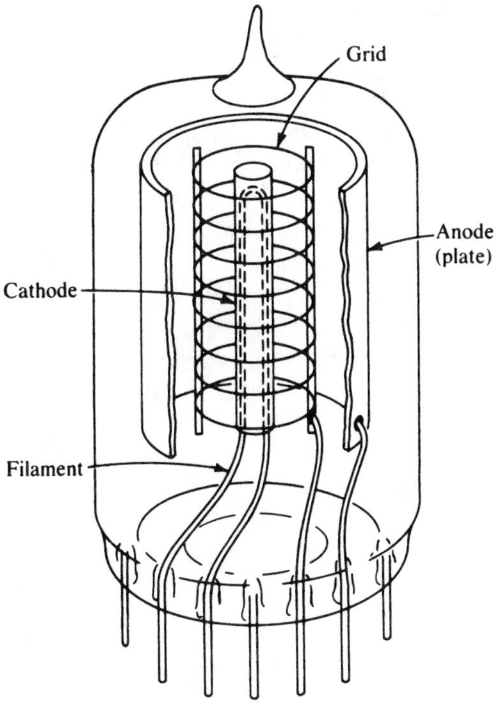

Grid

Anode
(plate)

Cathode

Filament

FIGURE 3–4
Vacuum tube

component, only rectified but did not regulate current flow.) Now, for the first time, electrons were all alone. These minute particles, traveling between the vacuum tube's cathode and plate, could be controlled by a tiny grid of wire with a negative charge placed on it. The result? Switching speeds in the hundreds of thousands of cycles per second and amplification factors on the order of a million or more.

When this first active component was combined in various ways, the electronic computer was born, as were radio, television, and a host of other twentieth-century electronic wonders. Clearly, the vacuum tube is one of the most important inventions of all time.

Amazing as it was, the vacuum tube had its shortcomings. It was relatively large, easily breakable, generally unreliable, hungry for power, and had a limited life expectancy. A more efficient replacement was needed.

In 1947 the **transistor** emerged from the laboratory, as a second-generation active component, to challenge the vacuum tube (Figure 3–5). The new transistor did everything its predecessor did, only better. It was smaller, highly reliable, consumed milliwatts instead of watts of power, and could be counted on to last at least as long as any other component in the circuit. In addition, once mass production opened up, transistor costs plummeted to only pennies per unit, while the average vacuum tube still sold for more than $2.

But even with the transistor, improvements were called for. Using basic transistor technology, engineers at Fairchild developed a tiny component, no bigger than an infant's fingernail, that contained dozens of transistor elements. These transistors, along with various resistors and capacitors fabricated in the same way, were combined to form entire circuits on a single

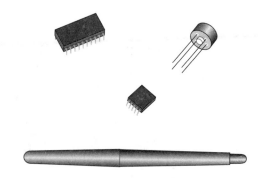

FIGURE 3–6
Integrated circuit

tiny sliver of silicon. The year was 1958, and the **integrated circuit** had arrived (Figure 3–6).

Today, integrated circuits (ICs) containing hundreds of thousands of transistor elements are being mass-produced for just a few dollars. This third-generation active component has all the advantages of discrete (individual) transistors, plus the added benefit of much greater circuit density and, surprisingly, even lower power consumption.

We are not through yet. Although the **microprocessor** is a type of integrated circuit, many feel that this device, developed in 1972, is so radically different from its predecessors that it deserves its own identity as an active component (Figure 3–7). The microprocessor, of which there are now more in use than there are people on Earth, is programmable. It is software that determines its function. This means that the same microprocessor can be used in a microwave oven, a DVD player, an automobile cruise control, or a personal computer, depending on the program that controls it. This component, which is the heart of all computer-based systems, has revolutionized the design of electronic products and services. Now, the electronics designer must know about computer software as well as computer hardware.

The passive components (Figure 3–8)—resistors, capacitors, inductors, and transformers—help the active components manipulate electrical current.

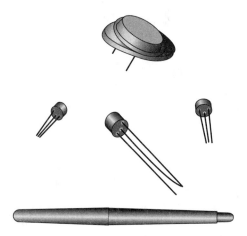

FIGURE 3–5
The transistor, an active component

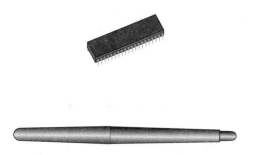

FIGURE 3–7
Microprocessor

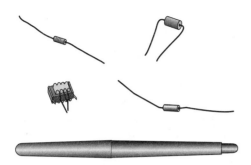

FIGURE 3–8
Passive components

The **resistor** limits current flow and drops voltage. It is probably the most abundant of all components; rarely does an electronic circuit not contain at least one resistor. The **capacitor** acts as a temporary battery to store an electrical charge. It is used in timing circuits, tuners, and filters. The **inductor** (coil), which stores electrical energy in a magnetic field, acts as a choke (a device used to limit or suppress the flow of alternating current) and, along with a capacitor, as a tuner. The **transformer** steps up or down incoming voltage. It is found in power supplies and as an energy-matching component in amplifiers and oscillators.

Electronic Component Description

Let's take a moment to examine more closely the most widely used active and passive electronic components. These components are found extensively in the projects discussed in Part III.

Diodes. A diode is a two-element (cathode and anode) component that allows current to flow through it in one direction only. The schematic symbol and pictorial presentation for a typical silicon diode are shown in Figure 3–9a. The diode conducts current when forward-biased (its cathode negative, its anode positive). It blocks current flow when reversed-biased (its cathode positive, its anode negative). Thus, the diode acts as a one-way valve. It is used primarily for switching, although it also operates as a rectifier to change ac to dc.

Diodes are rated with regard to current- and voltage-handling ability. For example, a 1N4001 diode is rated at 1 A/50 V. That is, it will carry *up to* 1 ampere while handling *up to* 50 volts. The question is, Could such a diode work in a circuit requiring only 500 mA at 25 V? Yes, it could.

As noted in Figure 3–9a, the cathode end of a diode is identified by an atypical marking, usually a band, a chamfer, or "-" symbol.

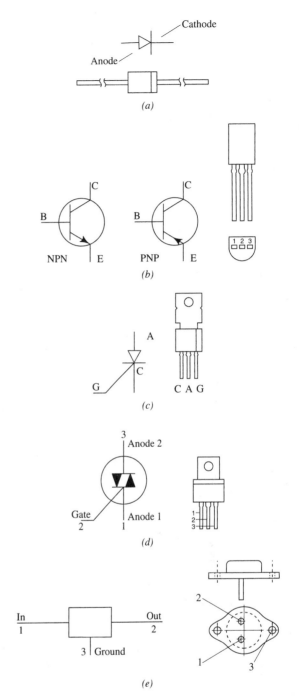

FIGURE 3–9
Active electronic components

Transistors. A transistor is a semiconductor component that can both switch and amplify. It consists of three terminals: emitter (E), base (B), and collector (C). There are two configurations, NPN and PNP. There are also two broad types, classified according to current-handling ability: general-purpose and power transistors. The schematic symbols for NPN and PNP transistors, along with a typical packaging style, are shown in Figure 3–9b.

A transistor is operated as an amplifier or as a switch. When a transistor is used as an amplifier, a small varying signal placed on the base causes a much larger varying signal to flow from collector to emitter. When a transistor is operated as a switch, it is either at cutoff (no collector–emitter current is allowed to flow) or saturation (maximum collector–emitter current flows).

The transistor as an amplifier finds extensive use in audio- and radio-frequency equipment. As a switch it is used to activate and deactivate a wide variety of circuit loads.

Silicon Controlled Rectifiers (SCRs). The SCR is a three-terminal solid-state latching switch. It conducts current from the cathode to the anode when the gate receives a positive voltage. Once the conduction takes place, it continues even when the control voltage to the gate is removed. The SCR ceases conducting only when the anode voltage is removed, reduced, or reversed. Thus, we have a latching switch, designed to operate primarily in a dc circuit.

In Figure 3–9c we see the schematic symbol and typical packaging for an SCR. SCRs are rated in current- and voltage-handling ability. For example, a 1-A/200-V SCR conducts *up to* 1 ampere while handling *up to* 200 volts. Will such an SCR work in a circuit delivering 500 mA at 6 V? Yes, of course. SCRs are used extensively in burglar alarms, color organs, strobe lights, and industrial control applications where dc is involved—anywhere, that is, where a solid-state latch is required.

Triacs. A triac is a back-to-back SCR (Figure 3–9d). It functions as an electrically controlled switch for ac loads. Like an SCR, the triac is rated in terms of current- and voltage-handling ability. The former ranges from 1 to over 40 A; the latter, from 200 to 600 V. Triacs are used to control high-current, high-voltage ac loads. In dc circuits we use the SCR; in ac circuits, where current and voltage tend to be higher, the triac is the choice.

Voltage Regulators. A voltage regulator is a circuit that holds an output voltage at a predetermined value regardless of normal input voltage changes or changes in the load impedance. There are fixed and variable voltage regulators. The former have a fixed output, such as 5.0, 6.0, 10.0, and 18.0 V. The latter hold their output voltage steady within a predetermined range. For example, the LM317 is a 1.2–17-V regulator. When set to any voltage within that range (usually with a potentiometer), it holds the output at the selected voltage.

Many voltage regulators come in a three-terminal package. They have only one input terminal, one output terminal, and one ground terminal, as shown in Figure 3–9e. Such regulators are chosen primarily to satisfy the voltage and current requirements of the load. For example, the LM7805 is a fixed 5-V/1-A voltage regulator. It holds the output voltage at 5 V while the load draws up to 1 A. If the load exceeds 1 A, the voltage regulator automatically shuts down. It resumes operation when the load requirement is reduced below 1 A.

Voltage regulators are a must for any dc power supply. Considering their simplicity and low cost (in many cases under $1), there is no reason why they shouldn't be designed into every power-supply project.

Resistors. A resistor is a passive electronic component that acts to oppose current flow and drop voltage. The schematic symbol and physical characteristics of a fixed carbon-composition resistor (the most popular type) are shown in Figure 3–10a. Note the bands, which we will discuss in a moment.

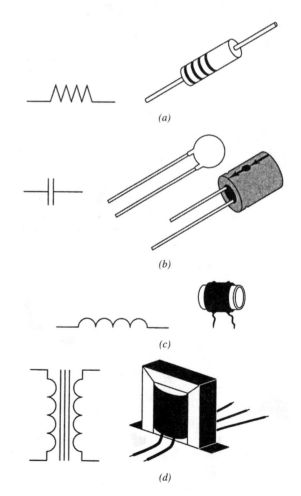

(a)

(b)

(c)

(d)

FIGURE 3–10
Passive electronic components

Resistors have two ratings. They have a resistance value, measured in *ohms* (Ω), and a heat-dissipation factor, measured in *watts*. The ohm is just a measurement of resistance, or opposition, to current flow. The more ohms, the more resistance. Standard carbon-composition resistors range in value from less than 1 ohm to over 100 million ohms.

A resistor's wattage rating is a measurement of how much heat it will get rid of. The larger its physical size, the greater its wattage rating. Common ratings are $\frac{1}{4}$, $\frac{1}{2}$, 1, and 2 W.

Carbon resistors use a color code to display their value in ohms. Four colored bands surround the resistor body. (See Appendix H for a discussion of five-band resistor color codes.) Each color is assigned a numeric value, as shown in Figure 3–11. Black is 0; brown, 1; red, 2; and so on. The first three bands (the first, or number one, band being closest to the edge) give the value in ohms, the fourth band the tolerance in percent of nominal value. Furthermore, the first two bands represent numerals, while the third band designates the number of zeros to be added to the numeric value.

It's really not that complicated. For example, a resistor color-coded *brown*, *red*, *orange*, *silver* would have an ohmic value of 12,000 ohms and a tolerance of plus or minus ($\pm$) 10%, meaning the resistor could be plus or minus 10% of the numeric value of 12,000 ohms. Its actual value could fall between 10,800 and 13,200 ohms. (For a more in-depth discussion of the resistor color code, consult a standard dc/ac textbook such as Thomas Floyd's *Electronics Fundamentals*, published by Prentice Hall Career and Technology.)

Resistors find application in virtually every circuit. They are the most popular of all electronic components, active or passive.

Capacitors. A capacitor is a passive component that stores electricity. It consists of two metal plates separated by a dielectric, or insulator. There are two types of fixed capacitors: polarized and nonpolarized. The former have a positive and a negative lead. Also known as electrolytic capacitors, polarized capacitors must be installed in a circuit in the correct direction. Nonpolarized capacitors can be installed in either direction.

FIGURE 3–11

Resistor color code

Resistor value, first three bands:

First band—first digit
Second band—second digit
Third band—multiplier (number of zeros following the second digit)

> *Note*: For resistance values less than 10 ohms, the third band is either gold or silver. Gold indicates a multiplier of 0.1, and silver indicates a multiplier of 0.01.

0	black
1	brown
2	red
3	orange
4	yellow
5	green
6	blue
7	violet
8	gray
9	white

Fourth band—tolerance

+/– 5%	gold
+/– 10%	silver

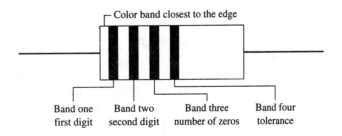

Color band closest to the edge

Band one — first digit
Band two — second digit
Band three — number of zeros
Band four — tolerance

The schematic symbols and body styles for two types of capacitors are shown in Figure 3–10b. Note that the positive lead of an electrolytic capacitor is identified by physical characteristics in one of two ways: In radial leads (those protruding from the same side), the positive lead is the longer of the two. In axial leads (those extending like the axle of an automobile), the positive lead emerges from the side where the cylinder indents.

Capacitor values are given in microfarads (μF) and picofarads (pF). Values up to 1 μF are nonpolarized; beyond 1 μF the capacitor is almost always polarized. Capacitors have a voltage rating, too, for example, 47 μF/50 V. This capacitor should not be placed in a circuit where more than 50 V would be impressed across its leads.

Capacitors are used as filters in power supplies, in tuning circuits for radios, and as part of resistance/capacitance (R/C) networks in timing circuits.

Inductors. An inductor, or coil, stores energy in a magnetic field. It consists of a coil of wire wound on an air or iron core. The schematic symbol and case styles for a typical coil are shown in Figure 3–10c. Inductors are nonpolarized; that is, they can be installed in either direction within the circuit. Inductor values are given in millihenries (mH) or microhenries (μH). They have no voltage rating. Coils are used in motors, antennas, tuners, relays, and solenoids.

Transformers. A transformer looks very much like a coil. That is because a transformer consists of two or more windings wrapped around a common core. It's really just two or more coils using the same core.

A transformer steps ac voltage up or down. Its primary coil receives the incoming voltage. Through magnetic induction, the voltage is induced into the secondary winding. If the secondary winding has more turns than the primary, the incoming voltage is stepped up. If the number of windings in the secondary coil is less than that of the primary, the incoming voltage is reduced.

Transformers are classified as power, audio, and radio frequency (RF). The former are the largest, the latter the smallest. The schematic symbol and physical characteristics of a typical transformer are shown in Figure 3–10d. Transformers are used in power supplies, audio and video equipment, and, in general, to isolate one ac circuit from another.

Component Packaging Styles

Most passive components, such as resistors, capacitors, and inductors, are identified as having either axial or radial leads (see Figure 3–8). If such components have polarity (positive and negative terminals), the positive terminal is usually specified with a longer lead, or the body of the component has a polarity marking clearly shown.

Leads for active components, integrated circuits, and components in the transistor family (diodes, diacs, SCRs, triacs, and so on) are identified differently. They are determined according to a packaging style that is usually associated with a specific lead configuration (see Appendix D), yet you can't always be sure that these configurations will match the particular components you are working with. To play it safe, look up the packaging style and lead pattern in Appendix D or in the **data sheet** for the appropriate component. This is the only way you can be absolutely sure of lead placement.

Electronic Component Selection and Acquisition

As an electronics technician, particularly one involved in prototype design and fabrication, you will be concerned with the selection and acquisition of electronic components, not just for the prototype, but for the mass-produced product should the prototype prove successful. With a prototype, it's simply not enough to get the circuit up and running. You must do so with electronic components and hardware that will be readily available in large quantities and at reasonable prices. In other words, the prototype technician must look ahead to the final, mass-produced product in determining component selection and acquisition.

The first step in learning to make such decisions is to gain as much knowledge as possible about electronic components: what's available, from what vendors, delivery schedules, and, of course, pricing. When you are out in industry, three sources will be of primary help to you: (1) your company's purchasing agent, (2) parts catalogs such as the EEM (Electronic Engineers Master catalog), and (3) the Internet.

When you are on the job, make it a point to get to know your company's purchasing agent. If you can, get on a first-name basis with that person. A serendipitous relationship may well develop: you will be helping each other do a better job.

Start to acquire now, while you are still in school, data manuals, line cards, and parts catalogs that will provide a valuable resource in selecting components and vendors. Such items are usually available free to those in the field. Call or write to suppliers today.

When searching the Internet for electronic components, you can take one of three approaches.

First, using your Internet provider's browser, simply name the component you want, say a 7402 IC. The trouble is, if you don't narrow the search by clearly delineating the name, you can wind up with thousands of results. A search on just the numbers 7402 turned up over 10,000 choices, including "Minnesota Rules Chapter 7402."

Second, you can simply use your preferred vendors' Web sites.

Third, you can use one of the many component search engines, such as PartMiner. Such services will list the components you are looking for by providing the part number, description, manufacturer, price, in-stock notice, quantity, and supplier. You then simply order on-line. With this approach, you have a choice of vendors and prices, both at one Internet location.

When selecting electronic components, quality and quantity must be uppermost in your mind. Although you do not want to "overdesign" (choose a Cadillac when a Chevy will do), selecting high-reliability components wherever possible can pay off and actually save money down the line. You are ahead if quality components reduce the demand for incoming inspection, if rejects are minimized, and if product failure out in the field is lessened. You may pay a little more on the front end for quality components, but in the long run, you'll often save money.

The ability to purchase in quantity doesn't just refer to the willingness of a given vendor to continue to supply parts at good prices in quantities you want when you want them—though that's important. It has more to do with the critical issue of *second sourcing*. Virtually all electronic components are "second-sourced," meaning they are available from more than one OEM (original equipment manufacturer). OEMs actually encourage this practice and often license the production of components they have designed (such as microprocessors and memory chips) to rival OEMs. They know, as you should, that manufacturers of electronic products are not going to "put all their eggs in one basket." For production houses to rely solely on one source for any item in their product lines puts them in a very vulnerable position should the source decide to discontinue the item or, worse, go out of business. The rule of thumb is simple: Never purchase any electronic component that is not second-sourced—period!

When you have selected components that might be used in your prototype, you or your company's purchasing agent can call the distributor and ask for samples. As a student, you can do the same. Don't be greedy. Ask only for what you need and be direct about who you are and why you require the item. If you are polite and positive, you'll usually get what you ask for.

What if vendors say they don't have the exact component you are looking for, but an "equivalent" substitute? Now, of course, you need a sample more than ever. When you get that sample, be sure it is not only electrically equivalent but physically the same as well.

Electronic Circuits

Active and passive components combine to make up **electronic circuits**. A schematic drawing for a three-stage audio amplifier circuit is shown in Figure 3–12. Notice the five active components—transistors Q_1–Q_5—and the passive components—resistors and capacitors—required to support them.

FIGURE 3–12
Three-stage audio amplifier circuit

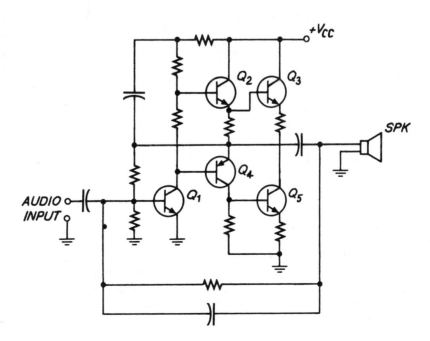

FIGURE 3–13
Digital burglar alarm

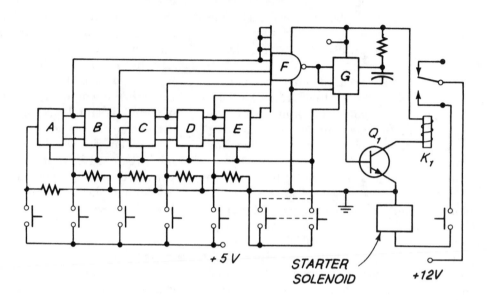

* G IS A MONOSTABLE MULTIVIBRATOR IC.

* GATE F IS AN 8 - INPUT NAND GATE

* A-E ARE J-K FLIP FLOPS

Though it may seem that there are hundreds, even thousands, of different types of circuits, such is not the case. All electronic circuits (just like the active components) can be grouped into two broad classifications: digital and analog circuits. **Digital** circuits switch currents on or off very rapidly; **analog** circuits use a small current to control a much larger current. All circuits, whether they are buried in a sophisticated radar system or are part of a household light dimmer, are analog or digital or a combination of both.

When we think of analog circuits, we imagine amplifiers and oscillators. Such circuits form the heart of practically all audio and video communications equipment. Chances are that your stereo, telephone, radio, and television systems are predominantly analog in design. Although the trend is clearly toward digital circuits, analog circuitry, and the designers who are familiar with it, will be needed for a long time to come.

When we turn to digital circuitry, it is computers that come quickly to mind. Virtually all the computers that you are ever likely to work with are digital, not analog. Remember, also, that if an electronic product uses a microprocessor, the device is essentially computer-based. Furthermore, with billions of microprocessors in the world, that means a large number of digital circuits to design and repair.

Computers, however, are not the only equipment to use digital circuitry. Many timers, counters,

and industrial control circuits can be built with a handful of inexpensive integrated circuit gates (switching circuits combined to make logic decisions). Figure 3–13 shows a simple burglar alarm circuit. Notice that although all the active components are digital, the circuit contains no microprocessor.

Electronic Systems

As discussed, adding input and output transducers to an electronic circuit creates an **electronic system**. Some people have found out about the need to fashion such a system the hard way. Many unsophisticated computer purchasers, lured by the "too good to be true" advertisements, bought computers at unbelievably low prices. When they got the computer home, however, and plugged it in, just like the amplifier mentioned earlier, it didn't do a thing. They forgot about the input and output transducers, known in computerese as input and output peripherals. Without a keyboard, light pen, joystick or mouse at the input and a video display unit, printer, or plotter at the output, the computer circuitry has no way of blossoming into a computer system (Figure 3–14).

Electronics designers spend a great deal of time thinking about transducers. Many of the basic circuit building blocks already exist, but determining how to sense a particular light or heat source, knowing how to actuate various valves and

FIGURE 3–14
Computer system

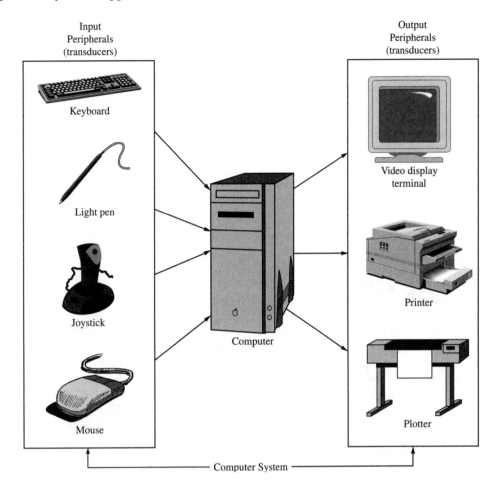

Input Peripherals (transducers)
Keyboard
Light pen
Joystick
Mouse

Computer

Output Peripherals (transducers)
Video display terminal
Printer
Plotter

Computer System

plungers—in other words, interfacing electronic circuitry to the outside world—that's where the real challenge often lies. For you as an electronics designer, knowing the makeup of a complete system is critical—and that means knowing about input and output transducers as well as digital and analog circuitry.

In this section we have emphasized the importance the designer must place on understanding the electronic system. Whenever you design an electronic product or service, think of how whatever it is you are doing fits into the complete system. Even though your design may seem like a small part of a much bigger picture, you will want to know where you are in that larger canvas.

Total Quality Management (TQM) and the Design Process

TQM is one of the hottest buzzwords in corporate America today. It's an umbrella term for the quality programs that have spread throughout U.S. industry in the past two decades. It developed in response to the Japanese emphasis on superior-quality products.

Total Quality Management started with large companies but has spread to smaller ones, as the former have insisted that the latter, who often act as suppliers to their larger kin, adopt quality programs of their own. Let's see just what TQM is and how it fits into electronic design and the systems approach.

The What and Why of TQM

Total Quality Management represents, as its name indicates, a company's total commitment to quality. TQM is essentially customer- and employee-driven. What does this mean? First, there is an intense focus on customer satisfaction. Customers, however, are internal as well as external. An assembler may be the "customer" of a technician who builds a prototype, for instance, just as the person who buys the final, mass-produced product is the customer of everyone in the company.

Second, TQM involves a new work relationship based on trust and teamwork. The idea centers around the term *empowerment*, which means that management gives employees wide latitude in how they go about achieving the company's goals. By getting employees involved at all levels from design to shipping, companies hope that a shift will occur from catching and correcting defects at the end of the process, to monitoring the process itself so that defects do not occur. Such an approach, the essence of TQM, represents a real change in philosophy for many U.S. companies.

The Electronics Technician's Role in TQM

Improving quality involves much more than beefing up the quality assurance (QA) department, increasing inspection and rejection at the back end—it's really about designing quality in at the front end. You'll come across this concept again and again out in industry. Obviously you, as an electronics technician dealing with prototype design and fabrication, are going to be a key player in TQM implementation, and the emphasis on "player" is significant. TQM makes considerable use of concepts such as the team method and quality circles. You might say it is the systems approach applied to people and organizations, not just an electronic circuit sandwiched between input and output transducers. In sum, you're going to be living and working with TQM for many years to come. Embrace it and find out all you can about it now, while you're still in school. It will be waiting for you out in the electronics workplace.

The Two-Step Design Process

The project design process involves two steps. We need to produce a **Concepts and Requirements Document** and a set of **design drawings** (Figure 1–3). The Concepts and Requirements Document, in addition to stating the project concept, spells out just what is to be done, by whom, and when. The design drawings consist of a system diagram, circuit design sketch (circuit schematic), and packaging plan. Let's examine both items in turn.

Concepts and Requirements Document

The Concepts and Requirements Document doesn't have to be a lengthy thesis; one or two typewritten pages will do. Its purpose is to state clearly why the project should be undertaken, what are the basic design requirements, how those requirements are to be met, and who is to do exactly what and when. First, a goals declaration is made, followed by a list of objectives, a clarification of responsibility statement, and a short discussion of the theory of operation. Here's what we mean:

Project Goals. The goals declaration simply states what the project is, why it is needed, and why it is better to build it rather than obtain it commercially. For example, suppose you want to solve the problem of break-ins in the electronics laboratory—an illustration of a need-in-search-of-a-product problem category. You conclude that by installing a burglar alarm system the incidences of theft will be reduced. The project, then, is a burglar alarm, and its purpose is to reduce break-ins. Also, because nothing is available commercially that will meet the company's particular needs, you are proposing that the company build the burglar alarm.

Project Objectives. First, you state the design requirements (objectives). For example, the system should be reliable, low in cost, safe to operate, have an entry delay, produce a loud siren sound when triggered, and protect all doors and windows. Next, you specify how those particular requirements, or objectives, will be met. Here you could say that reliability will be achieved by using solid-state components; low cost, by building the circuit on a PC board; safe operation, with the use of batteries; and so on. It's basically a question of stating specific design criteria and how they are to be met.

Project Responsibilities. It is very important, early on, to find out who is going to do what and set deadlines for the completion of the various tasks: Who will provide the design sketches? Who is going to make the complete drawing set? Who will build the prototype? Who will test and troubleshoot the project? and, of course, who will be responsible for the documentation? Adjustments can and will be made as the project progresses, but general commitments in terms of personnel and facilities need to be made now, in the Concepts and Requirements Document.

Theory of Operation. A brief explanation of how the circuit works should be included. Although marketing personnel, and even some production people, may wish to skip this section, it is useful for all those involved in the design and prototyping phases.

After the Concepts and Requirements Document is accepted, it is signed off by the appropriate personnel and appears later as part of the Project Report. We'll look more closely at the Concepts and Requirements Document in the next section, where the Sample Project, the Variable Power Supply, is also examined.

Design Drawings

There are three design drawings: a system diagram, a circuit design sketch, and a packaging plan. The **system diagram** is a block diagram illustrating the functional units that make up the system. Each

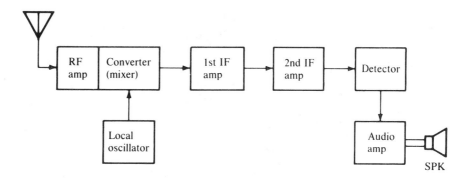

FIGURE 3–15
System diagram of an AM radio

transducer is represented by its own block, as are individual circuits within the complete system. A system diagram for a standard AM transistor radio receiver is shown in Figure 3–15. The diagram is often accompanied by a brief functional statement, the purpose of which is to explain in simple terms what each block does.

The **circuit design sketch** is the rough schematic drawing (drawn with traditional drafting tools or with CAD) of all circuits, along with input and output transducers. All components need to be identified and their values stated, either on the drawing or in a separate parts list. A circuit design sketch for the AM transistor radio is shown in Figure 3–16.

A **packaging plan** is a sketch showing what the final project might look like. Enclosures should be chosen with aesthetics, functionality, and safety in mind. The tentative locations of panel lights, knobs, dials, jacks, and meters are identified. The type of finish to be applied must also be stated. A packaging plan for the AM transistor radio is shown in Figure 3–17.

The two-step design process, consisting of a Concepts and Requirements Document as well as a three-part design drawing set, provides the necessary information to determine if the project is worth doing. If it is a "go," the proper personnel sign off, and we move on to producing a working schematic drawing. Before we examine that procedure (in Chapter 4), let's explore some basic sketching techniques and take a look at the actual two-step design process as it applies to the Sample and Exercise Projects.

Sketching Techniques

Sketching is an important skill that every technical person should develop. The ability to get your thoughts down on paper quickly, in graphic form, for someone to evaluate is an important asset. Let's examine a few techniques that will help you do just that. (Though you may choose to complete your Systems Diagram and Packaging Plan using CAD,

this section explores traditional methods of sketching—techniques valuable for anyone to know.)

Sketching System Diagrams

Sketching a system diagram (Figure 3–15) is really quite simple. Just be sure to draw a solid block for functional units that are part of the complete system and a broken block (dotted lines) for functional units that are part of the system but not the project you will be building. (See the system diagram for the Variable Power Supply Project in Part III.) Also, remember to include the lines, with arrows, that indicate signal flow between functional units.

Sketching Schematic Drawings

A schematic drawing consists of electronic component symbols with interconnecting conductors. These components are identified and their lead and pin numbers indicated.

A schematic sketch differs from a working schematic drawing (see Chapter 4) primarily in two ways. One, the sketch is often incomplete in detail. Integrated circuit pin numbers are usually omitted, supply voltages are "understood," and even the identity of major components may be missing (Figure 3–16). The assumption, made by the sketcher (most often the designer), is that someone else (most often the technician) will look up this missing information when the working schematic is produced.

The second way in which the sketch differs from the working schematic is that in the sketch, little if any attention is paid to such things as component layout, line work, and scaled lettering. The idea is to put down the basic circuit as quickly as possible.

Nonetheless, the schematic sketch is no excuse for sloppiness and unorganized work. The sketch must be understandable and correct. It is one thing to leave IC pin numbers off the sketch, to be looked up at a later time, but quite another to put the wrong pin numbers on the drawing.

Keep this in mind when you are called on to produce a quick schematic sketch. Because most

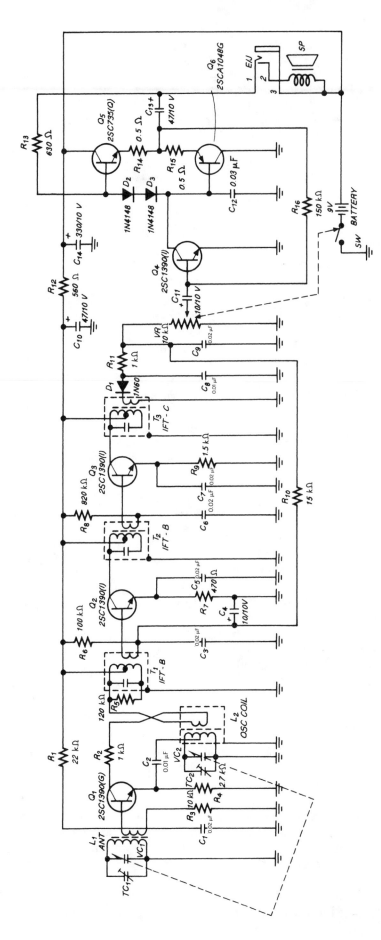

FIGURE 3–16
Circuit design sketch of an AM radio

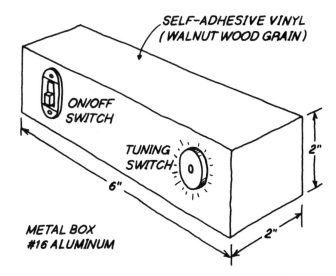

FIGURE 3–17
Packaging plan for an AM radio

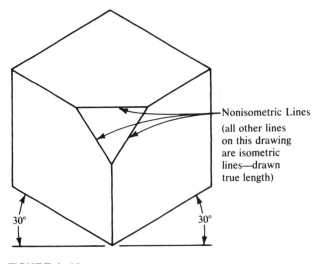

FIGURE 3–18
Isometric drawing

likely you will be sketching your schematic from one that is already drawn in books or magazines, your job will be fairly easy. Work in a logical progression, usually from left to right, and try not to forget any important details.

Sketching Pictorial Drawings

In order to sketch chassis, enclosures, electronic components, and the like, you need a basic knowledge of pictorial drawing (Figure 3–17). This often means drawing in **isometric**. In such an approach, objects are usually sketched by first developing an **isometric cube** and then "cutting the cube away" to get the final shape you want.

Isometric means "equal measure." In an isometric drawing, the three common planes showing height, width, and depth are drawn true length. On an isometric drawing, vertical lines remain vertical, but horizontal lines recede at 30° angles (Figure 3–18). Any nonisometric lines (those that are not vertical or in the horizontal plane) are drawn between isometric points.

To sketch a drawing like the one in Figure 4–1 using the cutaway method, you might proceed as shown in Figure 3–19. Start with the basic cube, drawn in isometric, and then in successive steps hack away at it until the final shape emerges. As the drawing progresses, you will probably need to erase unwanted guidelines. When the final shape has emerged, you can darken the remaining object lines.

Something to Think About

Searching the Internet is a great way to determine electronic component pricing and availability for your project. Why not pick a project now, then surf the Internet to find the best deals on what you will need?

3.3 ELECTRONICS DESIGN: TWO PRACTICAL PROJECTS

In this section we look at the two-step design process as it pertains to the Sample Project (the Variable Power Supply) and the Exercise Project (the 3-Channel Color Organ). For both projects, we present the complete Concepts and Requirements Document, as well as the three-part design drawing set.

Designing the Variable Power Supply

Here the Concepts and Requirements Document and design drawings—the two-step design process—for the Variable Power Supply are discussed.

Concepts and Requirements Document for the Variable Power Supply

The complete Concepts and Requirements Document for the Variable Power Supply is presented in Figure 3–20. The *project goals* state what is to be built, why it is a desirable project, and,

FIGURE 3–19
"Cutting away" an
isometric cube

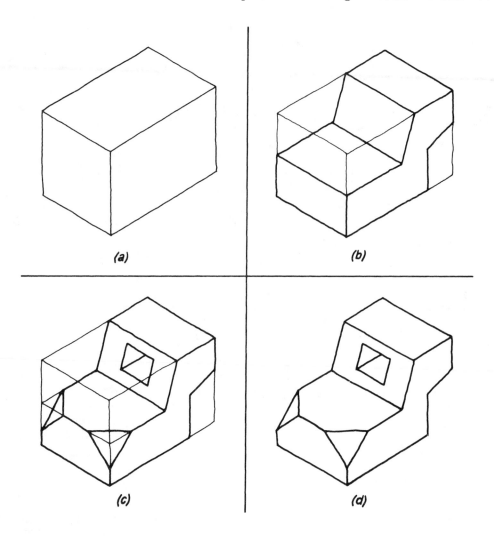

(a)

(b)

(c)

(d)

in this case, what is to be gained by having electronics students build the device. In industry, of course, having engineers and technicians work on a project just to gain experience may not be a reasonable justification for proceeding. Nonetheless, it is important to realize that the experience factor, even in the industrial setting, is not to be discounted.

Project objectives require achieving optimum benefits with regard to cost, reliability, safety, appearance, and ease of use. How these objectives are to be realized is also stated, in general terms.

Project responsibility rests solely with you, the student. You must do it all. Again, in industry, this would rarely be the case. Designers, technicians, and technical writers would be involved, and tentative due dates would be assigned; however, because this is also to be a complete learning experience, you'll carry forward every phase of project design and construction, under your instructor's supervision.

Project theory of operation, also shown in Figure 3–20, is discussed below, after the design drawings are examined.

Design Drawings for the Variable Power Supply

The system diagram for the Variable Power Supply is shown in Figure 3–21. Notice something very interesting: the power supply project itself is not a complete electronic system, only the circuit portion of it. A power supply converts alternating current (ac) into direct current (dc). It is an all-electronic device, taking one form of electricity at its input (ac) and supplying another form at its output (dc). In that sense it is similar to an amplifier, or any other electronic circuit, that "only" manipulates electrical current.

Nonetheless, the power supply is part of an electrical system. The input transducer is the generator at the power plant supplying ac to the wall outlet. The generator changes mechanical energy into electricity. The output transducer is the load connected to the dc output terminals of the power supply. Eventually, that load, be it a radio, television, computer, or alarm, will change electricity into another form of energy. Thus, there is indeed a complete electronic system.

Figure 3–22 shows the **circuit design sketch**, or rough schematic, for the Variable Power Supply. This is a sketch that designers, technicians, and other technical personnel use to evaluate how the power supply works. In this instance, note that the drawing is quite complete. After all, the project is a simple one and the circuit design is not hard to arrive at.

The **packaging plan** for the Sample Project is illustrated in Figure 3–23. It shows the proposed size and shape of the sheet metal enclosure, sug-

gested finishing treatment, and the locations of the on-off switch, panel light, voltage terminal posts, variable voltage control potentiometer, and dial face. Remember, this is not a final drawing, merely a first proposal showing how the complete prototype might look.

Variable Power Supply: Theory of Operation

As noted, a power supply converts ac to dc. All four active components (vacuum tubes, transistors, integrated circuits, and microprocessors) require dc.

FIGURE 3–20
Concepts and
Requirements Document:
Variable Power Supply

Concepts and Requirements Document
Variable Power Supply
by
Austin Babayan

Project Goals

This Concepts and Requirements Document proposes that students in Electronics 150 design and build a working prototype for a Variable Power Supply that may later be mass-produced. Electronics students need an inexpensive 0- to 15-V dc power supply to power such projects as transistor radios, burglar alarms, digital counters, and a host of solid-state-based experiments. A search has failed to uncover a low-cost commercial power supply of the type required. Furthermore, undertaking such a project will provide electronics students with a valuable electronics design and fabrication experience.

Project Objectives

The project is to have the following objectives: inexpensive to produce (less than $25), reliable, safe, attractive, and easy to use. These objectives will be met in the following ways:

* **Cost.** By using noncritical, readily available hardware and electronic components, bought in bulk quantities, the cost will be kept low.

* **Reliability.** By using solid-state active components and printed circuit board construction techniques, a high degree of reliability will be obtained.

* **Safety.** Use of an internal fuse, thermal shutdown voltage regulator, insulated output terminals, and a completely sealed metal enclosure, will assure safe operation.

3

FIGURE 3–20 (cont'd)
Concepts and
Requirements Document:
Variable Power Supply

* **Appearance.** A metal enclosure, either painted or covered with self-adhesive vinyl paper, will ensure an attractive project.

* **Ease of Use.** With the use of an on-off switch, a front panel-mounted LED indicator, and a panel-numbered voltage indicator, the project will have a high degree of functionality and be easy to use.

Project Responsibility

Because this is to be a learning experience for the electronics student, he or she will be responsible for the design (which, in this case, is being supplied), drawings, experimenting, prototyping, and testing, troubleshooting, and final documentation phases of the Variable Power Supply project. Due dates for each phase will be assigned by the electronics instructor.

Project Theory of Operation

The Variable Power Supply changes the 120 V ac from the wall outlet to a varying dc of 1.2 to 15 V. The ac line cord couples, via switch S_1 and internal fuse F_1, the 120 V ac to the primary of transformer T_1. The transformer steps down the 120 V ac to approximately 15.5 V ac. Diodes D_1–D_4 form a full-wave bridge to change (rectify) the ac to pulsating direct current (pdc). Light-emitting diode D_5 is used as a power "on" indicator; resistor R_1 limits current to the LED. Capacitor C_1 filters the pdc to smooth dc. Capacitor C_2 provides high-frequency bypassing. The variable voltage regulator U_1 sets its output voltage by a control loop consisting of resistor R_2 and potentiometer R_3. The output voltage can be varied over the 1.2- to 15-V range by adjusting the potentiometer. Finally, capacitor C_3 acts to swamp any ringing effect developed by the control loop and thereby ensures stability of the voltage regulator's output.

Electronics instructor's approval *Barb Jones* Date *4–11*

4

Some of the circuits using these components get their dc from batteries, but most will plug into a wall outlet, which, of course, supplies ac. The devices employing these circuits, such as televisions or stereos, will have power supplies built in as just one more circuit. There is, however, a need for what are known as "bench power supplies": separate units designed for use in experimenting, testing, and troubleshooting dc-powered projects. The Variable Power Supply fits into this category.

Specifically, the Variable Power Supply takes the 120 V ac at its input and supplies 1.2 to 15 V dc (at up to 1 A) at the output. The supply is fully regulated, which means it will maintain the set voltage even though load current requirements may fluctuate. It is also overload-protected. In case of a short circuit across the output terminals, the supply will shut down rather than burn out.

Let's look at how the circuit works. Figure 3–24 shows the project schematic, along with the appropriate voltage waveforms at important points. The ac line cord couples, via switch S_1, the 120 V ac to the primary of transformer T_1. The transformer steps down the 120 V ac to approximately 15.5 V ac. Diodes

FIGURE 3–21
System diagram: Variable
Power Supply

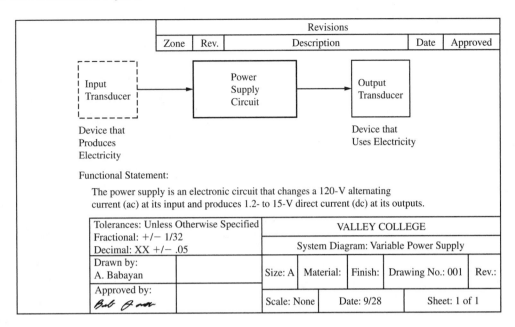

FIGURE 3–22
Circuit design sketch:
Variable Power Supply

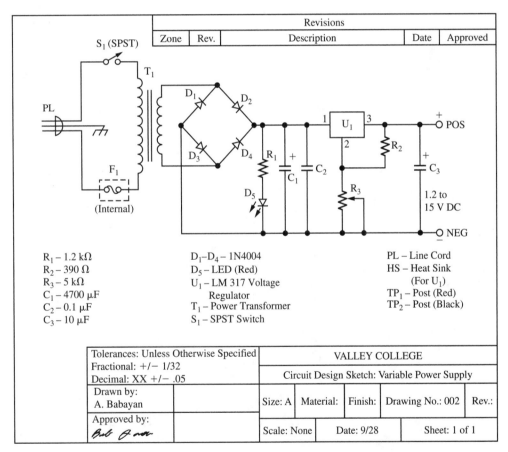

D_1–D_4 form a full-wave bridge to change (rectify) the ac to pulsating direct current (pdc). The light-emitting diode D_5, along with its current limiting resistor R_1, is used as a power "on" indicator. Capacitor C_1 filters the pdc to smooth dc. Capacitor C_2 provides high-frequency bypassing. The variable voltage regulator U_1 sets its output voltage by a control loop consisting of resistors R_2 and R_3. The output voltage can be varied over the 1.2- to 15-V range by adjusting potentiometer R_3. Finally, capacitor C_3 acts to swamp any ringing effect developed by the control loop and thereby ensures stability of the voltage regulator's output.

FIGURE 3–23
Packaging plan: Variable
Power Supply

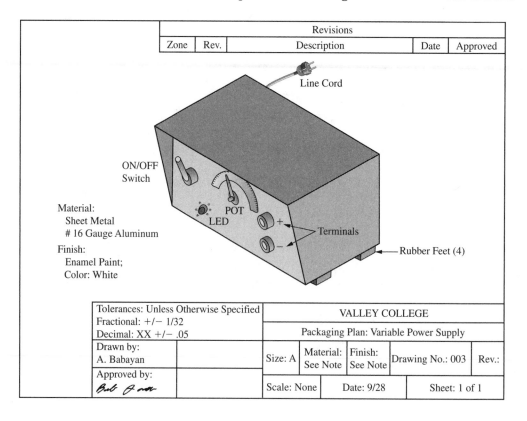

FIGURE 3–24
Variable Power Supply schematic with voltage waveforms

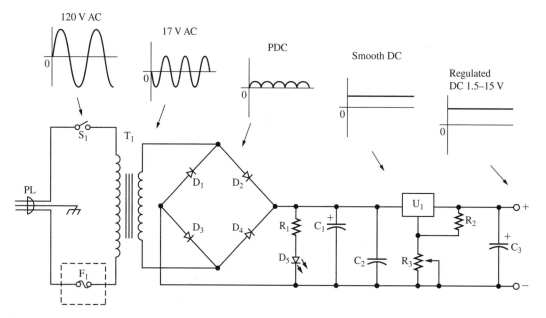

Designing the 3-Channel Color Organ

Here, as in the previous section, we review the two-step design process for the 3-Channel Color Organ Project. In Chapter 1 we said that the information supplied for the Exercise Project will not be as complete as that given for the Sample Project. At the design stage we make an exception, however. The Concepts and Requirements Document, as well as the three design drawings for the 3-Channel Color Organ, are presented and thoroughly dis-

cussed. In the following chapters, however, only hints and suggestions as to how to proceed will be provided.

Concepts and Requirements Document for the 3-Channel Color Organ

The Concepts and Requirements Document for the 3-Channel Color Organ (Figure 3–25) is similar in style and content to the document for the Variable Power Supply. The *project goals* are basically the

FIGURE 3–25
Concepts and
Requirements Document:
3-Channel Color Organ

Concepts and Requirements Document
3-Channel Color Organ
by
Larry Hyde

Project Goals

This Concepts and Requirements Document proposes that students of Electronics 150 design and build a working prototype for a 3-Channel Color Organ that may later be mass-produced. The color organ is a project that flashes three strings of Christmas lights to the sound of music produced by a stereo system or FM radio. Each string of lights responds to a range of frequencies: bass, midrange, and treble. When the lights are placed behind a piece of diffused plastic, the display can be quite dramatic.

   Commercial units, although generally available, are expensive, easily costing more than $50. Furthermore, undertaking the design and construction of such a project, rather than purchasing the item, will provide electronics students with a valuable electronics design and fabrication experience.

Project Objectives

The project is to have the following objectives: inexpensive to produce (less than $20), reliable, safe, attractive, and have good amplitude and frequency sensitivity. These objectives will be met in the following ways:

* **Cost.** By using noncritical, readily available hardware and electronic components, bought in bulk quantities, the cost will be kept low.

* **Reliability.** By using solid-state active components and printed circuit board construction techniques, a high degree of reliability will be obtained.

* **Safety.** The use of a fuse in the ac line and a completely sealed metal enclosure will assure safe operation.

* **Appearance.** A metal enclosure, either painted or covered with self-adhesive vinyl paper, will ensure an attractive project.

3

FIGURE 3–25 (cont'd)
Concepts and
Requirements Document:
3-Channel Color Organ

* **Sensitivity.** The color organ will control overall amplitude sensitivity with the use of a limiting resistor and a bypass switch. The frequency sensitivity of individual channels will be controlled by filter networks consisting of fixed resistors and capacitors, as well as a potentiometer.

Project Responsibility

Because this is to be a learning experience for the electronics student, he or she will be responsible for the design (which, in this case, is being supplied), drawings, experimenting, prototyping, and testing, troubleshooting, and final documentation phases of the 3-Channel Color Organ project. Due dates for each phase will be assigned by the electronics instructor.

Project Theory of Operation

The 3-Channel Color Organ gives a dramatic effect by flashing three strings of lights (each a different color) to the sound of music. Each string (with up to 200 W of light) responds to a particular range of audio tones: one for the bass, another for the midrange, and a third for the treble. Any audio source will provide the input signal, tapped directly off the speaker leads.

Resistor R_1 limits current to the indicator bulb I_1. The audio input signal, connected via jack J_1, is current-limited by resistor R_2. This resistor can be bypassed by slide switch S_2, applying the full audio input signal to transformer T_1. When the audio source provides a high input signal, S_2 must be open; when a low-level signal is present, S_2 is closed. T_1 is used as a step-up transformer to provide sufficient audio levels to trigger SCRs 1, 2, and 3. Potentiometers R_3, R_4, and R_5 are used as voltage dividers, allowing for variable adjustment of the SCR trigger levels. Resistor R_6 and capacitor C_1 form a low-pass filter for frequencies below 500 Hz for SCR_1 gate drive. Components R_7, R_8, C_2, and C_3 form a band-pass filter for frequencies between 500 and 3000 Hz for SCR_2 gate drive. Finally, capacitor C_4 and resistor R_9 form a high-pass filter for frequencies above 3000 Hz for SCR_3 gate drive.

Electronics instructor's approval *Bart Jones*   Date __*4-11*__

4

same, to give students experience in designing and building a working prototype of an inexpensive electronic project. *Project objectives* also involve such factors as cost, reliability, safety, and appearance. In addition, a sensitivity specification is presented in very general terms. Although "good" amplitude and frequency sensitivity are a bit hard to nail down, we can at least distinguish this color organ from those with virtually no sensitivity in either category. Finally, *project responsibility* for the entire project, as with the Variable Power Supply, rests completely

with the electronics student. Project *theory of operation*, shown in Figure 3–25, is discussed below, after the design drawings are studied.

Design Drawings for the 3-Channel Color Organ

The system diagram for the 3-Channel Color Organ is shown in Figure 3–26. Although this project does contain an output transducer (a series of lightbulbs that change electricity into light), it, like the Variable Power Supply, is still not a complete system by itself. The input to the color organ is an electrical

FIGURE 3–26
System diagram: 3-
Channel Color Organ

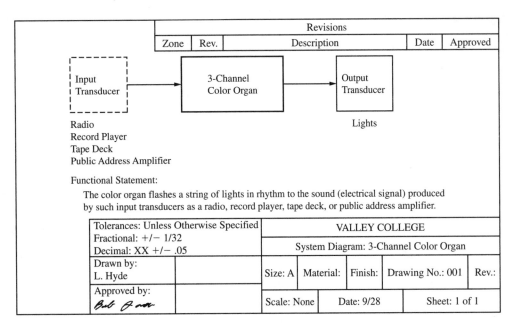

FIGURE 3–27
Circuit design sketch:
3-Channel Color Organ

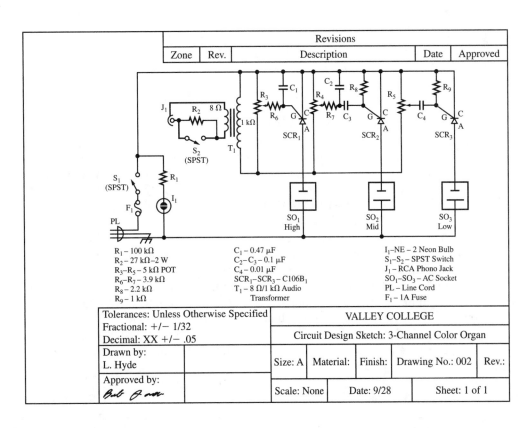

signal from a radio, tape player, or CD player. In a sense, these devices act as one complex input transducer. The radio changes electromagnetic energy (radio waves) into an electrical signal. The tape player "reads" information stored by magnetic particles and produces a corresponding electrical output. Together, the input device, color organ circuit, and lightbulbs make up the complete electronic system.

The **circuit design sketch** for the 3-Channel Color Organ is shown in Figure 3–27. Because the project is fairly simple, complete circuit design component specifications are obtainable at this stage.

The **project packaging plan** is shown in Figure 3–28. It should be pointed out that the enclosure is for the color organ circuitry only. The strings of lights are usually mounted in a large

FIGURE 3–28
Packaging plan:
3-Channel Color Organ

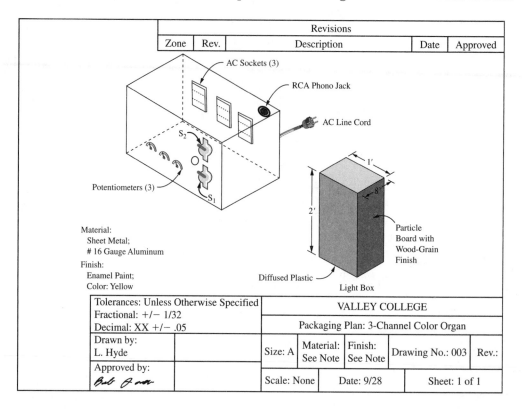

wooden box, one side of which is covered with a sheet of light-diffusing plastic. A sketch for such a light box is shown in the lower right-hand corner of the figure.

3-Channel Color Organ: Theory of Operation

The 3-Channel Color Organ gives a dramatic effect by flashing three strings of lights (each one a different color) to the sound of music. Each string responds to a particular range of audio tones: one for the bass, another for the midrange, and a third for the treble frequencies. Any FM radio, stereo tuner, phonograph, or CD or tape player can be used as a signal source. The electrical signal is tapped off from the speaker leads and fed to the color organ's input.

The color organ presented here will handle up to 200 W of light per channel. Since most standard Christmas tree bulbs are rated at 7 W, a string of 20 or more lights can be connected to each channel. You could, of course, just connect one 200-W light instead of a series of smaller bulbs. It all depends on the special effect you are trying to achieve.

Let's examine how the circuit works. Power is obtained from the 120-V ac wall outlet. Switch S_1 is the master on-off switch, while fuse F_1 provides overcurrent protection. Resistor R_1 limits current to the neon indicator bulb I_1. The audio input signal is fed, via jack J_1, to the matching audio transformer T_1. Resistor R_2 limits, or attenuates, the input signal strength. Notice that it can be bypassed by switch S_2. When high-level audio input signals are connected to J_1, S_2 must be in the "off" (open) position to prevent the transformer from saturating. When low-level input signals are used, S_2 must be in the "on" (closed) position. Then resistor R_2 is bypassed and the full audio signal is applied directly to T_1.

The audio transformer steps up the input signal to levels sufficient to trigger silicon-controlled rectifiers (SCRs) 1, 2, and 3. The SCRs are solid-state switches that turn on when a sufficient signal appears at their gate (G) electrode. Potentiometers R_3, R_4, and R_5 act as voltage dividers to allow variable adjustment of the SCR trigger levels.

Resistor R_6 and capacitor C_1 form a low-pass filter for frequencies below 500 Hz for SCR_1 gate drive. Components R_7, R_8, C_2, and C_3 form a band-pass filter for frequencies between 500 Hz and 3000 Hz for SCR_2 gate drive. Finally, capacitor C_4 and resistor R_9 form a high-pass filter for frequencies above 3000 Hz to drive the gate of SCR_3.

If you are building one of the elective projects or a project of your own design, now is the time to produce the Concepts and Requirements Document and the design drawing set for your project.

3.4   BEFORE BREADBOARDING: COMPUTER-BASED CIRCUIT DESIGN, SIMULATION, AND ANALYSIS

It is now possible to both design a circuit and simulate its operation on a computer. Actually, we've been doing so for some time. In the 1970s, powerful mini-computer systems, housed in large engineering laboratories, did just that. What's new, of course, is the ability to do such design and simulation on a personal computer, using cost-effective software selling for less than $300. There is little reason, therefore, or excuse, to jump from design sketch to breadboarding layout. Today, you can verify a design before breadboarding with computer-based design and analysis tools.

In this section we examine the elements of circuit design, simulation, and analysis. We next look at a typical software package, Multisim 7.0 from Electronics Workbench. Finally, we design and simulate our Variable Power Supply project using Multisim 7.0.

The Virtual Circuit

The virtual circuit is here. A circuit that appears to be, rather than actually is, can be represented on a computer screen. First, the *design*, in schematic form, is created. Next, the circuit is *simulated*, run, with LEDs flashing, speakers sounding, motors turning, and instruments displaying. Then, an *analysis* of its operation, using analysis graphs, is made. Thus the design, simulation, and analysis of a virtual circuit is complete. It's not the real thing,

of course, but the virtual circuit can get you closer to reality while saving considerable laboratory time.

Circuit Design

A circuit, real or virtual, consists of electronic components connected together with wires (or traces on a PC board). To design a circuit on a computer, you create a schematic consisting of components wired together.

Your components cannot just be drawn on the screen, however. First, they must be endowed with characteristics in the form of values and models. Then, when they are wired together and the resulting circuit is simulated, the circuit "works." If all you do is draw the components, as is done in a typical CAD package, you have only schematic symbols. Such symbols lack "intelligence": no circuit simulation or analysis is possible.

All circuit analysis programs have libraries chock full of components to be called up and placed in the drawing area. These components are predefined, having mathematical properties that allow them to function when the circuit is enabled.

The components, of course, can be moved, rotated, flipped, and copied. They can be assigned labels, values, and reference designations, and, as you would expect, they can be eliminated (see Figure 3–29).

Once components are placed on the computer screen, you wire them together by drawing lines from one terminal to another. It is simply a matter of pointing to a component terminal with the mouse and then dragging the mouse to a terminal on another component (see Figure 3–30). The wire is automatically routed at right angles, without overlapping other components.

With the circuit wired, your design is complete. Now comes simulation to see if it works.

Circuit Simulation

The circuit simulator calculates a numerical solution to a mathematical representation of a circuit you create. As mentioned earlier, for calculating to occur, each circuit component must be represented by a mathematical model. Mathematical models

FIGURE 3–29
Rotating and flipping components

  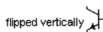 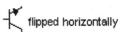

FIGURE 3–30
Wiring components

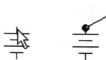 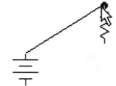 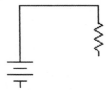

link the schematic in the circuit window with the mathematical representation for simulation. The accuracy of the component models determines the degree to which simulation results match real-world circuit performance.

Most general-purpose simulators have four main stages: *input*, *setup*, *analysis*, and *output*.

1. At the input stage, the simulator reads information about your circuit.
2. At the setup stage, the simulator constructs and checks a set of data structures containing a complete description of your circuit.
3. At the analysis stage, which is the core of the simulation, circuit analysis specified in the input stage is performed.
4. At the output stage, you view simulation results on instruments or on graphs appearing when you run an analysis.

Is simulation worthwhile? Do simulators work? Well, ask any engineer who worked on the Spirit and Opportunity Mars missions. Without simulation they wouldn't have been the roving success they were.

When it comes to circuit design, too, simulation is valuable. True, it's no substitute for the real thing. Nonetheless, running a simulation allows you to get to your prototype stage more quickly.

Circuit Analysis

Once your circuit is designed and operating, you can do an analysis on it and find out what's really going on. There are many types of analyses: dc operating point analysis, ac frequency analysis, transient analysis, Fourier analysis, noise analysis, distortion analysis, and parameter sweep analysis. The first three warrant our brief attention.

Dc operating point analysis determines the dc operating point of a circuit. The results are displayed on a chart that appears when the analysis has finished. The chart lists node dc voltages and branch currents.

In *ac frequency analysis*, the dc operating point is first calculated to obtain linear, small-signal models for all nonlinear components. The result is often displayed on two graphs: gain versus frequency and phase versus frequency.

In *transient analysis*, a circuit's response as a function of time is computed. Thus you wind up with a graph of voltage versus time.

Let's now turn to a widely used circuit design, simulation, and analysis software package that, as its promoters declare, "puts an electronics lab in a computer." Although many excellent schematic and simulation programs exist, here we examine Multisim 7.0, from Electronics Workbench.

Multisim 7.0

To give you a feel for what computer circuit design, simulation, and analysis is like, we look briefly at how Multisim builds circuits, works with instruments, and performs circuit simulation and analysis.

Building a Circuit

The user interface for Multisim is shown in Figure 3–31. It consists of the following elements:

☐ A system toolbar with buttons for commonly performed functions.
☐ The Multisim Design Bar, which allows you easy access to sophisticated functions offered by the program.
☐ The In Use list, which lists all the components used in the current circuit.

FIGURE 3–31
Electronics Workbench
Multisim user interface
Courtesy Multisim,
Electronics Workbench

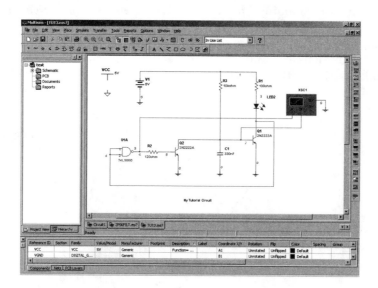

FIGURE 3–32
Toolbar Parts Bin Courtesy Multisim, Electronics Workbench

☐ The component toolbars containing Parts Bin buttons. These buttons let you open component family toolbars, which in turn contain buttons for each family of components in the Parts Bin.

☐ The Circuit Window, where circuit design takes place.

☐ The database selector, which allows you to choose which database levels are to be visible as component toolbars.

☐ The status line, which displays information about the current operation and a description of the item the cursor is currently pointing to.

In designing a circuit you first select, then drag various components from the tool bins to your circuit window, or drawing area (see Figure 3–32). Next, you select component values and reference IDs as needed.

With components in place, it's a simple matter to wire them together. Using a mouse, you point to a component's terminal (a short protruding line on the component) to highlight it, then you press and hold the mouse button and drag it so a wire appears. You continue to drag the wire to a terminal on another component, and when the new terminal is highlighted, you release the mouse button. The wire is automatically routed at a right angle.

With Multisim, you can choose to wire components either automatically or manually. With the former, Multisim automatically wires the connection for you, selecting the best path between your chosen pins. Automatic wiring avoids wiring through other components or overlapping wires. With manual wiring, you control the path of the wire on the circuit window. You can even combine the two methods in a single wire, for example, by starting wiring manually and then letting Multisim automatically complete the wire for you.

Working with Instruments

Multisim includes a number of virtual instruments (see Figure 3–33). Through the Instruments button

FIGURE 3–34
Virtual multimeter Courtesy Multisim, Electronics Workbench

on the Design Bar, you have access to a distortion analyzer, logic converter, multimeter, oscilloscope, wattmeter, word generator, spectrum analyzer, network analyzer, logic analyzer, function generator, and Bode plotter (not available in the real world).

With each virtual instrument you have two views: the instrument icon you attach to your circuit, and the opened instrument, where you set the instrument's controls and display options (see Figure 3–34).

Circuit Simulation and Analysis

To simulate a circuit with Multisim, you click the Simulate button in the Design Bar and from the pop-up menu choose Run/Stop.

When the simulation begins, you need some way of displaying the results. An oscilloscope will do nicely.

To see the results on the scope, you simply double-click on the oscilloscope icon to "open" the instrument display, if it isn't already open. You will see results similar to those shown in Figure 3–35.

Multisim provides many different types of analyses. When you perform an analysis, the results

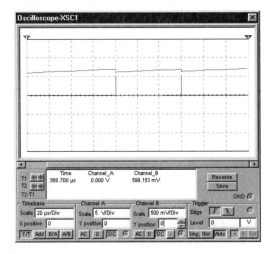

FIGURE 3–35
Circuit simulation Courtesy Multisim, Electronics Workbench

FIGURE 3–33
Instruments toolbar Courtesy Multisim, Electronics Workbench

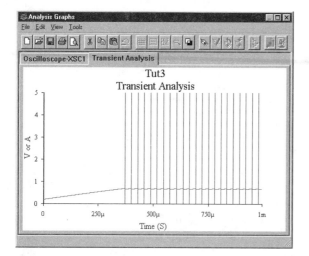

FIGURE 3–36
Circuit analysis Courtesy Multisim, Electronics Workbench

are displayed on a plot in the Multisim Grapher (unless you specify otherwise) and saved for use in the Postprocessor. An example of a transient analysis is shown in Figure 3–36.

As we have seen, computer-based circuit design and analysis tools are available to bring you quickly and effectively from concept to breadboarding. Investigate them, explore them, and use them to enhance the design process whenever you can.

Something to Think About

Numerous circuit design, simulation, and analysis software packages exist that sell for but a few hundred dollars. Why not gather information on as many as you can? Start building your information folder with a search on the Internet. Once you identify a product, send for a free test package.

Variable Power Supply

It's now time to design and simulate our Variable Power Supply project using what we have just learned. First, we place components in our circuit window. Then, we wire them together. Next, we simulate circuit operation. Finally, we create and print out a simple Bill of Materials (BOM) report.

Component Placement
After opening a new design window, we begin by dragging components from the Workbench Toolbar

onto the circuit window (Figure 3–37). Some components will be virtual, others real. The former, which exist in name only, cannot actually be purchased, but they have parameters that can easily be varied. In our circuit, the 120-V ac source and the transformer are virtual. All other components—the switch, the fuse, resistors, capacitors, and the voltage regulator—are actual components, with specific, unalterable specifications.

Note that Multisim treats virtual components slightly differently from real components. The default color of virtual components is different from that of real components on our schematic. This is to remind us that, since they are not real, these components will not be exported to PCB layout software, should we perform this step later.

Wiring the Circuit
We now simply wire the components together (Figure 3–38). Automatic wiring will take care of most of our needs in this circuit. We proceed as follows:

- [ ] Wire one end of V1 to one end of the switch.
- [ ] Wire the other end of V1 to one end of the fuse.
- [ ] Wire the other end of the switch to one end of the transformer's primary.
- [ ] Wire the other end of the fuse to the other end of the transformer's primary.

You get the picture.

Circuit Simulation
Once the Power Supply is designed, it's time to run a simulation (Figure 3–39). We are going to apply power and measure the output voltage with a voltmeter as we adjust the potentiometer through its range.

First, we choose the multimeter from the Multisim Instrument Toolbar and drag the instrument icon onto our work area. We then connect the multimeter's leads to the Power Supply's outputs, as shown. Double-clicking on the multimeter icon opens the meter for a full view. We select the voltmeter and DC settings.

Each press of letter *A* on our keyboard advances the potentiometer a given percentage of its full range. In our case, the increment setting is at 5%. With each press of key *A*, the potentiometer's resistance increases. That, of course, causes the output voltage to increase. The meter displays each new voltage reading. Pressing the Shift key and the letter *A* at the same time causes the potentiometer

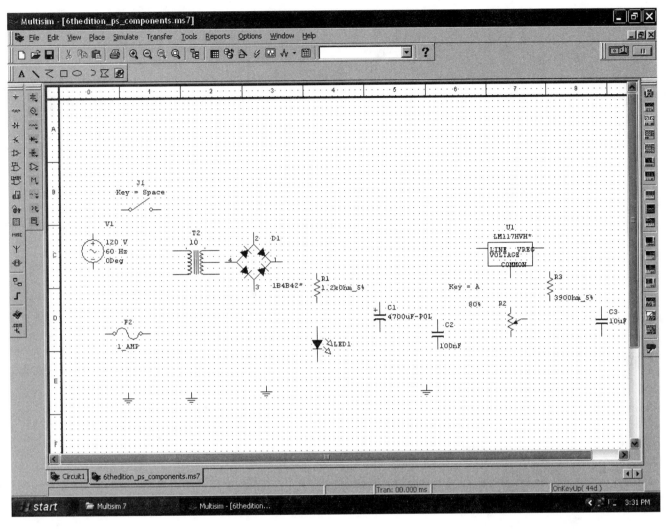

FIGURE 3–37

to decrease in resistance by 5%. The Power Supply's output voltage goes down.

Bill of Materials Report

A bill of materials lists the components used in our design and therefore provides a summary of the components needed to manufacture the circuit board (Figure 3–40). Information provided in the Bill of Materials includes

☐ the quantity of each component needed;
☐ a description, including the type of part (example: resistor) and value (example: 1.2 k ohms);
☐ the referenced ID of each component; and
☐ the package or footprint of each component.

Note that the voltage source and the power transformer are not listed in the Bill of Materials.

Because these are virtual components, and cannot be purchased, they are not included.

SUMMARY

In this chapter we first looked at design and designers, considering the types of electronic products and services created to meet particular needs and the kinds of people who do the creating and what type of training they need. We then examined a design methodology that identified problem and solution categories.

Next, we examined the fundamentals of electronic design and the two-step design process. We defined an electronic system as one that has three parts: input transducer, electronic circuit, and output transducer. We saw that circuits are made up of active and passive electronic components. The former control electrons by switching or regulating them; the latter

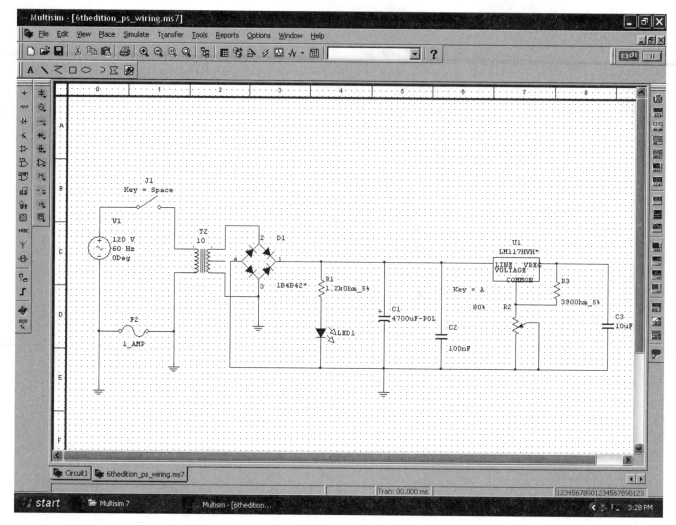

FIGURE 3–38

help the active components (vacuum tubes, transistors, integrated circuits, and microprocessors) do their job. We found out that circuits, too, can be classified as digital (switching) or analog (regulating). When the circuits are coupled with input and output transducers, components that change one form of energy into another, we have a complete electronic system.

The two-step design process requires a Concepts and Requirements Document and a set of design drawings. The former states the project goals, objectives, and responsibilities, as well as describes the project theory of operation. The design drawings consist of a system diagram, circuit design sketch, and a packaging plan.

Of course, we examined electronic design for the Variable Power Supply and the 3-Channel Color Organ Projects. We looked at the Concepts and Requirements Document, as well as the design drawings, for each project.

Finally, using **Multisim 7.0 (Electronics Workbench)** as an example, we explored the basics of computer-based circuit design, simulation, and analysis.

In Chapter 4 we turn to technical drawings, the language of engineering. Although we briefly discuss the entire 10-drawing set, our emphasis is on the schematic drawing, the "blueprint" for circuit understanding and assembly.

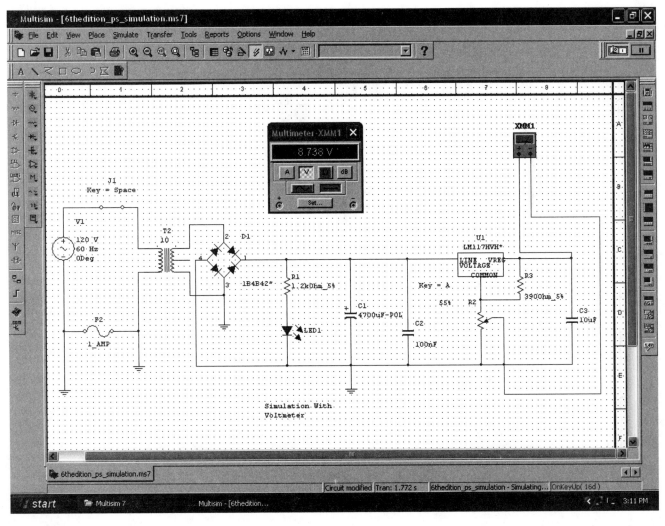

FIGURE 3–39

QUESTIONS

1. One type of design methodology involves identifying _____ and _____ categories.

2. An electronic system consists of an _____ transducer, electronic _____, and _____ transducer.

3. Electronic components can be classified as either _____ or _____.

4. Since the beginning of the electronic age, four active components have been developed: _____, _____, _____, and _____.

5. _____ electronic components help the active components do their job of switching or regulating current flow.

6. Electronic circuits can be classified as either _____ or _____.

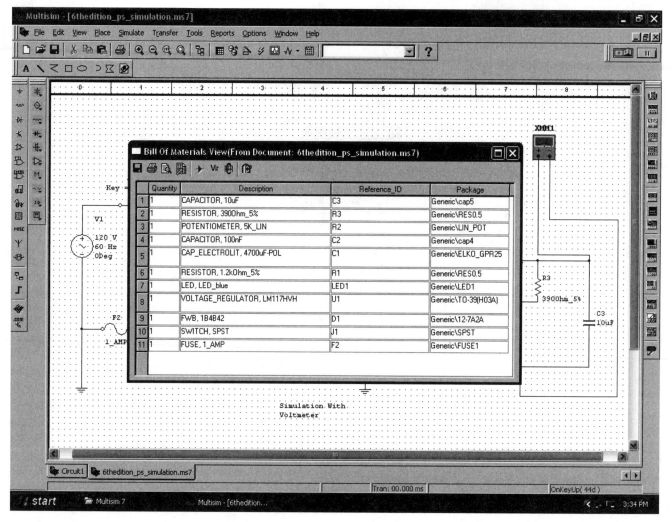

FIGURE 3–40

7. The Concepts and Requirements Document states the project's _____, _____, and _____.

8. The design drawing set consists of the _____ diagram, circuit _____ sketch, and a _____ plan.

9. Computer-based design tools allow you to do circuit _____, _____, and _____.

10. Whenever circuit analysis takes place in **Multisim 7.0**, an analysis _____ window automatically appears.

REVIEW EXERCISES

1. Write the Concepts and Requirements Document for a simple audio amplifier system, like the one shown in Figure 3–12. The amplifier is to be used as part of a small public address (PA) system. A microphone would be the input transducer and a speaker the output transducer.

2. Draw the system diagram for the audio amplifier in Exercise 1.

3. Draw the circuit design sketch (schematic) for a basic LED logic probe. It should indicate a high, low, and pulse condition. The probe should receive its power directly from the circuit "under test." If you are unable to design such a circuit yourself, obtain a schematic from an electronics magazine or project book. Redraw the schematic to answer this question.

4. List five electronic systems in your home, school, or workplace.

   a.

   b.

   c.

   d.

   e.

5. Name some possible input and output transducers for the following electronic systems:
   ☐ Stereo system
   ☐ Television system (remember, you should include the transmitter as well as the receiver to make up the complete system)
   ☐ Computer system
   ☐ Telephone system
   ☐ Thermostatic control system, such as the one found in most homes
   ☐ A bicycle speedometer (digital)

6. Sketch the system diagram for one of the electronic systems listed in Exercise 5.

7. Sketch the schematic drawing for an electronics project found in an electronics magazine.

8. Sketch the pictorial drawing for common cube-shaped objects found around the home or school laboratory.

9. Use a computer-based design software program to design a simple electronic circuit.

10. Use a computer-based design software program to simulate and analyze the circuit designed in Exercise 9.

4 Drawings—Show Me a Plan

OBJECTIVES

In this chapter you will learn

- [ ] Why drawings are indispensable to the art of technical communication.
- [ ] How mechanical, architectural, and electronic drawings differ from each other and how they are alike.
- [ ] What constitutes the 10-drawing set needed to design and fabricate the working prototype project.
- [ ] How to identify schematic symbols for most of the components used in modern electronics.
- [ ] How to read schematic drawings.
- [ ] How to draw schematic drawings.
- [ ] How to produce a working schematic drawing for the Variable Power Supply and the 3-Channel Color Organ.

No doubt you've heard it a thousand times, which may make it a bit stale but certainly no less true. Yes, a picture (or drawing) really is worth a thousand words. Maybe we could produce the required documentation for our projects without any drawings at all. Maybe. But the resulting Project Report would certainly be a great deal thicker and much, much more difficult to write and read. In fact, it's hard to conceive of doing such a thing. The effort required to put into words what can so easily be conveyed with special symbols and lines—well, one shudders just to think of it. Drawings not only are desirable in explaining something like project design and fabrication,

they are essential. For example, 10 drawings are required to clarify how a project works and how it is to be built. Anything less would make describing the project design and fabrication process in words too burdensome.

In this chapter we cover two major topics. First, we explain the role of the drawing in communicating a message or idea and introduce the 10-drawing set used throughout the book. Second, we zero in on the **working schematic**: the legible circuit design drawing that everyone can work from. We will find out how to read such a schematic and then how to draw one.

4.1 DRAWING: THE LANGUAGE OF ENGINEERING

Get two or more engineers together to solve a problem, and what do they do? Immediately they take out pencil and paper (envelope, napkin, or whatever else is handy) and start to draw. Like those individuals who can't seem to say anything without waving their hands, engineering types need props as well. To discuss any technical subject, they require that simple drawing equipment be within reach at all times.

In this section we find out why this is so, why the drawing is indispensable to technical communication. We start out by examining drawings in general and then those necessary to complete the design and fabrication process.

When Words Are Not Enough

Let's explore the basic need for drawings, then look at the three common types used in industry: mechanical, architectural, and electronic.

Why Draw?

Imagine trying to explain in words how to make the object shown in Figure 4–1. Notice the angled surfaces, holes, and rectangular slots. Could you do it? Probably, but it wouldn't be easy. What about the schematic drawing in Figure 4–2? Could you describe the function of each control component, as well as every resistor, capacitor, diode, and so on? Could you tie all these individual components together in explaining the entire circuit function? Again, the answer would likely be yes, but with a price to your mental stability.

The ancient Egyptians spared themselves such mental anguish when they built the pyramids and other architectural wonders. They used drawings sketched on papyrus to convey construction details.

Leonardo da Vinci was not only a superb artist but an excellent draftsman. He made detailed drawings of everything from a file-making device to a flying machine.

Today, the da Vincis—drafters, that is—of the world are still hard at work. They use computer-aided design (CAD) to assist in the drafting process, but in the end it is still a drawing that is produced. If the classical Chinese thought a picture was worth a thousand words, today's high-tech engineers and technicians couldn't agree more.

From the Realistic to the Abstract

Whether it be in manufacturing, construction, or assembly, the appropriate drawings are designed to show how to make or build something. In manufacturing, **mechanical drawings** based on orthographic (right-angle) projection illustrate three-dimensional objects on a two-dimensional surface (the drawing paper). Although it takes special training to read such drawings, they are, nevertheless, highly realistic, in that the objects as drawn look very much like the actual thing. **Architectural drawings** explain how a structure is erected. These drawings are also fairly realistic, especially with regard to building elevations, yet there is certainly an element of abstraction when symbols are used to represent everything from kitchen sinks to various types of concrete slabs. Finally, there are **electronic drawings**. Illustrating the "electronics" of a project, these drawings are almost entirely abstract. A pictorial drawing of a capacitor or diode, for example, is rarely illustrated, and then only when absolutely necessary.

You can see, then, that drawings vary in the degree of realism or abstraction. In the 10-drawing set we cover the entire gamut. Some will be very graphic (realistic), others extremely symbolic (abstract).

Electronic Fabrication Drawings

It's time, now, to outline the 10-drawing set that has been mentioned so often.

The 10-Drawing Set

Figure 4–3 shows a flowchart for the 10 drawings required to build a working prototype project. Don't be intimidated by the number of drawings. Remember, almost all can be drawn on an $8\frac{1}{2}'' \times 11''$ size A sheet of graph paper, and none is particularly complex or time consuming to produce. Furthermore, they are not all done at once but rather are spread out over a number of stages in the design and fabrication process.

The design drawings we have seen already. The working schematic we will review momentarily. The breadboard drawing is a simple layout of components as they will appear in an experiment layout. The next four drawings are concerned with printed circuit boards. The first is a PC design layout; the second, an artwork drawing developed from the layout; the third, a fabrication drawing of the PC board itself; and the fourth, an assembly drawing to show how components are installed on the PC board. The sheet metal drawing is basically a two-dimensional layout of a metal chassis or an enclosure. The wiring diagram illuminates how off-board components are connected. Finally, the packaging drawing shows how the finished project will look.

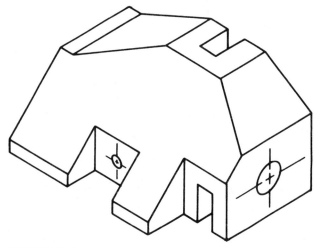

FIGURE 4–1
"A picture is worth a thousand words"

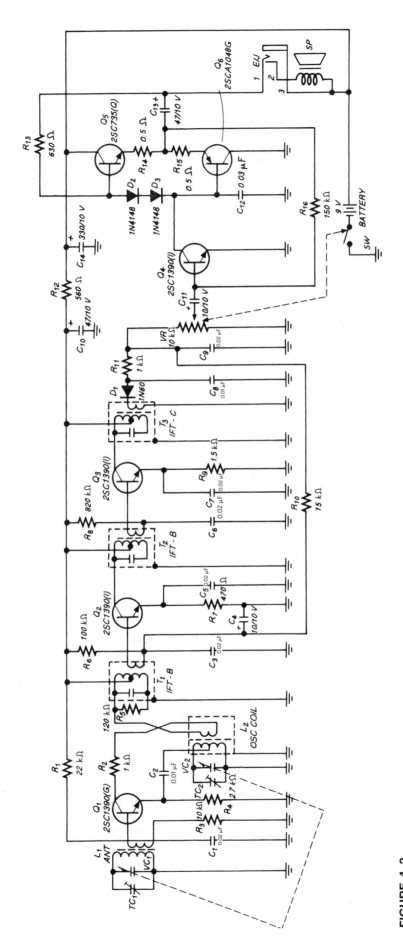

FIGURE 4–2
The schematic drawing

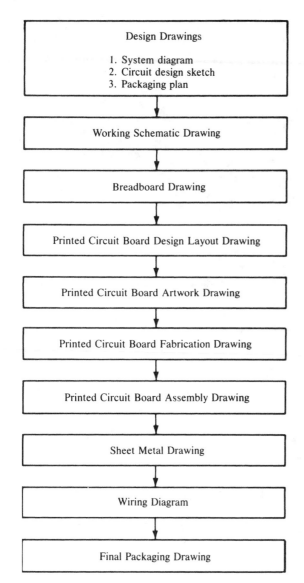

FIGURE 4–3
Ten-drawing set

The 10 Drawings: A Brief Explanation

Let's examine the purpose and characteristics of each drawing in more detail. As we do, take a moment to look at an example of each drawing in Chapters 15 and 16.

1. Design Drawings. The three design drawings—the system (functional) diagram, circuit design sketch, and packaging plan—are preliminary drawings presented with the Concepts and Requirements Document; however, the circuit design sketch and packaging plan will be used to develop the more formal working schematic and final packaging drawing.

2. Working Schematic Drawing. The single most important drawing in the entire package,

the working schematic illustrates in a precise and readable manner what the circuit consists of and how it is connected. It differs from the circuit design sketch in that the working schematic is clear, understandable, and complete enough for everyone to work from. If the circuit design sketch was made on a CAD system, updating to the working schematic drawing will be quick and easy. The drawing is made up of component symbols and interconnecting lines.

3. Breadboard Drawing. This drawing shows the layout of components on a breadboard. It is usually drawn to scale on graph paper. Although not always necessary, when experimental layouts are complex, the inclusion of a breadboard drawing can save valuable experimenting time.

4. PC Board Design Layout Drawing. Drawn on 0.10″ grid paper, often with colored pencils, this drawing shows a sketch of the component layout and trace patterns for the printed circuit board. Frequently drawn with the aid of a PC design layout template, the drawing may be produced at 1× or 2× scale. Again, if you are using a CAD system, this drawing may be skipped altogether, depending on the sophistication of your design software.

5. PC Board Artwork. This drawing consists of the actual traces and pads of the PC board design layout. It is from this artwork that a film negative will be produced to be used in etching the printed circuit board.

6. PC Board Fabrication Drawing. Basically a standard mechanical drawing, this illustration shows how the PC board is to be fabricated. All board dimensions, along with hole sizes and quantities, are given. In addition, specifications for PC board base material—copper-clad, solder coating, or silk screening (when applicable)—are provided on the drawing.

7. PC Assembly Drawing. Not unlike a mechanical assembly drawing, this type also shows where parts are placed and how they fit together. A template is used to lay out the electronic components over a screened image of the PC board traces. A parts list is included, often on the drawing itself.

8. Sheet Metal Drawing. The sheet metal drawing is a two-dimensional layout of the pattern required to produce a metal chassis or enclosure. Fold lines are indicated, as well as any holes, slots, and grooves. A pictorial illustration is often included to show how the finished object looks when "folded together."

9. Wiring Diagram. This diagram shows how any "off-board" wiring and components are connected to each other and to the printed circuit board. In complex projects the wiring plan may be pictorial. In simple layouts it is strictly schematic, with wires shown as horizontal and vertical lines.

10. Final Packaging Drawing. This drawing shows how the final project will look from the outside. The locations of all panel lights, knobs, switches, and so on are shown and dimensioned if necessary. Any surface-finishing treatment, such as paint or contact paper, is documented. Also, all appropriate lettering and other labeling are illustrated.

Let's now concentrate on the second, and most important of the drawings, the working schematic.

Something to Think About

Why not gather a set of professional drawings: everything from the working schematic to a packaging drawing? They don't have to be for the same project or device. Now, label and hang the drawings up in your fabrication laboratory.

4.2   WORKING SCHEMATIC: THE CIRCUIT SCHEME

A schematic drawing, as noted earlier, shows what's connected to what—in other words, how electronic components are tied together. We need to see how this is done. No electronics technician can get by without at least being able to read a schematic. Also, knowing how to draw one certainly wouldn't hurt. And because drawing a schematic is almost as easy to do as reading one, we'll show you how to do both.

Reading Schematic Drawings

Reading a schematic drawing involves knowing two things. You must recognize the schematic symbols for the components used, and you must understand how interconnections are represented on the drawing. That's really all there is to it.

Component Schematic Symbols

Figure 4–4 shows the schematic symbol and accompanying pictorial sketch for each of the major components used throughout the electronics industry.

FIGURE 4–4
Component schematic symbols

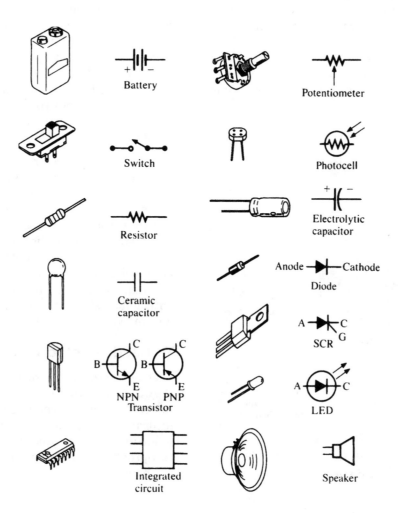

You simply memorize which symbol goes with which component. Here are a few points to keep in mind when studying the schematic symbols.

1. A component that is variable, or adjustable, includes an arrow as an integral part of its symbol. For example, a potentiometer is a variable resistor. Note the arrow representing one of its three leads.
2. Arrows are also used to indicate the emitting or receiving of various forms of energy. If an arrow is pointing away from the symbol, the component gives off energy. Note the light-emitting diode (LED). On the other hand, if the arrow is pointing inward, toward the symbol, it means the component receives energy from an external source. The photocell (light-sensitive resistor) is a good illustration.
3. Sometimes component leads are identified with letter symbols. For example, the transistor symbols shown in Figure 4–4 have their emitter (E), base (B), and collector (C) leads noted.

Finally, note the schematic symbols for ground (Figure 4–5a). The symbol on the left is the most widely used and denotes earth or chassis ground. If more than one such symbol is present at various points in a schematic, there is a common connection, or return point, for the circuit.

The ground symbol to the right is for chassis ground only. If it is drawn connected to the ground line of a three-prong ac line cord (see schematic for the variable power supply), the chassis is then at earth ground potential. The connection adds a measure of safety should the chassis become "hot."

Connecting Schematic Symbols

On a schematic drawing, lines indicate wires, traces, or component leads. No connection is indicated by letting the lines cross. If a connection is intended, a dot is placed at the junction. The half-loop to show no connection (Figure 4–5b) is no longer used.

Notice, also (Figure 4–5c), that a connection dot is used only when three or more lines (wires, traces, or component leads) come together. It is not used when only two such wires (or leads) are connected.

The Complete Schematic

Here are a few guidelines for reading the completed schematic (see Figure 4–2):

1. On the schematic drawing itself, each component is given its own letter–number reference designation to distinguish one from another. The letter identifies the class of component; the number, the sequence. For example, R_5 indicates the fifth resistor in the circuit.
2. Generally, circuit signals flow from left to right, with inputs on the left and outputs on the right.
3. Generally, voltage potentials are shown with the highest voltage at the top of the sheet and the lowest at the bottom. When differential supplies are used, the positive bus is usually on the top of the schematic, ground (or common) in the middle, and the negative bus on the bottom.

Now you are ready to practice reading schematics. There are essentially four ways to go about it. First, do the exercises at the end of this chapter; most ask you to complete the schematic drawing from the corresponding illustration. Then study simple schematics and sketch your own representation for them. Third, draw schematics from actual circuits. Finally, read, read, and read all the schematics you can find. Discuss the schematics with fellow students and dissect them together. In no time you will be reading schematics like a pro.

To Read Is Not to Write

Just as there are those who can read better than they can write, there are technicians who, while fully understanding a schematic, have some difficulty in drawing one. Yet, by using the proper tools and following correct procedures, practically anyone can produce a top-quality schematic drawing. Let's see how.

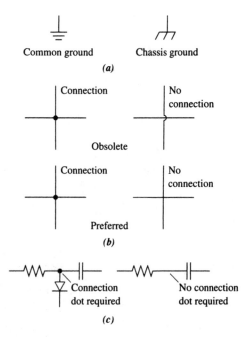

FIGURE 4–5
Connections on a schematic drawing

Tools for Drawing the Schematic

In Chapter 1 the equipment required for work on the board or at the computer was listed and described. If you are using traditional tools, you will need the following items to produce a schematic drawing:

- ☐ **Graph Paper.** $8\frac{1}{2}'' \times 11''$ will usually be large enough.
- ☐ **Pencil.** Preferably one with a lead hardness of H and a thickness of 7 mm or less.
- ☐ **Schematic Symbols Template.** At a minimum, one with resistor, capacitor, diode, and transistor symbols.
- ☐ **Straightedge.** A T-square, triangle, or mini-drafting machine will do.
- ☐ **Eraser.** The type known as a "Pink Pearl" should be used.

See Chapter 11 for a detailed discussion on CAD tools.

Drawing Procedures

In drawing schematics, problems arise most often when we forget to include something. A biasing resistor was not inserted, a reference designation was left off, or an IC pin number was not indicated. To avoid these and similar mistakes, and to progress in a logical manner, it's best to follow a step-by-step drawing procedure. Here, then, are 10 steps to completing a schematic drawing (see Figure 4–6):

1. Border and Title Block. Draw a border on the paper at least $\frac{1}{4}''$ from all outer edges, then include a title block that provides for, at a minimum, the name of the drawing, name of the drafter, approval signature, date, scale (if applicable), and drawing number. Make all lines wide and bold, but not uneven and smudgy. If you have trouble making your line work as even and sharp as it should be, consult the appropriate section of a good engineering drawing text.

2. Schematic Symbols. Draw the key (control) components first. Next draw the passive components, such as resistors and capacitors. Use the schematic template for both types.

3. Interconnections. Draw all interconnections with vertical or horizontal lines. Do not place a dot where lines cross but no connection is intended.

4. Connection Dots. Now "heavy up" all connection points (where three or more lines meet) with $\frac{1}{16}''$ to $\frac{1}{8}''$ dots. Such "dot holes" are usually on the schematic template.

5. Arrowheads. It's time to put in all arrowheads. They will be needed on energy-emitting and -receiving symbols, for example, LEDs and

FIGURE 4–6
Working Schematic:
Variable Power Supply

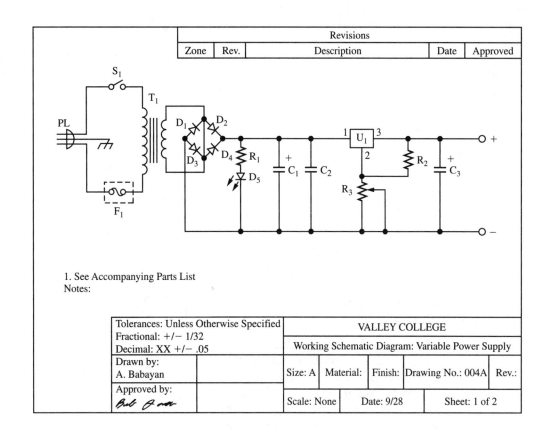

photodiodes, and on variable components, such as potentiometers and adjustable capacitors.

6. Reference Designations. Reference designations are very important and easy to mess up. Keep the letter–number combination on the same line. That is, don't put an R (for resistor) on one line and the 14 (the fourteenth resistor) on another. You should, however, place the resistance value, when shown (12 k, for example), on a different line from the reference designation.

7. Pin Numbers and Lead Identifications. This is another area where it's easy to forget something. When numbering the pins for ICs, voltage regulators, and the like, make sure you include numbers. When designating component leads, such as the emitter, base, and collector of a transistor, be equally vigilant in including all identifications.

8. Drawing Notes. One way to start your notes is from the bottom of the page and move up. The reason is obvious: you may want to include additional notes later on. Another approach is to start from the upper right corner and move down.

9. Parts List. On very simple schematics, the parts list can be included on the drawing. With more complex circuits, however, a separate sheet listing electronic components, connectors, and all hardware items is required. In using a separate sheet, it is a good idea to begin the listing at the top of the sheet so that new parts can be added. It's easy to forget small items, such as a screw or heat sink.

10. Drawing Check. The only way to reduce mistakes and present as complete a drawing as possible is to check it over carefully. Ask yourself these five simple questions.

☐ Does the drawing accurately reflect the design sketch?
☐ Is the line work sharp and even?
☐ Are all components properly identified?
☐ Are all connection dots in place?
☐ Is the drawing clean and neat?

Follow these basic steps and you will be not only reading schematics like an expert but drawing them like one as well.

Working Schematic: Two Practical Projects

Now it's time to turn to the Sample and Exercise Projects and examine their working schematics. Keep in mind that even though the working schematic is likely to undergo some revision if circuit changes are required in the breadboarding stage, such alterations are a small inconvenience

when compared with the value of a truly readable drawing. To have success at any circuit construction phase, even the breadboarding phase, you must possess a clear, accurate, and thorough working schematic.

Working Schematic for the Variable Power Supply

The complete working schematic for the Variable Power Supply is shown in Figure 4–6. Only a few points need to be noted:

☐ The power supply reads from left to right—that is, input power is at the left, output power, at the right.
☐ Connections are shown by dots, and nonconnections with crossed lines.
☐ Arrowheads are shown on the light-emitting diode symbol and on the potentiometer, R_3.
☐ All reference designations are shown, but, unlike on the circuit design sketch, component values are now placed on the parts list (Figure 4–7).
☐ Note the pin numbers for the voltage regulator U_1.
☐ Drawing notes start at the bottom of the page and move up.
☐ Finally, the parts list (Figure 4–7) is on a separate sheet and includes not only electronic components but some hardware items as well.

Working Schematic for the 3-Channel Color Organ

Again, only hints and suggestions on how to progress will be given for the Exercise Project.

When drawing the working schematic from the circuit design sketch of the 3-Channel Color Organ, simply follow the step-by-step drawing procedure discussed earlier. *Hint:* First draw four light horizontal lines that can be used as guides for bus lines and component placement, then set up three light vertical lines that will form the centerlines for each SCR. Put the parts list on a separate sheet.

Whether you are building one of the elective projects or a project of your own design, now is the time to draw the working schematic for your project.

Something to Think About

Find the most complex schematic drawing, then study it carefully. Note the many different schematic symbols. Now, count the total number of components. Are you surprised at how few unique components there are?

FIGURE 4–7
Schematic Drawing Parts
List: Variable Power
Supply

Revisions					
Zone	Rev.	Description		Date	Approved

Qty.	Symbol	Description	Part No.
1	R_1	Resistor, 1.2 Ω, ¼ W	
1	R_2	Resistor, 390 Ω, ¼ W	
1	R_3	Potentiometer, 5 kΩ	
1	C_1	Capacitor, 4700 μF	
1	C_2	Capacitor, 0.1 μF	
1	C_3	Capacitor, 10 μF	
4	D_1–D_4	Diode, 1N4004	
1	D_5	Diode, LED-Red	
1	U_1	Voltage Regulator, LM 317	
1	T_1	Transformer, 120/12 V AC	
1	S_1	Switch, S.P.S.T.	
1	PL	AC Line Cord	
1		Heat Sink (LM 317)	
1		Terminal Post-Red	
1		Terminal Post-Black	

Tolerances: Unless Otherwise Specified Fractional: +/− 1/32 Decimal: XX +/− .05		VALLEY COLLEGE					
		Schematic Drawing Parts List: Variable Power Supply					
Drawn by: A. Babayan		Size: A	Material:	Finish:	Drawing No.: 004B		Rev.:
Approved by: *Bob Gross*		Scale: None		Date: 9/28		Sheet: 2 of 2	

SUMMARY

In this chapter we looked at drawings, with particular emphasis on the working schematic. We examined the "language" of engineering, saw why it is indispensable to technical communication, and looked at how mechanical, architectural, and electronic drawings go from the realistic to the abstract. Next, we discussed the 10-drawing set required to design and fabricate a complete prototype project. Finally, we probed the working schematic, first learning how to read, then how to draw one. We concluded with a look at the working schematic for the Variable Power Supply, with some hints and suggestions on how to produce a similar drawing for the 3-Channel Color Organ.

Of course, when it comes to communications, drawings are rarely enough; words count, too. In Chapter 5 we examine the elements of technical writing, not to make you a full-time technical writer, though there's nothing wrong with that, but to help you become a strong technical communicator, as an adjunct to your duties as an electronics technician.

QUESTIONS

1. Of the three types of drawings found in industry, which one is most realistic? _____ Which one is most abstract? _____

2. Which drawing comes first, the PC board design layout or the PC board artwork? _____

3. The _____ _____ drawing is a two-dimensional layout of the pattern required to produce a metal chassis or an enclosure.

4. The _____ _____ drawing illustrates in a clear and readable manner how the circuit works.

5. An electronic component that is variable, or adjustable, includes an _____ as an integral part of its symbol.

6. On a schematic drawing, each component is given its own letter–number _____ designation.

7. Schematic symbols often are drawn with the aid of a schematic _____.

8. On the schematic drawing, a connection is usually indicated with the use of a small ($\frac{1}{16}$″ $-\frac{1}{8}$″) _____.

9. Drawing notes should begin at the _____ of the sheet and move _____.

10. The _____ _____ is sometimes placed on a separate sheet of paper.

REVIEW EXERCISES

1. Complete the schematic for the circuit shown.

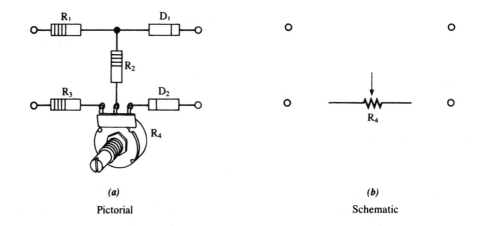

(a)

Pictorial

(b)

Schematic

2. Complete the schematic for the circuit shown.

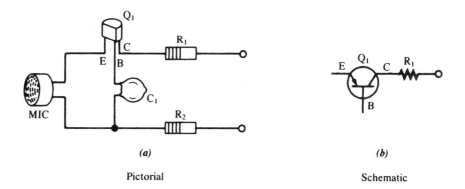

(a)

Pictorial

(b)

Schematic

3. Complete the schematic for the circuit shown.

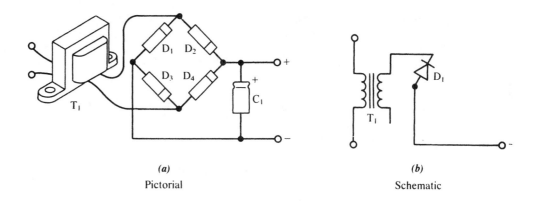

(a)

Pictorial

(b)

Schematic

4. Complete the schematic for the circuit shown.

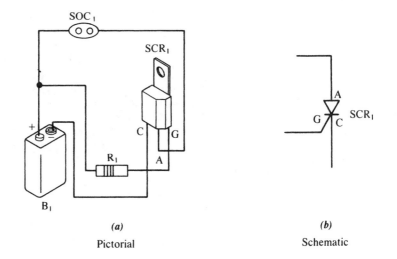

(a)

Pictorial

(b)

Schematic

5. Complete the schematic for the circuit shown.

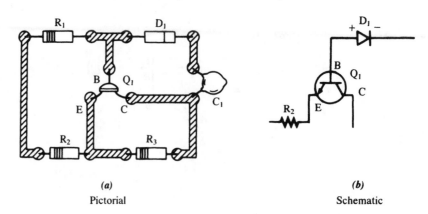

<table>
<tr><td align="center">*(a)*</td><td align="center">*(b)*</td></tr>
<tr><td align="center">Pictorial</td><td align="center">Schematic</td></tr>
</table>

6. Draw the schematic for the circuit shown.

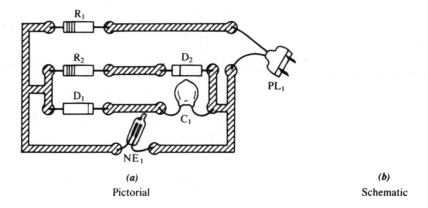

<table>
<tr><td align="center">*(a)*</td><td align="center">*(b)*</td></tr>
<tr><td align="center">Pictorial</td><td align="center">Schematic</td></tr>
</table>

7. Draw the schematic for the circuit shown.

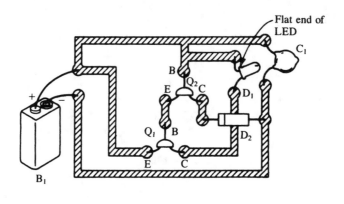

<table>
<tr><td align="center">*(a)* Pictorial</td><td align="center">*(b)* Schematic</td></tr>
</table>

8. Match the electronic components and their schematic symbols with their correct names.

a. Disc capacitor
b. Resistor
c. Diode
d. Speaker
e. Integrated circuit
f. Light-emitting diode (LED)
g. Transistor
h. Battery
i. Silicon-controlled rectifier (SCR)
j. Switch
k. Potentiometer
l. Electrolytic capacitor

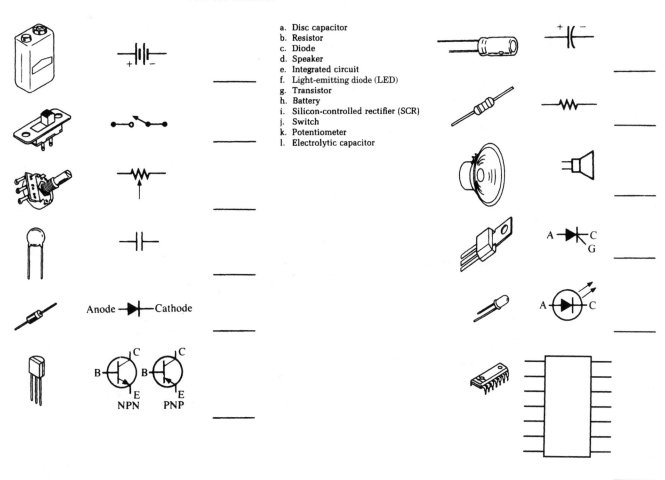

5 Technical Writing— Telling It Like It Is

OBJECTIVES

In this chapter you will learn

- [ ] What are the characteristics of technical writing.
- [ ] How technical writing differs from writing a novel or advertising brochure.
- [ ] About the career of technical writer.
- [ ] What a proposal is.
- [ ] About the different types of proposals.
- [ ] About the Concepts and Requirements Document as a proposal.
- [ ] How to identify your readers.
- [ ] How to outline a technical document.
- [ ] Why good writing is rewriting.
- [ ] About the technical writing style.
- [ ] How to write in the active voice.
- [ ] How to use the comma in the appropriate way.
- [ ] How to chop word fat from your final draft.

"A picture is worth a thousand words." We said it in the last chapter, in the introduction. True, drawings are vital to a full explanation of any technical subject, but they are not the whole story. Can a drawing tell you what an item is, what it is used for, what it does, how it works, or what it is made of? Often, not. To complete the "picture," a technical explanation, written clearly, precisely, and accurately, in a plain, impersonal, and factual style, is required. In this chapter we show you how to write that way. As a result, you'll produce well-written documents for all your Project Reports.

To begin, we examine technical writing: what it is and how it differs from creative writing. We discuss why you, a future electronics technician, should be interested in technical writing, how it may improve your job prospects or open up new ones.

Next, we study the proposal, what it is and how it is written. It's the proposal, more than any other document, that you'll create on the job.

Then, we explore the technical writing style, how you "tell it like it is" with clarity and accuracy. Finally, we examine simple self-editing techniques, punctuation and word elimination, that tighten up your writing.

This chapter is not a definitive work on technical writing—far from it. For something approaching that, see a listing of technical writing texts in Appendix G. At the conclusion of Chapter 5, however, you should feel more confident in your ability to state what you mean. Add to that the drawing talent you're now acquiring, and your Project Report, with its written and graphic documentation, will be all it can be.

5.1 TECH TALK, STRAIGHT TALK

Technical writing is writing about technical topics, anything to do with specialized areas of science and technology. Its purpose is to explain. Generally, technical writing deals with an object, process, system, or abstract idea. It involves facts and figures.

In this section we look more closely at technical writing and why it's important for you to learn its techniques.

Just the Facts, Ma'am

Exactly what is technical writing and how does it differ from writing a novel or an advertising brochure? Let's find out.

Plain Speaking

Technical writing, according to Herman M. Weisman, author of *Basic Technical Writing,* "is a specialized field of communication whose purpose is to convey technical and scientific information and ideas accurately and efficiently." Hence, it must be technically useful, concise, complete, clear, and consistent; correct in spelling, punctuation, and grammar; targeted, well organized, and interesting. Interesting? Yes! Technical writing doesn't have to be, indeed shouldn't be, boring.

Examples of technical writing are found in user guides and manuals, installation instructions, information brochures for high-tech equipment, technical or trade publications, textbooks, and online help. From a highly sophisticated medical journal to the instruction booklet accompanying your grandchild's new unassembled tricycle, technical writing, good and bad, permeates our technological age. Even if you're not producing it, you're reading it every day.

Objective Versus Subjective

Technical writing, unlike creative or advertising writing, is designed to explain rather than entertain or sell. (Of course, selling must be built into every proposal, as we will see shortly.) Technical writing concerns itself with factual information presented in a language that doesn't appeal to emotions or imagination, but to intellect. It is objective. Its words are exact and precise. It informs.

Literature, on the other hand, is subjective. It is an interpretive record of human progress and is based on imaginative and emotional experiences rather than a factual record of human achievement. Its purpose is to give us insight.

Sales material, of course, is created to sell us something. As such, it appeals to our emotions, with words and phrases conjuring up pleasant images of us and the products and services we are being besieged to purchase. In many ways, such writing, though legitimate in its purpose, is the antithesis of technical writing.

A Technical Writing Career?

You've got to be kidding! Me as a technical writer? Why not? There are excellent jobs for full-time technical writers, particularly those having a strong grounding in a technical subject, such as electronics and computer technology. Besides, even if you don't work at it 40 hours a week, your ability to pitch in from time to time, writing instructions and procedures for products and process you otherwise design, build, or repair, can only enhance your value to any company. Knowing how to write technically is always an asset.

The Technical Writing Life

Full-time technical writers write and edit materials for reports, manuals, briefs, proposals, instruction books, catalogs, and related technical and administrative publications. They also assist in planning and project coordination.

Technical writers work everywhere. They're employed by computer software firms or manufacturers of aircraft, chemicals, pharmaceutical products, and computers and other electronic equipment.

What do they get paid? It varies, depending on experience, geographic location, and type of employment, but in general, according to the Society for Technical Communication (STC), in 2000, the median annual salary for technical writers with seven years of experience was $51,000.

Although to become a technical writer you'll probably need more training in writing than you now possess, the biggest thing you have going is your technical background. Traditionally, technical writers have emerged from the ranks of college English majors. Although their writing skills are fine, these individuals often lack the technical savvy necessary to heighten their effectiveness. What is desperately needed is a person with a capital *T*—someone with strong technical skills first and writing ability second. Most firms will settle for a lowercase *w*. With enough exposure and encouragement, a good technical writer will surface.

Part of the Job

OK, so you're not an uppercase *T,* and unlikely to become one soon. Still, as an electronics technologist, you'll write often. Even if you don't write the manuals, instruction guides, and brochures accompanying the products and services your company provides, you'll most likely be asked to edit such materials, at least for technical accuracy. Your willingness and ability to do so will enhance both your job performance and your visibility.

However, all this professional writing aside, just communicating in writing with colleagues, engineers, supervisors, managers, and salespeople is a necessity. Even if you're never known as a technical writer, you are going to do some technical writing. Let's get started seeing how.

5.2 PROPOSAL WRITING

Although earlier we made a distinction between sales writing and technical writing, there is one form of the latter that incorporates elements of the former—proposal writing. Proposals are designed to sell something: an idea, a concept, a service, a piece of equipment, or a complex system. Not all proposals represent technical writing, of course. Yet, the kind you're likely to write or edit are sure to be technical documents. The Concepts and Requirements Document you will write is a proposal. It attempts to sell others on the building of a prototype project. In this section we examine the proposal and how to write it.

Sell Me On It

What does the proposal accomplish? What kinds of proposals are there? How is the Concepts and Requirements Document a proposal? Let's find out.

Problems and Solutions

All proposals, no matter how simple or complex, delineate a problem and lay out a solution. A memo written to your boss justifying the purchase of a new computer is as much a proposal as the 2000-page, six-volume tome a large aerospace company prepares for developing a light rail transit system. In either case, the problem is first spelled out, then a solution proposed. A proposal offers a plan to solve a problem.

In clarifying the solution, the proposal usually addresses the *how, who, where, when,* and *how much will it cost* questions. How will the solution be accomplished? Who will do it? Where and when will it be done? How much money is involved, and where will the money come from? Whether in one paragraph or one volume, these questions in most cases must be answered in the proposal.

Types of Proposals

When an external organization, particularly the government, requires a product or service your company can provide, it issues what is known as a *request for proposal* (RFP). Your proposal is prepared and presented in response to the RFP.

A proposal can be simple, only one or two pages written in letter form, or huge, requiring the efforts of personnel in sales and marketing, engineering, accounting, and publications. Regardless of length, a proposal should contain six elements, as Weisman discusses in *Basic Technical Writing:*

1. An *introduction,* explaining why the proposal is written and giving any background information
2. A *technical presentation,* describing the proposal plan for doing the work
3. A *technical description,* representing the creative contribution by the proposer for solving the problem
4. *Capabilities,* explaining how capable, in terms of facilities and personnel, the organization is of carrying out the work
5. *Programming,* clarifying who will do what, when
6. *Cost schedules,* delineating equipment, facilities, and salary costs

Often, if it's quite large, the final document may be published as three subproposals:

1. A *technical subproposal,* presenting the engineering solutions and technical specifications for the customer's problem independent of the cost.
2. A *contractual and cost subproposal,* covering the total project cost (broken down by phases). Cost can include materials, labor, research and development, and administration.
3. An *abstract,* highlighting the key points in a short, nontechnical document for the busy high-level executive who lacks the time to read the whole proposal.

The Concepts and Requirements Document as a Proposal

The Concepts and Requirements Documents in this text are short, usually two pages, yet each one contains most of the six proposal elements listed above. Under Project Goals we have an introduction and technical presentation. With Project Objectives and Theory of Operation, we see a technical description. Project Responsibility tells who will be doing what, when. Even capabilities are implied: a student will be building the prototype and preparing the Project Report in class.

Although the Concepts and Requirements Documents in the book do not follow, literally, the half-dozen steps outlined above, your document could. In preparing a Concepts and Requirements Document, pursue any approach emphasizing the document as a proposal.

Writing Your Proposal

Now, let's turn to proposal writing. We look first at the reader, then at outlining, and last at writing the first draft.

Getting to Know You

To write is to communicate. As a writer, you're sending a message to a recipient, the reader. For the reader to comprehend what you're saying, your written words must be presented within the framework and experience of the reader. Therefore, before you write anything, get to know your reader.

Most readers, particularly those in government and corporate America, are, as William Zinsser, author of *On Writing Well,* says, "Impatient birds, perched on the thin edge of distraction or sleep." Typically, they're busy, frequently interrupted, often lack knowledge of your product or service, and must share decision-making authority with others, most of whom have many different backgrounds and interests.

So, how do you get to know such readers? By listening to what they have to say. That means demonstrating an interest in their problems. William S. Pfeiffer, in his book *Proposal Writing,* suggests three approaches:

1. Spend most of your time asking questions in early conversations with the client or manager, rather than talking about what you have to offer.
2. Show patience in listening to responses, even when conversations depart from the immediate subject at hand.
3. Take careful notes, both to demonstrate the importance you place on their comments and to retrieve information to use later in writing the proposal.

Once you have a handle on your reader, or readers, it's time to begin an outline.

Planning Ahead—Outlining

Essay tests! You probably don't like them, but they're a part of college. The problem with such tests is that they seem open-ended. Hence, the test taker struggles, trying to figure out where to begin. Most instructors will tell you, "Don't dive right in when presented with an essay question." Instead, they suggest taking a few minutes to outline your answer, to develop a plan highlighting the main and subpoints you plan to make in response to the question. Then you can proceed with confidence, writing sentences that in effect expand your outline.

The same approach, with slight variation, works for any technical writing assignment. Outlining helps you organize information. It gives you a plan.

The first step in outlining really isn't outlining but brainstorming. The purpose is to get on paper, anywhere on paper, ideas and facts you may eventually include in your report. Don't concern yourself with wording, redundancy, sequence of ideas, or other fine points. Just get information on paper.

Now, highlight your main points. Connect subordinate points to the main points they support. Delete redundant or weak points. Then, number the circled main points according to their anticipated sequence in the proposal. According to Pfeiffer, "If at this point your paper is a mess, you're doing it right."

Next, establish order, a hierarchy of thought. Pfeiffer suggests:

1. Placing points in the order they will appear in your draft.
2. Clustering minor points below their respective major points.
3. Showing relationships with dashes, indentation, underlining, the familiar Roman numeral system, or any other arrangement that works for you.
4. Putting all points in either topic or sentence form.

The topic outline is preferred. You're familiar with it from your middle-school years. It reflects subordination and relationship, revealing a breakdown of greater detail. This chapter, indeed the entire book, was outlined in such a manner.

Every chapter is composed of first-, second-, and third-level headings, corresponding to Roman numerals, uppercase letters, and Arabic numerals, respectively. Your outline might not go to three levels, or it could go to four, even five. Regardless, keep an important point in mind. Since a topic is not divided unless there are two or more parts, there must be at least two subheadings under any division. Here is the typical pattern:

I.
  A.
    1.
      a.
      b.
    2.
  B.
II.

Finally, remember that the outline is only a tool. As with the document writing itself, the outline is subject to constant revision. Don't hesitate to make changes.

Writing Is Rewriting

"Why can't we do it right the first time?" How often have you heard that one? Whether repairing a VCR or installing a new computer network, we are constantly being implored to do it right, right away, so we don't have to revisit the problem over and over again.

When I was in college, a while back, I heard the same thing from my English teacher. The structure of the class and the writing technology of the time left her no choice. With pen or typewriter, but no word processor, revision of an essay in class or a term paper at home was close to impossible. Your first draft was your last.

Too bad, because that's no way to write. "Writing is rewriting," as anybody who does it for a living will attest. It's a process of constantly reworking your last draft: pruning, cutting, simplifying, correcting. How many such revisions are necessary? As many as it takes until you're satisfied with the result or simply run out of time.

Fortunately, today's word processing technology allows, even encourages, revision. Take advantage of it. As you will see in the next section, the editing process is a revising process—but one made easier with today's word processors.

> ### Something to Think About
>
> *Gather two or three good proposals. As you read them, break each one down into its outline. If you discover a topic outline hidden within, identify each first-, second-, and third-level heading.*

5.3 EDITING—CLEANING UP YOUR ACT

It is not necessary to be a writer to write well. Many technical communicators, from Charles Darwin to Paul Davies, were not writers first. They were biologists, inventors, and technologists who could write well. As an electronics technician, you may write only occasionally. When you do, you'll want it to be your best.

In this section we look at the technical writing style, with its emphasis on brevity, conciseness, and the active voice. Next, we examine the comma, the key element of punctuation. Finally, we investigate a technique for pruning every unnecessary word from your text. Making it short and sweet will be our goal.

The Technical Writing Style

First, we examine the value of "less is best." Then we distinguish the active from the passive voice.

Say No More

William Zinsser knows what good writing is. He sums it up beautifully in his book *On Writing Well:*

> Good writing is to strip every sentence to its cleanest components. Every word that serves no function, every long word that could be a short word, every adverb that carries the same meaning that's already in the verb, every passive construction that leaves the reader unsure of who is doing what—these are the thousand and one adulterants that weaken the strength of a sentence. And they usually occur in proportion to education and rank.

George Orwell said much the same thing:

> Never use a long word where a short one will do. If it is possible to cut a word out, always cut it out. Never use the passive where you can use the active. Never use a foreign phrase, a scientific word, or a jargon word if you can think of an everyday English equivalent.

Jargon is of particular importance to us in technical fields. We love our vocabulary. It separates us, makes us feel unique, superior. But we must shun it whenever possible. *Input, output, throughput, infrastructure, interface, articulate, prioritize—* these are the clichés of our field and should be avoided or used sparingly.

Furthermore, strive to shorten your words. "Small is beautiful." Replace the words in the left column with those in the right column whenever you can (William S. Pfeiffer):

advantageous	helpful
alleviate	lessen, lighten
commence	start, begin
discontinue	end, stop
endeavor	try
finalize	end, complete
initiate	start, begin
principal	chief, main
prioritize	rank, rate
procure	buy, get
subsequently	later
terminate	end
utilize	use

Do the same with phrases, especially trite ones. Again, replace what's in the left column with its one-word equivalent in the right column:

afford an opportunity to	permit
along the lines of	like
an additional	another
at a later date	later
by means of	by
come to an end	end
due to the fact that	because
during the course of	during
give consideration to	consider
in advance of	before
in the final analysis	finally
in the neighborhood of	about
in the proximity of	near

It all comes down to this: write like a person, not like a technician. Remember, simple writing is not the product of a simple mind—far from it. As Zinsser puts it: "Actually a simple style is the result of hard work and hard thinking; a muddled style reflects a muddled thinker or a person too arrogant or too dumb or too lazy to organize his thoughts."

The Active Voice

Most writing coaches tell you to use the active voice over the passive. They're right, and we'll see why in a moment. First, however, let's be sure we know the difference between active and passive verbs.

Active verbs usually transfer action from the subject to a direct object, whereas passive verbs describe an action performed on the subject. A sentence written in the active voice is about agents; one written in the passive voice is about actions.

For example, "We *will erect* six buildings at the site" is in the active voice. The emphasis is on the agent of the action. "Six buildings *will be erected* by us at the site" is in the passive voice. The emphasis is on the action being performed rather than on the agent.

Although some writers insist the passive voice is more objective, more "scientific," most readers view passive-voice writing as boring and often confusing. Write in the active voice if you want to shorten your sentences, make them more forceful, and make clear the agents of action. If the action is clearly more important than the agent, write in the passive voice, but do so only when there is no other choice.

Punctuate and Eliminate

Correct punctuation, even more than grammar per se, is the key to good writing, and of all the marks of punctuation, the comma is most critical. Therefore, in the limited space left, we will concentrate on the comma.

Word elimination is the road to concise writing. As we conclude our chapter on technical writing, we show you ways to trim the word fat.

Comma Studies

I once had a 12th-grade English teacher who said there are 22 uses for the comma. No wonder I was confused. Then I discovered Robert Brittain's *A Pocket Guide to Correct Punctuation,* and my outlook brightened. In his marvelous little book, still readily available, Brittain says there are only three uses for this indispensable mark of punctuation: as a single comma, a pair of commas, and as a comma-plus-coordinating-conjunction. Now *that* I can deal with. So can you.

The *single comma* is used in only one situation: in place of a word.

Bill wore a blue suit; Jim, a brown one.

A short, fat, giggling man walked quickly into the room.

He had red, yellow and blue, green, and gold balloons.

In the first example, the comma takes the place of the word *wore*. In the second, both single commas take the place of the word *and*. In the third sentence, the single commas also take the place of the word *and*. With commas inserted, we know there are only four balloons, not five. Even though the last comma is technically not required, inserting it eliminates any confusion.

A *pair of commas* is the most important single punctuational symbol we possess. It is composed of two symbols (commas) always separated from each other by a word or group of words. Yet, they are not two single commas.

What does a pair of commas mean? According to Brittain:

The pair of commas means, the element enclosed within this mark of punctuation is not essential to the grammatical structure of the sentence and is placed in such a position that it interrupts or changes the normal order.

It is important to note: the pair of commas must both enclose a nonessential grammatical structure and interrupt the normal order in a sentence.

Linda, when Ms. Smith corrected her, began to cry.

Before we decide what to do, however, let's eat lunch.

As I told you, the situation is not as bleak as it looks.

The president then recognized Mr. Edwards, the delegate from Los Angeles.

Each of the last two examples has a pair of commas, but in both cases, only one comma is visible. Where's the other comma? A more emphatic mark of punctuation absorbs it. Marks indicating the beginning and ending of sentences—a capital letter and period, exclamation mark, question mark—are more emphatic than the comma.

A coordinating conjunction, such as *and, but, or, nor, for, so, yet,* is often used to join two independent clauses, clauses that can stand alone as sentences. But when they do so, a comma must accompany them in the form of a *comma-plus-coordinating-conjunction.* As Brittain states:

> The comma-plus-coordinating-conjunction means, at this point in the sentence, one independent clause has been completely stated, and the second one is about to begin.

It is true she needs to rest, but I think we should wake her before 3 p.m.

Either you will take the package to Larry's house, or I shall have to go myself.

Mr. Smith, the new captain, read the first statement, but the second statement was delivered by the private.

Word Chop

After you've completed your first draft, go through it, cutting and trimming wherever you can. Now, grab a felt-tip highlighter, say a yellow one. With pen in hand, read through your document from beginning to end, highlighting every word from the list below:

and

or

it

here

fact

kind

sort

type

that

of

How many of the words you have just highlighted are absolutely necessary? I'll bet you can eliminate at least 50% of them.

Next, go through your document again, with a different color, highlighting words found in this second list:

by

not

is

are

was

were

have

exist

words ending in "tion"

words ending in "ly"

Often these words signal the passive voice. Sometimes they can be changed.

Finally, with yet another color, highlight these words:

by

to

for

toward

on

at

from

in

with

These words indicate prepositional phrases. Rephrase such sentences and you'll shorten your document even further.

That will have to do it—we're out of space. We hope you've gained a few insights on technical writing and the technical writing process. As mentioned earlier, for a more in-depth analysis, see the listings in Appendix G.

Something to Think About

Why not "play" editor? Volunteer to read over a proposal written by a fellow student. Using the editing techniques learned here, dive in, edit the document. You and your colleague will be better for it.

SUMMARY

In this chapter we began by examining technical writing, what it is and how it differs from creative writing. We looked at the career of technical writer and why you, a future electronics technologist, should want to become proficient in technical writing.

Next, we explored the proposal, what it is and how it is written. We reviewed types of proposals and examined how the Concepts and Requirements Document is, in effect, a miniproposal. Finally, we saw how to write a proposal by looking first at the reader, next at outlining, and then at the rewriting process.

It was then on to editing, or cleaning up our act. We saw ways to cut jargon and long phrases from our drafts and how to write in the active voice. We concluded with a study of the comma and how to chop, even further, word fat from our technical documents.

In Chapter 6 we get on to experimenting, breadboarding, to prove out our design. We discuss soldering and different breadboarding approaches, and we examine troubleshooting techniques relevant to the breadboarding stage.

QUESTIONS

1. Technical writing, unlike creative or advertising writing, is designed to _____ rather than to entertain or sell.

2. There are excellent jobs for full-time technical writers, particularly those having a strong grounding in a _____ subject.

3. All proposals, no matter how simple or complex, delineate a _____ and lay out a _____.

4. When an external organization, particularly the government, requires a product or service, it issues what is known as a _____ _____ _____ (RFP).

5. The best way to get to know your reader is by listening to what he or she has to _____.

6. The first step in outlining really isn't outlining but _____.

7. If it is possible to cut a word out, _____ cut it out.

8. Never use a long word where a _____ one will do.

9. A sentence written in the active voice is about _____; one written in the passive voice is about _____.

10. Basically, there are only _____ uses for the comma.

REVIEW EXERCISES

Suppose that you're an electronics technician working at a small electronics contract manufacturing firm. The company employs 25–30 assemblers to stuff printed circuit boards with electronic components.

The personnel turnover rate in the assembly area is high, and new hires tend to be inexperienced and lacking in electronic component familiarity. An assembler needs to have a series of **Electronic Component Physical Description Sheets** depicting and describing the physical characteristics of each component to be installed. The purpose of each sheet is to allow the assembler to quickly and accurately identify the

Electronic Component Physical Description Sheet
Electrolytic Capacitor

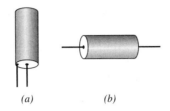

(a)   *(b)*

* **Function.** An electrolytic capacitor is a polarized electronic component used to store electricity.

* **Shape.** Electrolytic capacitors are usually cylindrical in shape.

* **Size.** Electrolytic capacitors range in size from 1/4" to 3" in diameter, and 1/2" to 6" in length.

* **Material.** Most electrolytic capacitors have an outer casing made of aluminum. The casing may or may not be coated with a plastic sleeve. Some electrolytic capacitors have an outer coating of ceramic.

* **Lead Configuration.** Electrolytic capacitors come in two lead configurations: axial and radial (see figure above). Axial leads protrude from the sides of the cylinder, radial leads from one end.

* **Color.** Electrolytic capacitors come in a variety of colors: black, blue, and silver are the most popular.

* **Polarity Designations.** In addition to printed markings, discussed below, the positive lead of an electrolytic capacitor is identified by physical characteristics in one of two ways: (1) For radial leads, the positive lead is the longer of the two. (2) For axial leads, the positive lead emerges from the side where the cylinder indents.

* **Markings.** Most electrolytic capacitors have numerous markings, only three of which are of concern to the assembler. First, the *value* of the capacitor is indicated in microfarads (µF). For example, 1 µF, 220 µF. Second, the *voltage* of the capacitor is stated in volts. For example, 10 V, 35 V. Third, a *polarity* indicator will be printed near the positive lead, or negative arrows will point to the negative terminal.

* **Special Considerations.** Some very large electrolytic capacitors, known as *cans,* have screw terminals instead of leads. A red dot is usually placed next to the positive terminal.

correct electronic component, determine its electrical value, know if it has polarity (and if it does, what polarity), and recognize the special physical characteristics affecting its selection and proper installation. With a three-ring binder of individual **Electronic Component Physical Description Sheets** close at hand, the assembler will be greatly aided in performing his or her assembly job.

Your assignment is to prepare an **Electronic Component Physical Description Sheet,** similar to the sample shown on page 95. Pick any component you like. Consider the sample sheet as a guide only. Feel free to structure your sheet in any way you determine is most effective in helping the assembler identify an electronic component and install it in the correct manner.

6 Experimenting—Breadboarding to Prove Out the Design

OBJECTIVES

In this chapter you will learn

- [ ] The differences among project breadboarding, prototyping, and production.
- [ ] About solder, soldering irons, and how to solder and desolder.
- [ ] How to breadboard using the cut-slash-and-hook (CSH) method.
- [ ] How to breadboard with solderless wiring terminals and solderless circuit board.
- [ ] How to breadboard using wire wrapping.
- [ ] How to breadboard with the universal printed circuit board and etchless printed circuit board methods.
- [ ] How to breadboard the Variable Power Supply and 3-Channel Color Organ Projects.
- [ ] How to troubleshoot breadboarded projects.
- [ ] How to document the breadboarding stage.

It's time to move off the board, or away from the computer, and onto the bench, to put down the pencil, template, and mouse (for a moment) and pick up the pliers and diagonal cutters. It's time to grab a handful of electronic components and start to build. You are ready to move on to the experimenting, or breadboarding, stage. You're ready to prove out your electronic design—to see if what's on paper can really be made to work.

In this chapter we examine the differences between breadboarding and prototyping; we discuss solder, soldering irons, soldering, and desoldering;

and we explore five popular methods of breadboarding electronic projects. We will go on to breadboard the Variable Power Supply and 3-Channel Color Organ Projects. Finally, we'll investigate some common troubleshooting techniques used at this, the experimental, stage.

6.1 MAKING IT, STEP BY STEP

Before clipping components together and powering up circuits, let's examine the reasons for breadboarding and the skills necessary to do it.

Trying It Out Before Going All Out

Why breadboard? Why not go all out, go for broke—now? After all, you have a working schematic, it looks good, and that means the project is sure to work; and, most importantly, you've used computer-based circuit design and analysis tools to build a virtual circuit. Call up production and tell them to tool up. Tell them to order the components and PC boards for 10,000 units. Why waste time? No need to experiment or build prototypes. Who knows where the competition may be? We've got to take a chance.

But hold on a moment. True, many a company has had just this kind of attitude. But notice the word *had*. The surest route to catastrophic failure is to skip the breadboarding and prototyping stages and jump right into production. Software circuit design, simulation, and analysis tools aside, every project, no matter how simple in design or concept, must be thoroughly analyzed and tested through the building of at least one breadboard and one prototype unit. Only then can we consider mass production.

Proving Out the Design

Electrons don't run around on the schematic drawing. A design may work in theory, but that simply isn't good enough. We need to see it work in practice. We need to prove out the design.

At this point we are not so much concerned with project layout and packaging but with whether or not the circuit functions—that is, if the circuit design is valid. To determine that, we need a quick and easy method of assembling the components into a functioning unit. Such a technique must allow us to change or substitute components quickly and easily. Minor and even major circuit modifications will often be required. We need to experiment with new designs; in other words, we must breadboard.

Breadboarding is a circuit assembly system that allows components and interconnections to be assembled and changed in their design stage easily. There are many breadboarding systems to choose from, and shortly we'll be examining the five techniques most widely used. But first, let's see just what the differences are among breadboarding, prototyping, and production.

Breadboarding, Prototyping, Production: Where to Draw the Line

There is some disagreement as to where to draw the line between the breadboarding and prototyping stages. Is wire wrapping, for example, a breadboarding or prototyping method? Actually, wire wrapping has been used not only in both of these stages but in the production stage as well. Solderless circuit board is definitely part of breadboarding, but many prototype projects also use it. To understand what methods work best at each stage, you need to remember some definitions.

In breadboarding, it is essential to produce easily temporary circuits using available parts. The prototype, although also easy to produce, must be rugged and capable of operating under severe conditions. The prototype should be suitable for complete evaluation of mechanical as well as electrical form, design, and performance. Approved parts must be used. In other words, the prototype must be completely representative of the final, mass-produced equipment.

Figure 6–1 illustrates the characteristics and the construction methods used in the breadboarding, prototyping, and production stages. Although some of the approaches identified with the breadboarding stage could easily move down a notch or two (and certainly would have a decade or more ago), in this book we stick with the breakdown as

---

Breadboarding Stage

Construction method that results in a temporary, easily changeable circuit

- ☐ Cut-slash-and-hook
- ☐ Solderless wiring terminals
- ☐ Solderless circuit board
- ☐ Wire wrap
- ☐ Universal printed circuit board

Prototyping Stage

Construction method that results in a rugged circuit that is suitable for complete evaluation

- ☐ One-of-a-kind printed circuit board

Production Stage

Construction method that results in a permanent, reliable, and easily built mass-produced product

- ☐ Printed circuit board

FIGURE 6–1
Project construction methods

---

shown. For us, a dedicated, one-of-a-kind printed circuit board is the way to go at the prototyping stage. All the other nonproduction methods will find their place at the breadboarding level. Which one is actually used in a given situation depends on the type of circuit being breadboarded.

Breadboard Documentation

You can't spend every minute of the breadboarding stage on the bench. A little time is required on the board and at the word processor. Some documentation will be needed.

At this stage a breadboard drawing and an Experiment Results Document are produced. The **breadboard drawing** is nothing more than a simple component layout to be used as a guide in assembling the breadboarded circuit. The **Experiment Results Document** is a statement about what happened during the experimenting, or breadboarding, stage. It can be as short as one paragraph or run into many pages of explanation.

Breadboard Drawing

To be perfectly fair, the breadboard drawing is not always required. If the project is extremely simple and the breadboarding construction technique very familiar, you can probably skip this drawing altogether. Yet, for the most part, making such a drawing is well worth the little effort required. Having

FIGURE 6–2
Breadboard Drawing:
Variable Power Supply

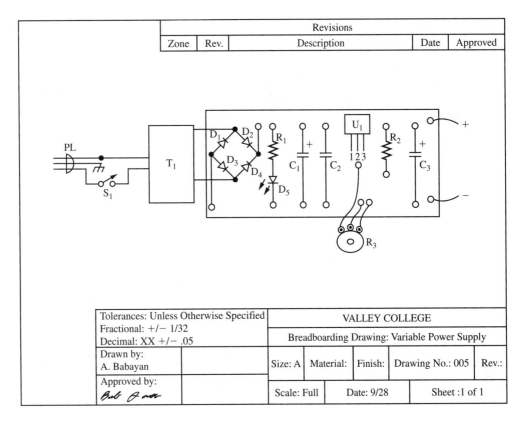

some idea as to where the components should go on the breadboard before beginning construction is always a good idea. Breadboarding can save more than minutes if things go astray simply because a component placement was not considered.

Figure 6–2 shows the breadboard drawing for the Variable Power Supply Project, to be constructed on solderless circuit board. The main idea is to account for all major components. With this drawing, sketched on $8\frac{1}{2}'' \times 11''$ graph paper, you can see where components will go. You can proceed with the confidence that there is room for your components and wires and have some idea as to how it should all be laid out.

Experiment Results Document

The Experiment Results Document tells what happened when the circuit was built. Did it work? Did it blow up? Did it work—but? What changes had to be made to get the project to do what the designer intended?

In an extreme case, the Experiment Results Document could become an obituary—the last thing said about a failed design and a failed project. It is just possible that the original plan is so flawed that the project cannot be saved. The circuit simply doesn't work, and it is best to cut our losses—now. If that is the case, this is the time to say so. We will have to explain why nothing more can or should be done with the project.

However, in most cases, circuit modification, not circuit abandonment, is the way to proceed. We need to explain what changes were required, and why. The Experiment Results Document is a brief statement about where we are now. It begins by stating a conclusion and then details the test results that support that conclusion. Of course, the circuit hasn't been tested under all user or environmental conditions. That will come with the prototype, but now, with a working circuit and the appropriate documentation in hand, we are ready to move on to that next fabrication step.

Something to Think About

Breadboarding is an odd word to describe a circuit assembly system allowing for quick connection and disconnection of components and wiring. Why not investigate the term's origin? The result may surprise you.

6.2 SOLDERING: THE TIE THAT BINDS

For well over 75 years, using basically the same materials and essentially the same procedure, technicians have employed **soldering** as the "tie that binds" in electronic circuits. For a secure electrical and mechanical joint between components and wires, nothing has ever replaced it, or is likely to in the foreseeable future. As an electronics technician you will need to know how to solder and desolder, and how to do both well. At the production stage it is mandatory; it is a rare prototype project that does not contain at least a few solder joints. Furthermore, in many cases even breadboarding requires some soldering here and there.

In this section we begin by examining soldering characteristics. We also look at **flux** and its role in the soldering process. Next, we investigate the **soldering iron**, looking at basic construction, various types, temperature characteristics, and **soldering iron tips**.

We then discuss the soldering process itself. We probe basic methods, the soldering on printed circuit boards, point-to-point soldering, the characteristics of good and bad solder joints, and the need for and means of flux removal.

Finally, we examine desoldering equipment and techniques. Throughout, the discussion will be confined to hand soldering and desoldering. We'll explore "automatic" wave and dip soldering in Part II, which deals with surface mount technology.

Solder: A Liquid Bond

Before we can learn to solder and desolder, we must investigate solder itself: what it is and what it does. Then we need to examine the all-important role of flux in the soldering process.

Solder Characteristics

Solder is an alloy (mixture) of tin and lead. The exact ratio of tin to lead determines its strength, hardness, and melting point. Solder can also be successfully alloyed with other metals such as copper. When the solder applied to copper melts, a thin film of copper is actually dissolved from the surface, forming an alloy that is part solder and part copper, with characteristics all its own. An intermetallic bond is formed between the parts. Thus, solder is not a glue: it does not stick component leads and copper traces together. It doesn't fuse them, as in welding, either. A new electrically conductive and mechanically strong material (alloy) is actually formed when the soldering process takes place.

Solder is available in various mixtures of tin and lead. A 63% tin and 37% lead combination (63/37) forms what is known as a **eutectic** (ideal melting point) composition. Such a mixture has two important characteristics. One, it has the lowest melting point of any tin/lead combination—370°F. Two, when going from solid to liquid, a 63/37 solder

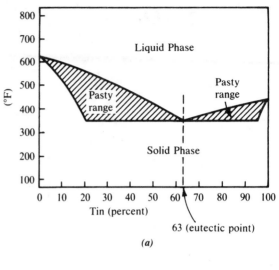

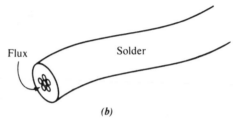

FIGURE 6–3
Solder

does not go through a pasty stage (Figure 6–3a); it goes straight from solid to liquid. Why are these characteristics important? First, a low temperature is obviously desirable because if a lower temperature is sufficient to melt the solder, less component damage will occur due to excess heat. Second, when component leads are accidentally moved during the pasty phase, cold (bad) solder joints tend to form. If there is no pasty stage to begin with, the formation of cold solder joints is minimized.

Given the obvious advantages of 63/37 solder, why, one might ask, isn't all solder 63/37? There are at least two reasons. One, tin is more expensive than lead. Therefore, the higher the tin-to-lead ratio the more expensive the roll of solder. If you are buying hundreds of pounds of solder a year, the difference can be significant. Two, slightly less tin but more lead will result in better wetting action—the ability to spread an alloy over the entire metal surface. Given these two factors, a good compromise for hand soldering is a solder consisting of 60% tin and 40% lead (60/40). Such solder costs less, has excellent wetting action, and only a slight pasty stage (Figure 6–3a). Also available is a solder with 62% tin, 28% lead, and 2% silver. It is used extensively with surface mount devices. We will examine it more closely in Part II.

Solder is available in both bars and wire. The former is melted down to be used with wave and dip soldering processes. Wire solder is the type we use in hand soldering.

Wire sizes range from 0.090″ to 0.015″ in diameter. The size you choose depends on the application. A 0.031″ diameter solder (#21 gauge) is the largest you will want to work with. Solder 0.025″ in diameter (#23 gauge) is better for most applications; however, 0.020″ diameter solder (#24 gauge) is excellent for work with integrated circuits. Finally, 0.015″ diameter solder (#28 gauge) is the choice for use with surface mount components.

Virtually all wire solder comes with a core of rosin **flux** (Figure 6–3b). As the solder melts, the flux flows out onto the connection being joined. It makes little difference if the core of flux is solid or stranded (multicore).

Soldering Flux

Plain solder, that is, solder without any flux in its core, cannot adhere to metal surfaces. This is because such surfaces are covered with *oxide films*. These films are created when oxygen combines with metal and forms an oxide layer. Rust is a good example.

Oxide films interfere with the solvent, or alloying, action that must take place for a good solder joint to form. The problem is that oxidation occurs rapidly when the temperature of metal is raised, as when applying the tip of a hot soldering iron to the joint to be soldered. The oxides formed must be removed immediately before soldering takes place.

Fluxes can dissolve oxides that have developed and prevent new ones from forming when heat is applied. (They also lower surface tension and aid its wetting action.) When rosin-core flux is used, oxides are brought to the surface while solder is molten. The leftover oxides and flux form a slag around the joint, which is later cleaned off with a suitable flux remover.

Fluxes used in soldering are almost always rosin-based. Rosin is a mixture of organic acids extracted from pine trees. At room temperature, these fluxes are noncorrosive. But as their temperature rises, they become "active," raising their corrosive characteristics. It is this activity, however, that causes the deoxidizing action.

The corrosive nature of rosin-based flux is not a problem, since its corrosive action takes place for only a short period of time as the solder joint is heated. When the joint is returned to room temperature, corrosion ceases. Nonetheless, flux residue should be removed from a printed circuit board after soldering for at least three reasons. One, the flux is sticky and may cause foreign particles to adhere to the board and create bridges and shorts. Two, with

the flux removed, the printed circuit board is easier to inspect for defects. And, three, the board looks better when the flux is eliminated. We will look at how flux is removed shortly.

Flux, as we have seen, is embedded in solder (rosin-core solder). It is also available separately in paste or liquid form. The former is used when cleaning and tinning a soldering iron, the latter to add flux when needed in soldering and desoldering.

Soldering Irons

We briefly discussed soldering irons in Chapter 1. Here we will look at this essential tool in more detail. We'll examine soldering iron types, the temperature factor, and the all-important soldering iron tip.

Types of Soldering Irons

The conductive soldering iron is used to administer heat to the joint to be soldered. Solder is then applied to the heated connection (not to the iron), which melts the solder. In its simplest form, such an iron consists of a *resistance heating unit* (coil connected to the ac line), a *heater block* to act as a heat reservoir, and the metal **soldering tip**, which is a pipeline for heat flowing to the work (Figure 6–4).

There are two types of soldering irons that are of particular interest to those in prototype work. The least-expensive all-around iron for electronic assembly is a 30-W pencil-type with easily interchangeable tips (Figure 1–22). Such an iron is easy to handle and, with proper care, can last many years. The iron should have a grounded tip to prevent electrostatic discharge (ESD). Prices range from under $10 to over $50.

A temperature-controlled soldering iron "station" (Figure 1–22) is very handy because it provides precise control of tip temperature. Some types

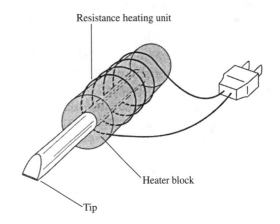

FIGURE 6–4
Soldering iron elements

consist of a closed-loop circuit with a temperature-sensitive tip. They are designed to maintain a set tip temperature throughout the soldering process.

Soldering Iron Temperature

Controlling and maintaining the tip temperature of your soldering iron is important, though not critical. A good iron will maintain its idle temperature with little variation, but when the soldering iron tip contacts the joint to be soldered, a myriad of factors act to cause temperature to vary. *Relative thermal mass,* determined by the mass of the metal to be heated, *surface condition,* oxides or contaminants covering the pads or leads, and *thermal linkage,* the area of contact between the metal tip and the work area, all play an important role, especially in production work. In prototype construction, however, where soldering is done at a more leisurely pace, these factors are of less concern. Using a quality iron with a proper tip is the best way to ensure excellent heat transfer to the solder joint.

Soldering Iron Tips

The purpose of the soldering iron core, we must remember, is to transfer heat to the all-important tip. Let's see what these tips are made of, what different shapes are available, and how such tips should be maintained.

Tips may be either *unplated copper* or *iron-plated* (clad). While the former transfer the maximum amount of heat, they require frequent dressing and tinning, since the tip wears away with use as the solder dissolves some of the copper. Unplated copper tips are rarely used in prototype work.

Iron-plated tips demand only occasional surface cleaning and no traditional tinning, whereby the tip is filed clean before heating. With this type of tip, the entire surface is protected against scaling. Though iron-plated tips are more expensive than the unplated variety, their added cost is justified by lower maintenance. They are the choice for all prototype and most production work.

Tips come in a wide choice of shapes, the most popular of which are shown in Figure 6–5.

☐ The *conical-* and *pyramid-*style tips are used for general assembly and repair work.
☐ *Bevel* designs allow for rapid heat transfer and are used for soldering terminal pad connections on single-sided PC boards.
☐ *Chisel-*style tips allow for large areas to be heated rapidly. They work well where point-to-point soldering is called for.

To some extent tip style is a matter of individual preference. The best approach is to maintain a vari-

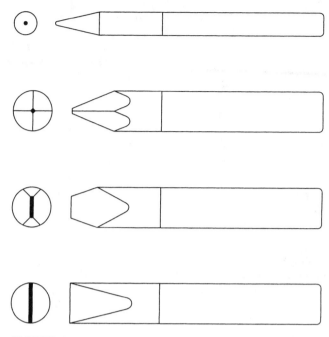

FIGURE 6–5
Soldering iron tips

ety of styles and experiment with each under varying soldering conditions.

Soldering iron tip maintenance is essential. A poorly maintained tip installed in the best soldering iron will render the entire tool useless. Here are some tips to follow in maintaining your soldering iron tips:

☐ Make sure the tip is fully seated (installed) in the heating element and tightly attached to the iron.
☐ Remove the tip daily to prevent an oxidation scale from accumulating between the heating element and the tip and setscrew.
☐ First, clean a plated tip while cold with a fine grade of steel wool until the surface is bright. Then, heat the iron and apply solder as the tip warms up. This is known as *surface tinning.*
☐ Before using the iron, wipe the tip lightly and quickly across a damp sponge to remove any oxides.

Your soldering iron is perhaps the most important hand tool you will use in electronic project fabrication. Treat it at least as well as you do your expensive digital meter or oscilloscope, and it will perform well for you day after day.

Soldering

We have looked at solder and soldering irons; now it's time to turn to the deed itself—soldering. We examine the basic soldering procedure, soldering on printed circuit boards, soldering point-to-point, and flux removal.

Soldering: The Basic Procedure

To solder an electronic project, you will need, at a minimum, three basic items: a soldering iron, solder, and a damp sponge. Recall, the purpose of the iron is to heat the joint. It is the heat from the joint that melts the solder. To rid the iron of accumulated gook (contaminants), wipe it frequently on a damp sponge.

After cleaning the joint to be soldered of all dirt and grease, follow these four steps in completing any solder connection:

1. With the iron, apply heat to the connection.
2. Apply solder to the heated connection.
3. Remove the solder from the heated connection.
4. Remove the source of heat (the soldering iron).

Of the four steps, which should take a total of 2 to 3 seconds, the last one is perhaps the most critical. In order to ensure a smooth, even wetting action of the solder, keep the iron on the joint a "long instant" after the solder has been removed. If the iron is taken away too quickly, the solder connection will not form well.

Soldering on Printed Circuit Boards

Soldering on a printed circuit board, particularly a single-sided board, requires special caution due to the heat sensitivity of the board pads and traces. Too much heat, applied with too much pressure, can quickly lift a pad or trace from the board laminate. To avoid this catastrophe, apply the iron tip with a light touch for no more than 3 seconds. The iron should rest on the pad rather than be pressed to it. (*Note:* For details on hand-soldering surface mount devices to a printed circuit board, see Chapter 13.)

After the component lead has been inserted and clinched (see Chapter 9) and the joint area cleaned of all contaminants, proceed as follows to create a good solder connection:

1. Apply the soldering iron to *both* the component lead (or wire) and copper foil pad (Figure 6–6a-1).
2. Next, apply solder first to the same side as the tip, then quickly bring it around to the opposite side (Figure 6–6a-2). The tip should always remain in contact with the lead and pad, as should the solder wire.
3. After the joint cools (give it about 10 seconds), clip the excess lead off as close to the board as possible.

FIGURE 6–6
Soldering on a printed circuit board

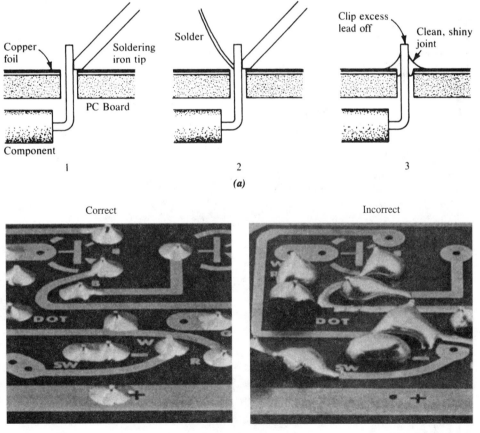

Now, inspect the solder joint. The surface of the solder should be smooth, well feathered at the edges, and nice and shiny. It should also have a slightly concave configuration (Figure 6–6b).

Poor solder connections (Figure 6–6b) are evidenced by *cold* and *disturbed joints*. The former is a result of withdrawing heat too soon; the solder does not become liquid. In appearance, the joint is dull gray. A disturbed joint, the result of movement of leads or wires during solder solidification, is characterized by a granulated and frosty appearance, often with many cracks showing.

In soldering integrated circuits to a printed circuit board, a few specific steps are worth noting:

☐ Heat damage to the component must be avoided. Try to complete the soldering of a given pin in no more than 3 seconds.

☐ After inserting the IC, clinch two leads (pins) on opposite corners to hold the IC in place while completing the soldering.

☐ When soldering the lead to the pad, touch the iron tip to one side of the joint while applying solder to the opposite side.

☐ To avoid heat buildup when soldering a row of pins, alternate so that leads next to each other are not soldered consecutively.

Soldering Point-to-Point

When soldering point-to-point, follow the steps shown in Figure 6–7a:

1. Make a good physical connection with the component lead or wire and the terminal post.
2. Heat both the lead (or wire) and the terminal.
3. Apply solder to the connection and then remove the heat.

Allow the connection to harden for about 10 seconds before moving any leads or wires.

The correct and incorrect ways to solder point-to-point are illustrated in Figure 6–7b. A good connection is shiny, has a well-contoured fillet, and has just the right amount of solder to form a good electrical connection. It is not necessary to fill the entire lug full of solder. A poor solder connection does not create an electrically sound connection. The cold solder joint in the figure resulted from insufficient heat. To correct it, just reheat the connection.

When connecting stranded wire, either point-to-point or to a PC board, first tin (coat) the wires. To do so, twist the strands together and apply a thin coat of solder along the length. Handling the wire will now be much easier because the strands will stay together.

FIGURE 6–7
Soldering point-to-point.
Courtesy Graymark
International, Inc.

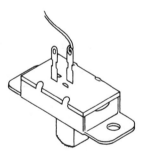
1. Make a good physical connection.

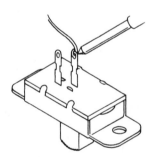
2. Apply heat.

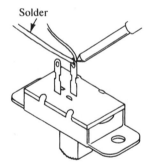
Solder
3. Apply solder.

(a)

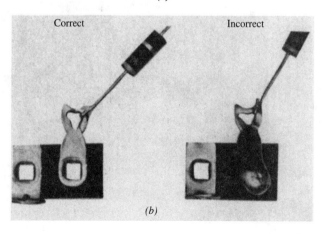

Correct   Incorrect

(b)

Flux Removal

As mentioned earlier, flux residue forms during the soldering process. Even though such residue is, in most cases, noncorrosive, it should be removed because it is sticky and can attract foreign particles, its removal will aid in board inspection, and a "flux-less" board has a neater, more professional-looking appearance.

There are many commercial "defluxers" on the market that can be purchased at your local electronics store. Most, such as Tech-Spray and Flux Stripper W, come in aerosol cans. They are not, unfortunately, environmentally friendly, since trichlorotrifluoroethane is the main ingredient.

To use, simply aim the spray head toward the surface to be cleaned and push the spray head cap down. All such spraying should be done in a well-ventilated area, preferably outdoors. When cleaning hard-to-reach areas, you can insert an extension tube into the spray head to direct the spray. Allow the chemicals to saturate the board for 10 to 15 seconds, then clean it with a medium-bristle brush (a fingernail-cleaning brush works well) or a clean cloth. Of course, allow components to dry completely before applying current to the circuit.

Desoldering

Regrettably, there comes a time when it is necessary to desolder. When components fail, if you have installed the wrong component, installed the right component but incorrectly, or when you just want to make changes or upgrades to a circuit, desoldering is unavoidable. When it becomes necessary to do so, it is worth noting that the skills required in desoldering are every bit as demanding as those for soldering, especially when you are trying to save both the PC board and the electronic component.

To remove electronic components from a circuit is not, however, the only reason for desoldering. Desoldering is also required to eliminate potential shorts caused by solder balls, globs, bridges, and icicles. Furthermore, poor-quality connections resulting from fractured, overheated, or contaminated solder may necessitate desoldering. Even removing excess solder on script that prevents legible PC board identification is often dictated. Desoldering, as you can see, is serious business. It demands your full study and attention.

In Chapter 1 we briefly looked at four tools/materials used in the desoldering process. We spoke of the *desoldering braid (wick), desoldering bulb, desoldering pump,* and *combination soldering iron–sucker.* All have their advantages and disadvantages. Which one you use in a given setting depends on the condition of the solder joint and what exactly you want to accomplish; e.g., component removal or elimination of shorts. We'll look at each approach and when it is best used.

Desoldering Braid

Desoldering braid is a loosely woven, flux-impregnated, stranded braid of copper wire (Figure 6–8a). A strip of the braid is placed over the solder joint and a soldering iron placed on top of the braid. The heat from the braid not only melts the solder but causes it to travel up the braid through capillary action. When a portion of the braid becomes clogged with solder, it is simply snipped off with diagonal cutters.

The techniques for removing solder from PC boards using desoldering braid are simple and straightforward. To begin with, always saturate the braid with liquid flux. Either dip the braid in a jar of flux or brush the flux over the braid after you have laid it in place. To remove a solder bridge, place the braid first on one trace and heat, then on the opposite trace and heat again. Repeat the process as necessary until the solder is completely removed. To remove solder from PC board script, just place the braid over the solder blob and heat.

To remove an electronic component from a PC board using desoldering braid, you must first decide which is more appropriate: to heat and lift the component leads off the board one at a time and then use the braid to remove solder from the pads, or first to remove all solder from each lead before extracting the component. The first approach is applicable with resistors, capacitors, diodes, and similar two-lead components, but with ICs and other multilead components, you must first remove solder from all component leads or pins.

To desolder a connection from a terminal post, it is necessary to distribute heat from the soldering iron throughout the connection. Rock the soldering iron back and forth over the braid to gain maximum solder absorption.

Although desoldering braid can be used to remove solder from almost any type of joint, it is best used on printed circuit boards, and single-sided boards at that. In reality, it works best on flat surfaces, where the capillary force created by the braid is stronger than the surface tension forces of the solder at the joint.

Desoldering Bulb

The desoldering bulb operates on the vacuum-pulse principle. The bulb is simply a hollow rubber "ball" with a heat-resistant Teflon tip (Figure 6–8b). It is designed to be used with any soldering iron.

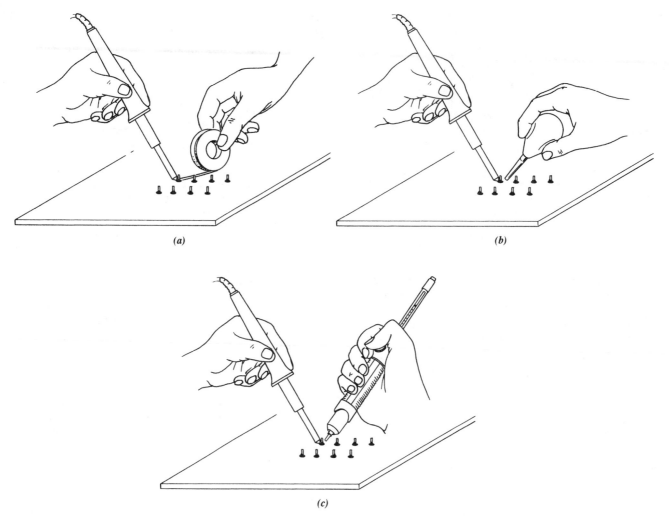

(a)   *(b)*

(c)

FIGURE 6–8
Desoldering methods

First, the iron is held in one hand and used to melt the solder. The bulb, held in the other hand, is then squeezed to create a vacuum. The tip is next applied directly to the reheated solder connection and the pressure released, creating a vacuum action that sucks the solder up through the hollow tip into the bulb. The bulb can be disassembled and emptied at any time.

Although the desoldering bulb does work, its application is limited to printed circuit board solder joints that are large. On smaller connections, the simultaneous action of soldering iron pressure and vacuum suction can cause pad lifting, particularly on single-sided PC boards.

Desoldering Pump

The desoldering pump (Figure 6–8c) is a mechanically operated, handheld desoldering tool that also functions on the vacuum-pulse principle. Various

trade names are given to the tool, Solder Extractor, Soldapullt, and Solder Plunger being the most popular.

The tool, in its basic form, consists of an outer shaft, piston handle, release pin, and Teflon tip. It is designed to be held and operated with one hand. The pump can be disassembled and cleaned as required.

In operation, suction is produced by a spring-loaded piston. The piston handle is first pushed downward. The Teflon tip is then held against the solder joint. As the solder begins to melt (by application of a soldering iron), the pin is disengaged and the solder is drawn up through the hollow tip as the piston snaps upward inside the tubular handle. The tool is easily disassembled to remove the accumulated solder. A step-by-step solder removal procedure is as follows:

1. Load the tool.
2. Heat the solder connection with the soldering iron until solder melts.

3. Hold the desoldering pump lightly against the melted solder.
4. Press the pin to vacuum solder from connection.
5. Reload and repeat as necessary to clean the area of solder.

Certain safety precautions should be observed when working with the desoldering pump:

☐ Never point the tool at another person.
☐ Use caution to prevent ejected solder from falling on the circuit board, possibly causing a short circuit.

Although the desoldering pump works on the same principle as the desoldering bulb, it is actually more effective, since the sucking action created is stronger. With that in mind, extra caution should be observed in deciding on which solder joints it should be used. Repeated application of the pump, along with soldering iron pressure to the same connection, may cause the pads to lift on single-sided PC boards.

Combination Soldering Iron–Sucker

This tool (Figure 1–24) is a combination soldering iron and desoldering bulb. The iron, however, does not have a traditional soldering tip, but rather a hollow tip. In operation, the tip is placed over the joint to be desoldered and rocked slightly in a circular motion. At the same time, the bulb is squeezed to create a vacuum. After 2 or 3 seconds, the bulb is released, creating the action necessary to suck the molten solder through the hollow tip into the bulb. The combination soldering iron–sucker has the same limitations, in terms of application, as the desoldering bulb and pump.

Something to Think About

The National Aeronautics and Space Administration (NASA) offers a highly regarded soldering certification program. To be "NASA Certified" is an accomplishment. Why not surf the Internet to find out more about such certification?

6.3 BREADBOARDING: FIVE WAYS TO GO

As we examine the five common methods used in breadboarding it is important to remember two points: First, there is no right or wrong procedure; one particular technique is not necessarily better than any other. Which method you choose depends on the type of circuit you are building. Second, you may combine different methods to meet particular needs. This is often done with the wire-wrapping and universal printed circuit board approaches, for example.

Cut-Slash-and-Hook (CSH): Tried and True

CSH has been around for decades and is used at the breadboarding, prototyping, and production stages. In fact, there was a time, not that long ago, when the *only* way to build anything in electronics required cut-slash-and-hook, or what is generally referred to as *point-to-point* wiring. Today, CSH is used at the breadboarding stage only with circuits requiring large amounts of current, such as big power supplies and audio amplifiers. In such cases, it is a method that still offers many advantages.

An Oldie but Goodie

CSH is a breadboarding procedure in which you take a piece of wire, cut it to length, strip it on both ends, and hook it around a terminal post before soldering. You do the same thing with component leads, but of course you do not strip the lead of insulation.

With this method, terminals are attached at various locations to a breadboard material usually made of perforated phenolic (plastic) $\frac{1}{16}''$ or $\frac{3}{32}''$ thick (Figure 6–9a). The terminals are either fastened with screw and nut or pressed into place.

Two basic types of terminals are used: the terminal strip and the push-in terminal (Figure 6–9b). The **terminal strip** consists of a varying number of lugs spaced $\frac{3}{8}''$ apart on a phenolic or Bakelite strip $\frac{3}{8}''$ wide. In some units, one or more lugs may be grounded. In assembly, a wire or component lead is simply hooked around the terminal and soldered in place. With the **push-in terminal**, a perforated board with hole sizes of 0.093″ must be used. The terminal is pressed into the board with a pliers or hand tool designed for the purpose. Wire and component leads are squeezed into two serrated slots on the top and one on the bottom, then soldered in place.

Electronic components too heavy or bulky to tie directly to a terminal can be screwed or strapped down onto the breadboard (Figure 6–10a). Wire of almost any gauge is then used to connect the component terminals to the terminal strip.

When using the CSH method, begin by sketching a breadboard drawing showing the major component layout. Insert terminals at convenient locations. Remember, additional terminals can easily be added, or those not used, removed. All component and wire connections must come together at a

FIGURE 6–9
CSH breadboarding
materials

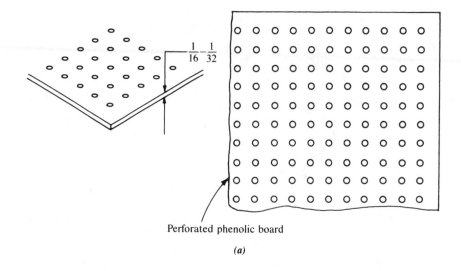

Perforated phenolic board

(a)

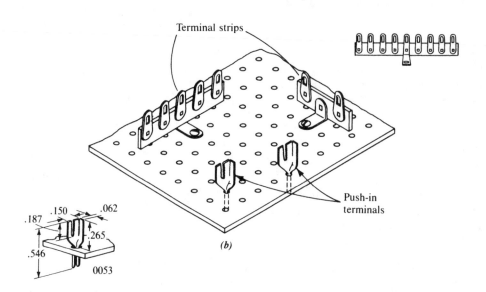

(b)

terminal tie point or post. No connection should "float" in the air (Figure 6–10b).

Advantages and Disadvantages

The CSH method of breadboarding still has a number of things going for it. It's easy to do, and the skills required to perform cut-slash-and-hook are quickly learned. You can use practically any gauge of wire. Circuits built using the CSH method are easy to trace and troubleshoot. Such a construction is stable; you don't have to worry about wires and leads falling off the breadboard.

However, the CSH technique has some major drawbacks, too. It is very time consuming. The process of cutting, stripping, hooking, and soldering all component leads and wires is very tedious. Also, this approach takes up a great deal of space.

Breadboards with individual terminals tend to be big and spread out. Furthermore, it is difficult to change things. The process of desoldering from a wrapped terminal strip can be downright frustrating, but perhaps the biggest disadvantage of all has to do with compatibility. The CSH procedure of breadboarding is totally unsuitable for the assembly of many modern electronic circuits, especially those using integrated circuits. Digital and many analog circuits simply cannot be breadboarded with this method.

Solderless Wiring Terminals: Point to Point the Easy Way

Are you ready for a cut-slash-and-hook breadboarding method without the hook and soldering? Solderless wiring terminals may be the answer.

FIGURE 6–10
Mounting components
using the CSH method

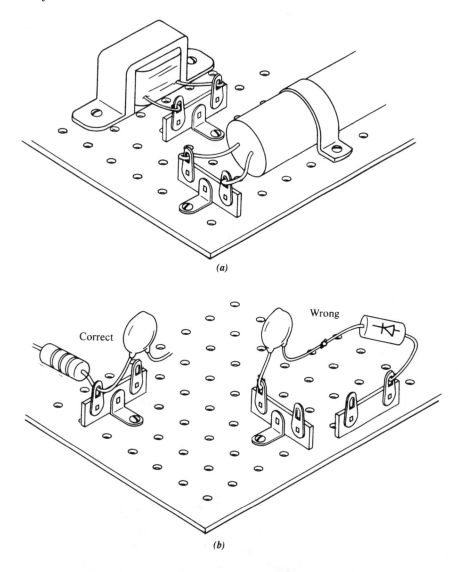

(a)

(b)

"Look! No Solder"

Solderless wiring terminals provide for quick connection or disconnection of wires and component
leads without hooking or soldering. The terminal
presses into a phenolic board with 0.062″ holes
(Figure 6–11). When a spring or plunger is depressed, the component lead or wire is simply inserted. Release the tension, and the spring holds the
leads in a reliable solderless connection. Like the
CSH terminals, these solderless springs can be
placed anywhere on the phenolic breadboard.

Advantages and Disadvantages

Solderless wiring terminals are used with the same
types of circuits as those employing the CSH method.
The advantage, of course, is that no hooking or soldering is required. This means that circuits can be
assembled and disassembled much more quickly and
easily. Circuit changes, the important characteristic
of breadboarding, can easily be accomplished.

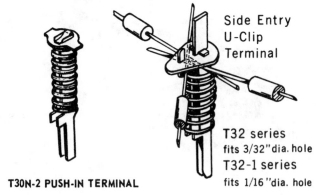

T30N-2 PUSH-IN TERMINAL

Side Entry
U-Clip
Terminal

T32 series
fits 3/32″dia. hole
T32-1 series
fits 1/16 ″dia. hole

FIGURE 6–11
Solderless wiring terminals

There are disadvantages, to be sure. Like CSH,
this method is incompatible with modern circuit
design. It's just as hard to spring-clip a 16-pin IC as
it is to hook and solder it to a terminal lug.
Furthermore, solderless wiring terminals are very

expensive and increasingly more difficult to find. Yet, if you want an easy CSH-type method without the hook and solder, and you can find some clips, solderless wiring terminals may be the way to go.

Solderless Circuit Boards: Breadboarding Modern Circuits

Finally—a breadboarding method that is compatible with modern components and circuits. In fact, the solderless circuit board was designed specifically with integrated circuits in mind. A plug-in technique had to be developed that could easily accommodate ICs and other components with 0.10″ lead spacing. The solderless circuit board satisfies this need.

A Holey System

Solderless circuit board is a plastic breadboard containing hundreds of holes (Figure 6–12). The boards come in many different sizes, $\frac{1}{4}″ \times 2″ \times 5″$ being the most popular. They can be snapped or butted together to increase the working surface. The holes are placed in a rectangular grid at regular intervals, 0.10″ apart. Some boards have number-and-letter designations for each column and row of holes to aid in component lead and wire placement.

Each hole has a tiny metal lug inside. The lugs are connected underneath in small and large groups. The usual pattern is to connect 5 lugs in a vertical column of small groups and 25 to 40 or more lugs in large horizontal groups known as *buses*.

The system works like this: Two or more component leads or wires inserted in the same group are "shorted" together by the lug-connecting strip underneath. The wire (or component lead) can usually be pushed in or pulled out with your fingers, although a needle nose pliers is handy.

One important feature in every solderless circuit board, regardless of the manufacturer, is the center channel. Integrated circuits are straddled across the channel so that one row of IC pins does not "short" to the row on the opposite side.

Easy as Said and Done

In addition to the solderless circuit board itself, the only other material used with this system is hookup wire. You can buy a kit of precut, prestripped single-strand #22 gauge wire from a number of distributors, or you can use commonly available #22–30 gauge telephone wire and do the cutting and stripping yourself.

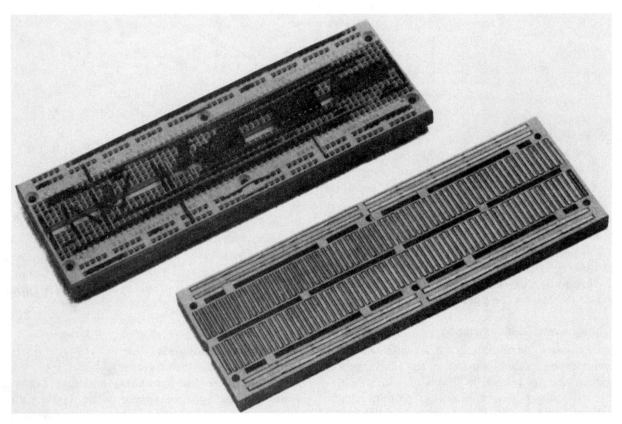

FIGURE 6–12
Solderless circuit board construction

FIGURE 6–13
Assembling on a
solderless circuit board

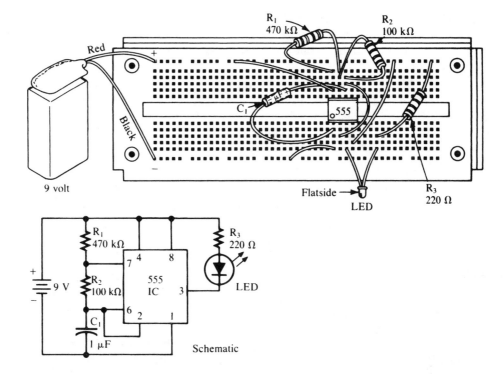

Schematic

Circuit assembly, as shown in Figure 6–13, is very quick and easy. Note, for example, how resistor R_1 is connected between pins 7 and 8 on the 555 IC by simply inserting each lead in the vertical column of holes aligned with those respective pins. If two components must come together at a location other than at an IC pin, simply pick a free row of pins and insert. This was done with the junction of the LED and resistor R_3. Where many component leads or wires come together at a single point, the bus strips can be used.

In many respects, the actual layout of the circuit follows the schematic drawing. As with the schematic, the supply bus is at the top, the signal bus in the middle, and the ground bus on the bottom. What could be easier?

When breadboarding a circuit, keep the following tips in mind:

☐ Keep wires as close to the board as possible. Avoid the "rat's nest" approach. A neat layout is always much easier to trace.

☐ Keep component leads short, for the same reason.

☐ It is easy to misalign IC pins, component leads, and wires. Check carefully to be sure you haven't inadvertently moved a connection over a hole or two.

☐ It is very easy to tuck an IC pin under a chip without knowing it. From the outside it looks fine; the pin seems to be inserted into the hole, yet it may be bent underneath the IC. To avoid

this problem, press evenly on all pins at the same time.

☐ If component leads are too big to insert into the board, simply solder a smaller wire to the component lead and press the other end into the board.

☐ When possible, use different colored wires for different electrical functions. For example, use black wire for ground connections and red wire for positive voltage lines.

Advantages and Disadvantages

The advantages to breadboarding with solderless circuit board are numerous. The method is a great time-saver. Circuits can be assembled in a fraction of the time it takes with any other breadboarding technique. Circuit changes are quick and easy. You just pull out one component and pop in another. There are few cold solder joints to worry about because there are so few solder connections (none on the board itself, of course). You can use components over and over again because they are not consumed by the breadboarding process. Furthermore, everything is laid out in front of you. It's hard to miss a connection. Finally, all types of components can be used. If the component is large or bulky, like a power transformer, mount it to the side of the board and extend its leads to the appropriate holes.

There aren't many disadvantages to breadboarding with solderless circuit board. It's true that some of the lugs will go bad from time to time, but when that happens, just mark the defective ones. Some technicians who do a great deal of radio-frequency work point out that their circuits need large amounts of shielding, which is difficult to create with this type of breadboard, and, of course, the circuit assembly itself can be somewhat flimsy. You do have to be careful how you handle the finished breadboard. Yet when you compare the important advantages against these limited drawbacks, you quickly see that solderless circuit board makes for an ideal breadboarding approach to be used with modern electronic circuits. (*Note:* See Chapter 13 for a discussion on how to breadboard surface mount components with solderless circuit boards.)

Wire Wrap: Twisting the Night Away

Wire wrapping is a breadboarding method that is not only compatible with modern electronic circuits but also provides a strong and secure bond between components—without the use of solder. Although circuit construction is not as fast as with solderless circuit board, wire wrapping takes considerably less time than would be required to design and produce a dedicated printed circuit board.

Stronger Than a Solder Joint

In the wire-wrap technique, connections are made by simply wrapping wire ends around rectangular or square terminals. The key is to use terminals with right-angle corners. Using a special (though inexpensive) wrapping tool, tightly wrap a piece of #28 or #30 gauge wire around a terminal six to eight times (Figure 6–14). The wire "bites" into the corners of the terminal, and the two materials are actually "welded" together to form a solid connection. The connection is even better than solder because no cold joints are possible. Unwrapping is as easy as wrapping: Just twist the tool in the direction opposite that when wrapping the wire.

For those who have grown up on solder, the idea, if not the reality, of wire wrapping is a bit hard to accept. Can such a thing really work, and can it actually be better than solder? Well, ask NASA or the military. They have been doing it for years, and not just in breadboarding. Many of their mass-produced circuits, operating under extreme environmental conditions, are wire wrapped.

Materials and Tools

We talked about wire-wrap equipment and tools in Chapter 1. Now is a good time to review that section and glance at Figure 1–25 again. You are going to need perforated phenolic board with holes spaced 0.1″ apart. Into these holes will go wire-wrap IC sockets and various wire-wrap posts (Figure 6–15). The sockets have long rectangular pins to take two or three levels of wrapped wire. The posts are square throughout their length and have a trifurcated clip on the top. Electronic components with round leads are soldered to the clip, and the posts are joined by wire wrapping. Remember, you cannot wire-wrap round component leads. Another approach to handling small components, such as transistors, low-wattage resistors, and disc capacitors, is to solder them directly to a **parts carrier**, or *header,* as shown in Figure 6–15 (lower right). The header is pressed into a wire-wrap IC socket, which is then wrapped in the usual way to make the circuit interconnections.

The sockets are held in place by a dab of glue or a press-fit cardboard retainer (Figure 6–16a). Another way to hold the sockets in place is to wrap a few strands of wire on the pins at two opposite corners. Press the wire down slightly, and the socket will be held securely.

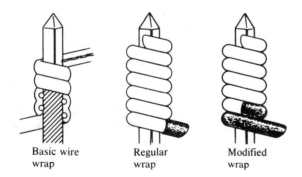

FIGURE 6–14
Wire-wrap techniques

Basic wire wrap   Regular wrap   Modified wrap

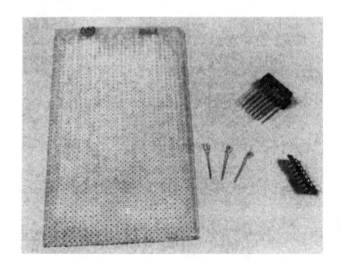

FIGURE 6–15
Basic wire-wrap materials

FIGURE 6–16
Installing sockets and terminals

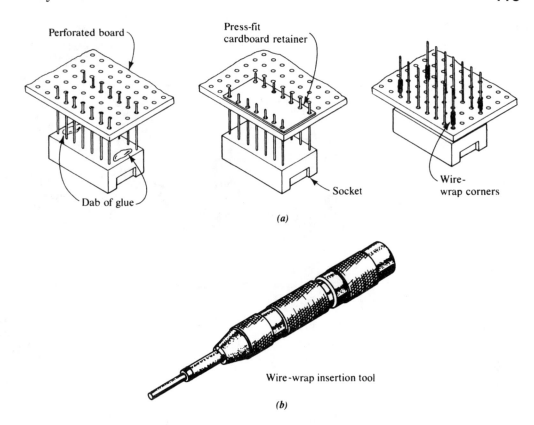

Perforated board

Press-fit cardboard retainer

Dab of glue

Socket

Wire-wrap corners

(a)

Wire-wrap insertion tool

(b)

Terminals can be pressed into place with pliers. A better way is to use an inexpensive insertion tool designed for the purpose (Figure 6–16b).

You can't use just any wire; it has to be #28 or #30 gauge, bare or insulated (known as Knyar-insulated wire), and if insulated, stripped back on both ends. The wire comes in either rolls or precut/pre-stripped lengths. Many colors are available, the most popular being red, blue, white, black, and yellow.

The wire-wrap tool itself can be had in a budget, standard, or deluxe model. Hobby-Wrap-30, a budget tool from OK Industries, works great, and is ideal for anyone just getting started in wire wrapping (Figure 6–17a). It will strip wire and wrap and unwrap it; best of all, it sells for less than $20. A standard model from Vector Electronic Company, known as the "Slit-N-Wrap," eliminates precutting and prestripping (Figure 6–17b). As with the budget version, your wrist has to supply the twisting action. This model runs about $25. If you wish to go all out, battery-powered units are available that do practically everything (Figure 6–17c). You just slip the end of the tool over the post to be wrapped, push the trigger, and that's it. They sell for over $50, however.

Wrap Away

In the following discussion, we assume you are using the budget tool from OK Industries. (Other low-cost wire-wrap tools work in a similar manner.)

To prepare the wire, first cut it to length. Run a strand from terminal to terminal, keeping the wire low to the board. Add about an extra $1\frac{1}{2}''$ because you will need to strip approximately $\frac{3}{4}''$ of insulation from each end. Now, using the little stripper on the tool, strip the wire.

The tool has a center hole that fits over the terminal (Figure 6–18). The end of the wire is inserted into a smaller off-center hole, or tunnel, near an index marker.

Here are the eight steps required in wire wrapping:

1. Hold the tool horizontally with the index mark up.
2. Insert the wire in the tunnel up to the insulation.
3. Bend the wire at an angle to hold it in place.
4. Place the tool over a terminal.
5. Hold the tool to the bottom of the terminal, but do not press down too hard.
6. Start a clockwise twist.
7. Wrap six to eight turns. You don't have to count; only the stripped wire will wrap around the post.
8. Remove the tool from the terminal.

If you have done a good job, the wire should climb up the terminal with one layer neatly placed on top of the other.

To remove a wire, just turn the tool over, insert it on the terminal, and twist in a counterclockwise

FIGURE 6–17
Wire-wrap tools

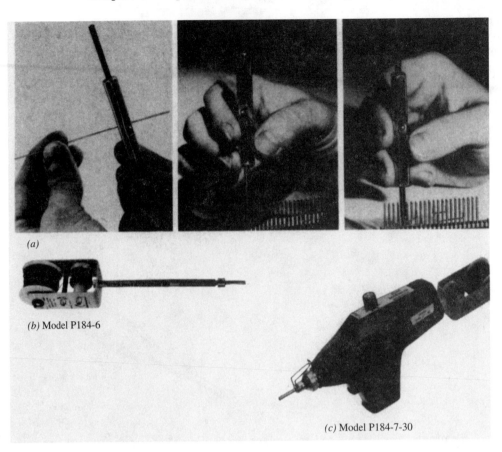

(a)

(b) Model P184-6

(c) Model P184-7-30

direction. Do not attempt to reuse the removed wire. If you try to insert the now-crooked wire into the tunnel of your tool, the wire will probably jam and clog the slot. Use a new piece of wire.

With a little practice and a little patience, you'll have the technique down in no time.

Advantages and Disadvantages

Wire wrapping has two main advantages: It is, like solderless circuit board, compatible with modern circuits, and it is relatively solder-free. As you have undoubtedly noticed, although the wire wrap itself

requires no solder, solder is needed when fastening electronic component leads to parts carriers. But even with that limitation, wire wrapping has much going for it. Connections can be made in tight spots without the danger of shorts from solder bridges or the nicking of insulation with a hot soldering iron. Wire-wrapped circuits are also easy to inspect. If it looks good, it usually is. And this circuit construction method results in solid, secure connections. It definitely won't come apart in your hand.

Like all breadboarding systems, wire wrapping does have one or two disadvantages. Wire-wrapping

FIGURE 6–18
"OK" wire-wrap tool

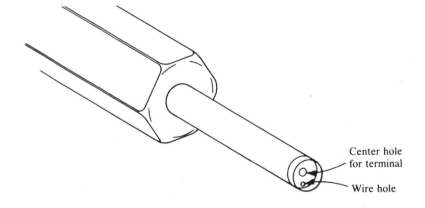

Center hole
for terminal

Wire hole

materials tend to be a bit expensive. Sockets, posts, wire, and the perforated board can strain your budget, costing as much as $8 to $10 for a project with just a few ICs. Also, the process is time-consuming, especially when compared with breadboarding on solderless circuit board. But if you are looking for a stable, more or less solderless circuit compatible with today's components, then you should definitely give wire wrapping a try.

Universal Printed Circuit Boards

Universal printed circuit boards are not what you may think. They are actually more like a wire-wrap board than a printed circuit board and are designed to make wire wrapping more convenient. Why they are called universal printed circuit boards nobody seems to have quite figured out.

Perforated Board with Solder Pads

Universal printed circuit board is basically wire-wrap board in which rings of tinned copper surround holes on one side of the board (Figure 6–19). On some boards, strips of copper tie two or more holes together. Also included are bus strips, which can run the entire length of the board, at the edges, or anywhere in the center. Even boards with odd-shaped patterns are available, where holes are connected with copper strips in various square, triangular, or rectangular configurations. The variety is endless.

The purpose for adding all this copper to an otherwise "clean" perforated phenolic board is twofold. First, the individual copper pads allow for soldering wire-wrap sockets and terminals in place. Just a dab of solder secures the pins and posts. Second, the multipad connections can be used to tie components together. In some cases you don't even need wire-wrap terminals. Just insert the component leads from the plain side of the board and solder them to the common pads (Figure 6–20). Bus strips, of course, allow you to connect many wires and component leads to a widely used common circuit point.

Advantages and Disadvantages

Universal printed circuit boards take the wire-wrapping approach one step further. With the copper pads and strips laid out in a universal pattern, you can decide which ones to use as you assemble the project. What you don't need, you simply leave alone. The system allows you to avoid the use of parts carriers (or headers) and extensive wire wrapping for buses. It makes wire wrapping a great deal more flexible.

But don't get too excited. You're in for a big shock when you price these universal wonders. An 8″ × 8″ board can cost over—are you ready?—$25. Obviously, this method is strictly for professional breadboarders. If you can afford it, the universal printed circuit board is a great way to go.

FIGURE 6–19
Universal printed circuit board

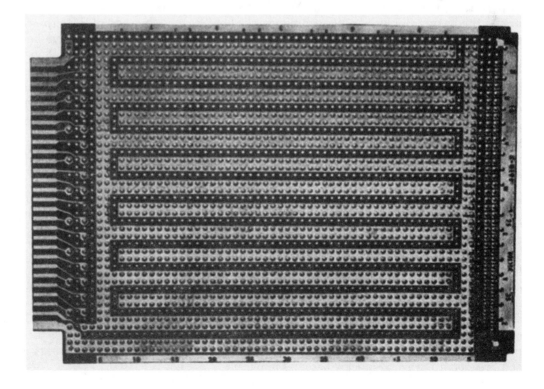

FIGURE 6–20
Installing components on
a universal circuit board

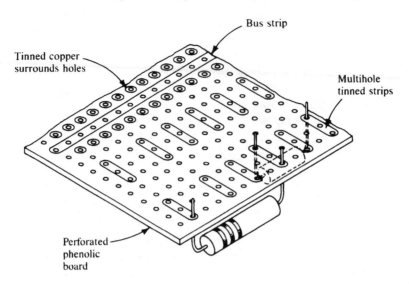

Labels in figure:
- Bus strip
- Tinned copper surrounds holes
- Multihole tinned strips
- Perforated phenolic board

Something to Think About

Wouldn't examples of the five breadboarding approaches make a great display in the laboratory? Why not gather such breadboards, mount them on a Peg-Board or similar surface, label each one, and display the entire unit in the lab? Talk about a "class" project!

6.4   PROJECT BREADBOARDING AND TROUBLESHOOTING

It's time now to breadboard the Variable Power Supply and 3-Channel Color Organ. We will also examine the basic troubleshooting techniques used at this, the breadboarding (or experimenting) stage.

Breadboarding the Projects

The Variable Power Supply and 3-Channel Color Organ will be breadboarded using the solderless circuit board approach; however, any of the methods discussed could be used, although the wire-wrapping and the universal printed circuit board systems are not particularly applicable with these projects.

Breadboarding the Variable Power Supply

Working from a breadboard drawing and a copy of the working schematic, we start to build the circuit. The breadboard drawing (Figure 6–2) takes only a few minutes to prepare. The tentative locations of all components are shown. Note how much alike the breadboard and schematic drawings are. This is a characteristic of breadboarding with solderless circuit board. Compare the breadboard drawing with a picture of the final layout (Figure 6–21) and the value of a breadboard drawing quickly becomes apparent.

Regardless of the breadboarding method chosen, it is an excellent idea to trace over the schematic drawing with a colored pencil as you proceed in the actual project construction. As a component or wire is installed, color over the schematic symbol or line on a copy of the working schematic. This will allow you to see in an instant not only what you have completed but what you still have to do.

The following points are worth noting:

1. The transformer, line cord, and switch are not mounted directly on the solderless circuit board. Three wire nuts are used to connect these components: one to connect the line cord to one primary lead on T_1, another to connect the line cord to one end of S_1, and a third to connect the other end of S_1 to the remaining primary lead of T_1. Two wires have been soldered onto the terminals of S_1 to make the switch connections.
2. Diodes D_1–D_4 are arranged in a bridge, as shown in the schematic drawing.
3. One lead of components D_5, C_1–C_3, and R_3 is connected to a common bus line, in this case the negative bus.
4. Some soldering is required when wires have to be connected to components, such as switch S_1 and potentiometer R_3.
5. The leads of all multistrand wires must be tinned before being inserted into solderless circuit board.

Before plugging in the project, place an ohmmeter across the output terminals (+ and −). Adjust potentiometer R_3 throughout its range. As you do so, the meter should indicate approximately 390 to 5000 ohms. If there is no variation in resistance, or if the reading falls below 300 ohms, check for faulty wiring. Finally, double-check diode polarity and that the transformer primary and secondary leads are not reversed.

FIGURE 6-21
Breadboard for Variable
Power Supply

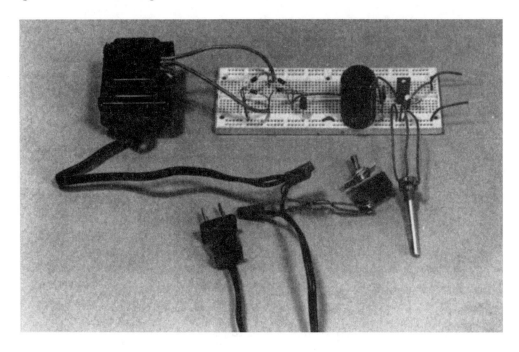

If everything checks out, plug in the line cord and turn on switch S_1. Does LED D_5 light? If it does, with a voltmeter measure the output voltage across the negative and positive terminals while adjusting potentiometer R_3. You should get 1.2 to 15 V dc. If the LED does not light, or you do not get the correct voltage output, immediately unplug the project and read the section on troubleshooting.

Breadboarding the 3-Channel Color Organ

It is suggested that the 3-Channel Color Organ be breadboarded with solderless circuit board. However, the solderless wiring terminal method will work just as well for this project.

As with the Variable Power Supply, lay out the color organ with the aid of breadboard and schematic drawings. Be sure to use a colored pencil to mark your progress.

Here are a few hints to follow in your breadboarding:

1. Some soldering will be required. You will need to attach wires to sockets $S0_1$–$S0_3$, as well as to potentiometers R_3–R_5 and switches S_1 and S_2. The line cord will not fit directly into the solderless circuit board. Therefore, you will have to solder single-strand wire to it and then insert the other end of each wire into the board.
2. It is best to build one channel at a time rather than to connect all capacitors, then all resistors, and so forth.
3. All components except the line cord and S_1 should mount directly on the solderless circuit board. You may have to trim the leads of the

SCRs to fit them in the board. Use diagonal cutters to narrow the lead slightly.

To check out the color organ, first plug three table lamps or three strings of Christmas lights into each ac socket, $S0_1$, $S0_2$, and $S0_3$. Be sure that no socket is fitted with more than 200 W of light. Next, connect a sound source to the input jack J_1. If you are going to use a radio, make it an FM rather than an AM receiver. Now, plug in the line cord and turn on switch S_1. Turn the sound source on. Adjusting potentiometers R_3–R_5 should cause the lights to flash to the sound from the radio or tape player. Turn down the volume of the sound source and place switch S_2 in the "on" position. The color organ should respond. When the volume of the sound source is normal to high, S_2 should be in the "off" position.

If the 3-Channel Color Organ does not function correctly, immediately unplug the line cord and proceed to the troubleshooting instructions.

Troubleshooting the Projects

First, we look at troubleshooting during the breadboarding stage in general, then at how to deal with specific projects. Troubleshooting is examined in more detail in Chapter 10.

Troubleshooting at the Breadboarding Stage

Because we breadboarded the Variable Power Supply and the 3-Channel Color Organ on solderless circuit board, we confine the troubleshooting analysis to this type of breadboard assembly.

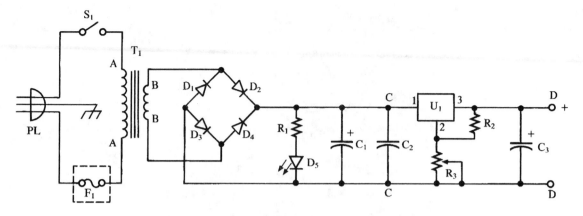

Points	Voltage reading	Possible component problem
A–A	110–120 V ac	S_1, T_1
B–B	15–20 V ac	T_1
C–C	15–20 V dc	D_{1-4}, C_1, C_2
D–D	1, 2–15 V dc variable	U_1, R_2, R_3, C_3

FIGURE 6–22
Troubleshooting the Variable Power Supply

The biggest problem encountered in breadboarding on solderless circuit board is the misalignment of wires and component leads. It is easy to get things moved over one or two holes to the right or left of where they should be. Check the layout very carefully. Also, as mentioned earlier, IC pins frequently get tucked under the body of the integrated circuit. Double-check the chips. Another common problem develops from inserting ICs, diodes, transistors, SCRs, voltage regulators, and polarized capacitors in the wrong direction, or backward. It is also possible, especially with resistors, to pick up the wrong value. Only the third-band color separates a 47 k from a 470 k resistor—black versus brown—for example. And don't forget the fragility of single-strand wires. They can break inside the insulation without your knowing it. Finally, the solderless circuit board itself can give you headaches. Some pins may go bad, in which case you must simply move the circuit over and start again.

Troubleshooting the Variable Power Supply

If the LED in the power supply does not light, it may be in backward or it may be burned out. Check for both possibilities. If the LED now lights, check the output voltage, as discussed earlier. If the LED still does not light, or the output voltage is incorrect, measure the voltages at the points indicated on the schematic and chart shown in Figure 6–22. Work forward, starting with points A–A. If you encounter an incorrect reading (or more likely, no reading at all), examine the components noted on the chart; they may need to be replaced.

Troubleshooting the 3-Channel Color Organ

Four types of problems tend to occur with the 3-Channel Color Organ. None of the lights for any channel responds to the sound source, or some channels work as they should and others do not. Perhaps lights are on for all three channels, but they remain on and do not respond to the audio signals. Finally, the color organ does not work at low audio levels. Let's look at each problem in turn.

☐ *Problem 1.* Check lamp I_1. If it is off, the problem could be with switch S_1. Shunt the switch terminals with a clip lead. *CAUTION:* Unplug the project, and after the shunt is complete, plug it back in. Never shunt, or short, a component

on an ac-operated project without first discon-
necting it from the wall outlet.

☐ *Problem 2.* If one or two channels are working,
it means that the audio coupling circuit consisting of R_1, S_2, and T_1 is working. Examine carefully the components associated with the
nonworking channel. The problem is most likely
in the SCR. Switch it with one in a channel you
know is working.

☐ *Problem 3.* Here the problem is most likely in
the audio coupling circuit. Check transformer T_1
carefully; it may be in backward.

☐ *Problem 4.* In this situation the problem is almost certainly with switch S_2. Short it out with
a clip lead. If the circuit now works on low audio
input, replace the switch.

The Experiment Results Document

It's time now to write the Experiment Results
Document. Such a document for the Variable Power
Supply is shown in Figure 6–23. Look it over carefully.
Remember, the purpose of this document is to tell
what happened when you breadboarded the project. If
all went well, say so and support your conclusion. If

FIGURE 6–23
Experiment Results
Document: Variable Power
Supply

Experiment Results Document
Variable Power Supply
by
Austin Babayan

Conclusions

The Variable Power Supply project performs as it was designed to perform, under
breadboard conditions. Output voltage is adjustable from 1.2 to 17.2 V, however, it
should be noted that at the breadboarding stage, the Variable Power Supply was
tested under a no-load condition. This is because at this time voltage regulator U_1
was not equipped with a heat sink.

Test Results

1. Output resistance test OK. Measured 290 to 5.2 k ohms.

2. Project was plugged into the wall outlet and switch S_1 closed.

3. LED D_5 did not light. Close examination revealed that the LED had been installed
   backward. When the LED was reversed, it lit as it should.

4. A voltmeter was placed across the output terminals. While potentiometer R_3 was
   being adjusted through its range the output voltage reading recorded was 1.2 to
   17.2 V.

5. The project was left on for 1 hour, after which output voltage readings were again
   taken. They were identical with those taken in step 4.

5

you encountered problems that required a modification in project design, those problems need to be documented at this time.

If you are building one of the elective projects or a project of your own design, now is the time to breadboard and troubleshoot your project. You must also produce the breadboard drawing and the Experiment Results Document.

Something to Think About

Each of the five breadboarding approaches has its own troubleshooting characteristics. Why not develop a list of such factors for each method and then post the list next to the appropriate breadboarding method? Another class project!

SUMMARY

In this chapter we began by looking at the need for breadboarding. We examined the differences among project breadboarding, prototyping, and construc-

tion and discussed the characteristics of the breadboard drawing and Experiment Results Document. We then went on to look at soldering: exploring what it is, how it is done, and the various ways to desolder.

Next we investigated five ways to breadboard. We saw how the cut-slash-and-hook (CSH) method is still used when assembling high-current circuits, such as large power supplies and power amplifiers. We examined how solderless wiring terminals can do the same thing, only without the need to solder. We explored ways to assemble modern circuits using solderless circuit board and wire wrapping and looked at the universal printed circuit board as a way to enhance the wire-wrapping technique.

We then breadboarded the Variable Power Supply and 3-Channel Color Organ. We also discussed breadboard troubleshooting in general and specific troubleshooting procedures with regard to the two projects.

In Chapter 7 we look at printed circuit boards, what they are, and how they are fabricated. Yes, it's time to prototype.

QUESTIONS

1. The breadboarding technique must allow us to _____ or substitute components quickly and easily.

2. Solder is an alloy of _____ and _____.

3. _____ is a substance that helps solder flow around the joint being soldered.

4. List the four steps in soldering any connection: _____, _____, _____, _____.

5. Desoldering is often required to eliminate potential shorts caused by _____ balls, _____, _____, and _____.

6. The cut-slash-and-hook and solderless wiring terminal methods of breadboarding are used with circuits requiring large amounts of _____.

7. The solderless circuit board and wire-wrap approaches to breadboarding are compatible with _____ electronic circuits.

8. The _____ printed circuit board method of breadboarding is really designed to facilitate the wire-wrapping technique.

9. When breadboarding with solderless circuit board, _____-_____ of wires and component leads can be the biggest problem.

10. The _____ _____ Document tells what happened during the breadboarding stage and, specifically, if any project design changes were required.

REVIEW EXERCISES

1. Make a breadboard drawing for the circuit shown. Choose a breadboarding method you think appropriate for this type of circuit.

Light-wave receiver

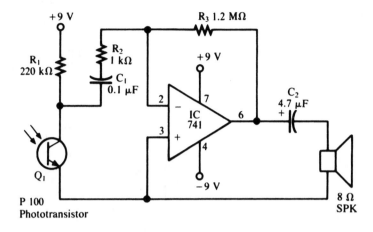

2. Breadboard the circuit shown in Problem 1.

3. Make a breadboard drawing for the circuit shown. Choose a breadboarding method you think appropriate for this type of circuit.

Bar-graph light meter

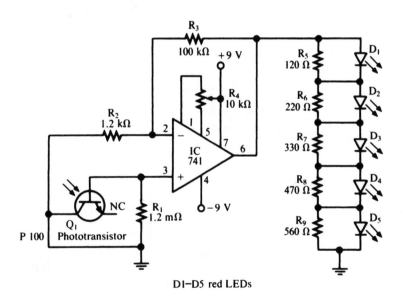

D1–D5 red LEDs

4. Breadboard the circuit shown in Problem 3.

5. Make a breadboard drawing for the circuit shown. Choose a breadboarding method you think appropriate for this type of circuit.

Toy organ

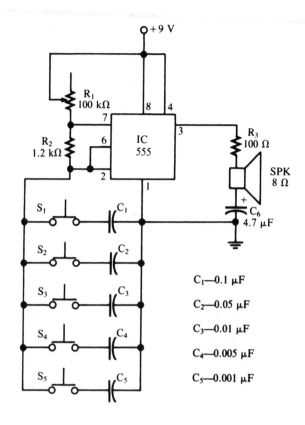

6. Breadboard the circuit shown in Problem 5.

7. Make a breadboard drawing for the circuit shown. Choose a breadboarding method you think appropriate for this type of circuit.

Neon flasher

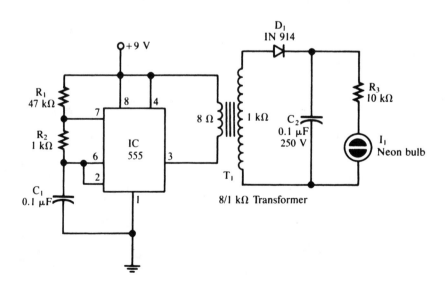

8. Breadboard the circuit shown in Problem 7.

9. Make a breadboard drawing for the circuit shown. Choose a breadboarding method you think appropriate for this type of circuit.

Stepped-tone generator

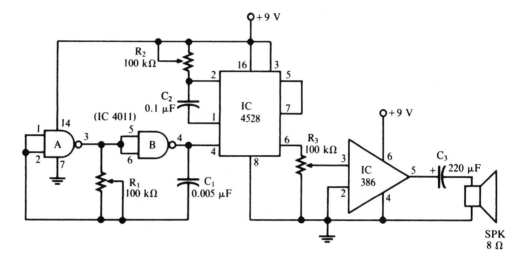

10. Breadboard the circuit shown in Problem 9.

11. Make a breadboard drawing for the circuit shown. Choose a breadboarding method you think appropriate for this type of circuit.

Test-bench amplifier

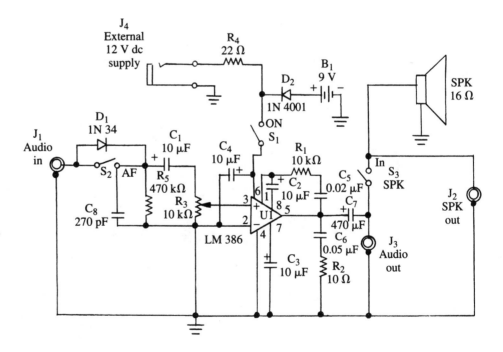

12. Breadboard the circuit shown in Problem 11.

7 Prototyping—Printed Circuit Board Design

OBJECTIVES

In this chapter you will learn

☐ What printed circuit boards are and why they are so widely used.

☐ What preliminary considerations are necessary before beginning the actual PC design layout.

☐ The basic techniques for doing a PC design layout.

☐ The two-step procedure for producing a PC design layout.

☐ Factors to consider in creating computer-generated PC board artwork.

☐ How to produce the printed circuit board design layout and artwork drawings for the Variable Power Supply Project.

☐ How to proceed with the printed circuit board design layout and artwork drawings for the 3-Channel Color Organ Project.

With the completion of the experimenting (breadboarding) stage, we now have a working circuit, one that does what its designers intended it to do; but although the circuit itself may be operating just fine, we still do not have a practical, functioning **prototype:** a device that is suitable for complete evaluation of electrical as well as mechanical form, design, and performance. What we need is a prototype that can be shown around, banged around, and played with, without falling apart in our hands. To ensure that it will "hang together," endure, we must construct the circuit portion of the project on a printed circuit (PC) board of our own design and fabrication.

In this chapter we examine printed circuit boards to see just what they are and how they are designed and laid out. In Chapter 8 we learn how the boards are fabricated, and in Chapter 9, how they are assembled, or "stuffed," with electronic components. Remember, however, that what is to follow is not a treatise on the potentially complex subject of printed circuit board design and fabrication. We are concerned here only with the basics. Nevertheless, our coverage will be extensive enough to allow anyone to design and fabricate PC boards for the projects offered in this book, as well as those found in hobbyists' electronics magazines.

7.1 THE PRINTED CIRCUIT BOARD

In this section we examine printed circuit boards to see what they are, the types that are in wide use, and the many advantages they offer for both mass production and prototype construction. We then explore the overall PC board design and fabrication process, as it pertains primarily to single-sided boards.

Circuits without Wires

Since its development shortly after World War II the printed circuit board (also known as a *printed wiring board*) has revolutionized the way electronic circuits are assembled. Let's see what these boards are and examine the basic type: the single-sided printed circuit board.

125

The Basic PC Board

In its basic form, a printed circuit board is a circuit in which the interconnecting wires have been replaced by conductive strips (traces) of copper left on an etched board (Figure 7–1). The interconnections terminate at copper pads that have holes drilled in them. From the nontrace side of the board, components are installed and their leads brought through the holes, to be soldered to the pads.

The PC board fabrication process is such that duplicate boards—that is, those with identical trace and pad patterns—can be fabricated easily by the hundreds or thousands. Once the accuracy of a layout is verified, every identical circuit is error-free. No wiring mistakes are possible.

Three types of printed circuit boards are widely used in electronics manufacturing: single-sided, double-sided, and multilayer. First let's look at a single-sided printed circuit board.

Single-Sided Printed Circuit Boards

The single-sided board, as its name implies, has circuit traces and pads etched on only one side. Components are mounted on the other (nontrace) side, with leads protruding through drilled holes to the copper pads, where they are soldered in place. This type of board is shown in Figure 7–1.

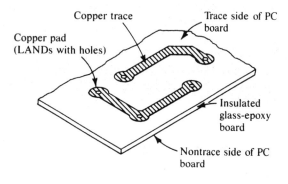

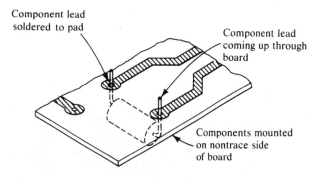

FIGURE 7–1
The printed circuit board

Single-sided PC boards were the first to be developed and are still the most widely used. They find application in relatively uncomplicated and inexpensive circuitry. Low-cost, mass-produced consumer electronics products make extensive use of single-sided boards. They are also the logical choice to get started with in a school laboratory setting, since they are the easiest of the three types to fabricate.

Double-Sided and Multilayer Printed Circuit Boards

Although this book does not deal with double-sided or multilayer printed circuit board design and manufacture, you will encounter such boards extensively in industry. Therefore, a brief discussion of each type is in order.

Double-Sided Printed Circuit Boards

The double-sided board has traces and pads on both sides. Components are mounted on only one side, however, and leads are brought through and soldered as with single-sided boards. A double-sided board is shown in Figure 7–2.

Interestingly, a double-sided board is easier to design but more difficult to manufacture than a single-sided board. The design process is facilitated by the fact that traces are run on both sides of the board. This makes it much easier to get from one pad to another without using jumper wires. Fabrication, however, often requires precise alignment between pads on both sides and the use of plated-through holes. In industrial manufacture this is no problem. In a typical school electronics laboratory, it is often an insurmountable challenge.

Multilayer Printed Circuit Boards

The use of multilayer printed circuit boards has expanded greatly since their beginnings in the mid-1970s with the advent of the microcomputer. Even PC boards used in those early machines required so much circuitry that designers searched for ways to reduce the number of boards crammed into a limited-size personal computer case. The multilayer board was the answer.

Such a board consists, at a minimum, of two outer circuit layers and two inner layers (see Figure 7–3). Some multilayer boards, however, often costing hundreds of dollars, have a dozen or more layers. Except for its thickness, a multilayer board looks almost identical to a double-sided board. Each outer layer contains circuit traces and pads. Components are mounted on one side, as with double-sided circuit boards.

The inner layers on a multilayer board usually serve a special purpose. One inner layer consists of a ground plane and is reserved exclusively for

FIGURE 7–2
Double-sided PC board

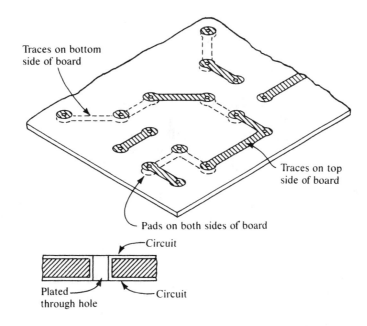

Traces on bottom side of board

Traces on top side of board

Pads on both sides of board

Circuit

Plated through hole

Circuit

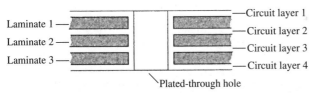

Laminate 1 ——
Laminate 2 ——
Laminate 3 ——

—— Circuit layer 1
—— Circuit layer 2
—— Circuit layer 3
—— Circuit layer 4

Plated-through hole

FIGURE 7–3
Multilayer board

ground interconnections. The other inner layer is used for making all the positive voltage connections. If more than four layers are involved, the additional layers consist of traces for circuit interconnections.

As we have indicated, multilayer boards are complex and relatively expensive. It takes advanced design software and specialized equipment to create such boards. We can offer only an introduction to the subject here.

The PC Board Advantage

Which type of printed circuit board do we use and when? And what are the advantages of PC boards in general? Let's briefly examine both issues.

Which Board Type to Use?

Each board type has its advantages. The single-sided board is simple to design and relatively easy and inexpensive to fabricate. The double-sided board, although more costly to make, can accommodate more complex circuitry than a single-sided board of the same physical dimensions. The multilayer board requires sophisticated design and fabrication techniques, yet very dense (crowded) circuit boards are possible with this approach.

In this book we concentrate on single-sided printed circuit boards. The knowledge and skills acquired in designing and fabricating this type of board contribute to an understanding of double-sided and multilayer boards.

Advantages of Printed Circuit Boards

Printed circuit board construction offers many advantages for circuit assembly. Let's examine a few of the more obvious reasons for using PC boards.

☐ *Less Room for Error.* Because all the wiring is already determined and in place, no wiring mistakes are possible, assuming that the circuit layout is correct.

☐ *Quicker Assembly Time.* With all wiring in and completed, all you need to do is insert components and solder. The time saved in not having to strip wires and mount terminal lugs, as with the cut-slash-and-hook (CSH) method, is tremendous.

☐ *Greater Circuit Density.* Much less space is required to produce a circuit using PC board techniques because long wires and mounting terminals are not needed. In addition, a great deal of care is given in the design stage to component layout and placement. The result is much more efficient use of board space.

☐ *Ease in Troubleshooting.* Because PC boards are laid out in a neat and orderly manner, locating components is easy and quick. More accurate and faster troubleshooting is the result.

Here are four more advantages that may be less obvious, but no less important.

☐ *Less Skill Required.* The skills required to assemble printed circuit boards, while not to be dismissed as trivial, are, nonetheless, not as advanced as those needed in most other types of electronic assembly. Little, if any, knowledge of electronic theory is necessary. One just stuffs the correct component in the right place and solders.

☐ *Use of Automatic Assembly Equipment.* Circuit assembly with PC boards lends itself to the use of automatic insertion equipment. Here, electronic components, from ICs to resistors, are automatically inserted into the board without ever being touched by human hands. Furthermore, even the soldering is often automatic. With the use of a flow-soldering machine, the board is passed over a bath of molten solder. All components are quickly and effortlessly soldered in place.

☐ *Less Prone to Vibration Problems.* The use of a printed circuit board results in a circuit assembly that is of high quality and sturdy construction.

☐ *Fewer Gremlins.* The many gremlins—that is, difficult-to-find spurious radiations, oscillations, and electrical noise caused by crossed wires and poor component placement—tend to diminish with printed circuit board construction.

Thus, printed circuit board assemblies offer many benefits over traditional circuit construction methods. This is true not only for mass production but for prototype work as well, as you will see when you design and build your own PC boards.

Overview of the Design and Fabrication Processes

Producing a PC board that is ready to be stuffed with electronic components requires two stages of development: design and fabrication. In the *design* stage, you determine where the circuit traces and pads are to go. The *fabrication* stage involves transforming a blank copper-clad board (an insulated board with a thin sheet of copper on one side) into one with etched traces and drilled pads. We'll begin with an overview of both stages, starting first, strange as it might seem, with the fabrication stage; but before we do even that, let's take a moment to examine the need to understand basic design principles in the age of computer-based drafting.

But the Computer Does It All!

In Chapter 11 you'll see how schematic capture and PC board design and layout is accomplished with a computer. If you are familiar with how this is accomplished, you may be tempted to skip the following material and jump straight to Chapter 11. That would be a mistake.

Anyone successful in a design field today, be it animation, packaging, fashion, or graphics, is well versed in computer-based design tools. Nonetheless, all, without exception, will tell you the same thing: "The computer is just a tool—first you must understand basic design principles."

Animators, skilled in the latest animation software, are no good without an art background, and graphic designers, sitting at their G4s, running Photoshop, will never "get it right" without a knowledge of the fundamentals of design. The same is true for PC board designers. First, you must understand the principles of board design and layout: what components go where, whether it should be a single- or double-sided board, and so on. Only then can you take full advantage of the computer-based design and layout tools discussed in Chapter 11.

PC Board Fabrication

The fabrication process begins with a copper-clad board consisting of a glass–epoxy laminated base (usually $\frac{1}{16}''$ to $\frac{1}{8}''$ thick) coated with a thin sheet of copper. Phenolic or Melamine is also used as a base, but both lack the superior electrical and mechanical characteristics of glass–epoxy. The former are found primarily in low-cost consumer electronics products—the latter in just about everything else.

To get from the laminated base to a board with copper traces and pads, all unwanted copper clad, or copper that *will not* be left to form traces and pads, must be removed (Figure 7–4). This is done by applying a chemical **etchant (acid)** to the copper. If the entire board was immersed in an acid bath,

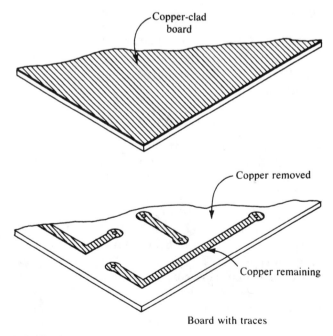

FIGURE 7–4
The etched PC board

all the copper clad would be eaten away. To protect the traces and pads from the etchant solution, a protective coating, known as a **resist** (because it resists the action of acid), is deposited over the copper board *in the pattern of the circuit traces and pads*. Then the board is put in an acid solution where the copper is removed, except for that protected by the resist coating. After the board is withdrawn from the acid bath, the resist material is scraped off (it takes only a little rubbing with fine steel wool), revealing the shiny copper traces and pads. Finally, holes are drilled in the pads to accept component leads.

PC Board Design Concept

To determine where the various traces and pads are to go, two stages are required: the printed circuit board **design layout** and the printed circuit board **artwork.** (The latter, once done by hand with black tape, is now accomplished almost exclusively with computer-aided drafting.)

With graph paper, colored pencils, PC board drawing template, and straightedge, a penciled layout of where the traces and pads will go is created (Figure 7–5). It usually takes a number of tries, or passes, to arrive at the final design layout. First, a very rough sketch of component placement and trace patterns is produced without the use of a template or straightedge. In successive steps more accurate layouts are developed, until a neat, well-defined drawing, as shown in Figure 7–5, is created.

If the artwork (final trace and pad layout) is done on a computer, as is typical, and the opaque

FIGURE 7–5

The PC board design layout

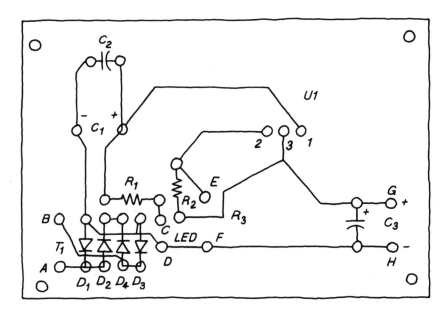

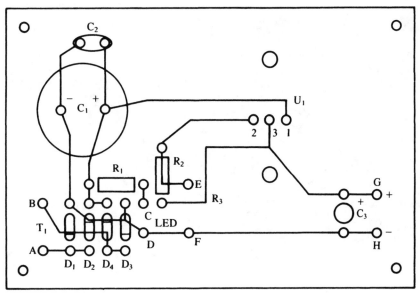

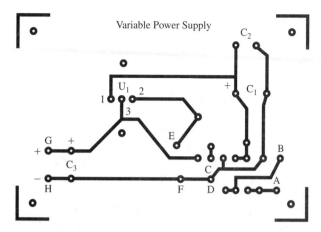

FIGURE 7–6
PC board artwork positive

printout produced with a laser printer, an acetate **positive** is made using a photocopy or thermofax machine (Figure 7–6). An **artwork negative** is produced from this positive using the negative PC design system. The pattern on the negative is reversed; that is, the traces and pads are now transparent, and the background is opaque (black) (Figure 7–7). The negative is positioned over a blank PC board that has been coated with a photosensitive resist material. The board, with the negative over it, is then placed under ultraviolet light. The light shines through only the transparent areas (the traces and pads), "baking on," or activating the photoresist. The photoresist material unexposed to light will simply wash away in the developing operation. Thus, a circuit resist pattern is created, and the board is ready to be immersed in an acid solution.

You may be wondering why we don't just cover the copper-clad board itself with tape-up artwork materials and drop the board in an acid bath. The artwork stick-ons would act as a resist and protect the selected areas from the etching solution. This so-called direct method is possible, but not desirable, for two reasons. First, the stick-ons do not act as a good resist. Often, they work loose in the acid bath, and once that happens, all is lost; the board is ruined. Second, and even more important, this method is a one-shot approach. The artwork produces only one board. All that time laying down the donuts, component-mounting configurations, and tape is for only a single board. What a waste of time! With the indirect method, outlined a moment ago, we have the added step of producing a negative, but we can now make as many boards as we like from only one negative (and one tape-up effort). Clearly, this is the much better procedure.

It's time now for a detailed explanation of the PC board design process. The result will be two drawings: (1) the PC board design layout, and (2) with the aid of the design layout, and the use of computer-based PC board design tools, the final PC board artwork drawing.

Something to Think About

Maybe you have already noticed that scrap printed circuit boards are now used for all manner of contemporary products, from business card holders, to corporate lunch boxes, to clipboards. Why not begin your own collection of PC boards, single-sided, double-sided, and multilayer? Create a montage and hang the finished product on your wall. PC boards as art!

7.2   THE PRINTED CIRCUIT BOARD DESIGN LAYOUT AND ARTWORK GENERATION

In this section we examine what it takes to produce a printed circuit board design layout. Specifically, we discuss preliminary considerations, factors that must be taken into account before the design is actually begun. Next, we review design layout basics, the nuts and bolts of laying out a PC board. Then, we go through the two-step procedure required to produce the design layout drawing. Finally, we examine factors to note in creating the computer-generated artwork.

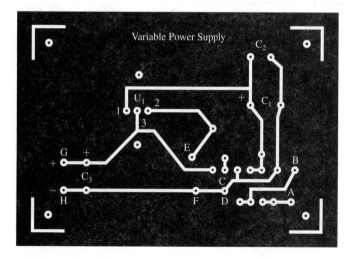

FIGURE 7–7
PC board artwork negative

Preliminary Considerations

There are a half-dozen factors that must be considered, or decisions that must be made, before the PC design layout can begin. Let's discuss each factor, in the order that you should deal with them.

Gathering the Circuit Components

No matter how experienced you may be at PC design, it is always a good idea to have in front of you the electronic components needed to build the prototype project. This does not mean every $\frac{1}{2}$-W resistor that will be used, but you should have on hand at least one of every type of component (capacitors, coils, diodes, and the like). It is particularly important to be able to examine the odd components you may be using, such as unusually packaged SCRs, triacs, transformers, xenon tubes, switches, and lamps. The idea is to have the physical package in front of you for "sizing" when planning your design layout. The PC design template gives you the configurations for most standard components, but it's still a good idea to have the real thing within easy reach at all times.

Where should you get these components? The breadboarded project, right? Wrong! It's best to leave that circuit alone—and functioning. It is always desirable to have at least one working version of the project around at all times. Don't start tearing down the breadboard until the prototype is up and going. Get your components from somewhere else. If that simply isn't possible, and you must examine what's on the breadboard, take only one component off at a time and put it back before you proceed to the next one.

On-Board and Off-Board Components

The next decision is to determine which components will be mounted on the PC board and which ones off the board. Of the off-board components, some will be attached inside the box and others to the outer surfaces (Figure 7–8).

If a component is quite bulky or heavy, or if it is likely to generate excessive heat while in operation, it should not be mounted on the PC board. Power transformers, very large electrolytic capacitors, and some solid-state components with heavy heat sinks fall into this category. These components, referred to as *internal components* (because they are inside the enclosure), will be mounted inside the box, on the chassis.

There is another group of components that we must be able to observe or have frequent access to. These components are the lamps, switches, potentiometers, terminals, jacks, meters, and so forth that will be mounted on the cabinet enclosure. If we have to look at it, touch it, or plug something into it, that component belongs on the outer surfaces of the cabinet. (The distinction between enclosure and chassis is not always clear-cut. Generally, when we refer to enclosure we mean the outer surfaces of a box used to house the project. The chassis portion is either a separate internal mounting surface or the bottom of the box.)

FIGURE 7–8
On-board and off-board components

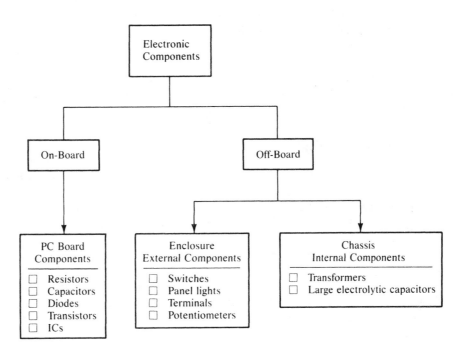

Essentially, then, there are three types of components to identify: those that go inside the box, those that will be attached to the outer surfaces, and the rest, the components that will be mounted on the PC board. As you decide which ones should go where, it is a good idea to trace over the component on a copy of the schematic drawing. Use one color for all components that go inside the cabinet, another for those that will be on the outer surfaces, and a third color for the components to be mounted on the PC board.

Single- or Double-sided Board?

As we learned earlier, a single-sided board has circuit traces on one side and components mounted on the other. A double-sided board has circuit traces on both sides but components placed on only one side, usually the surface with the fewer traces and pads. Single-sided boards are easier to design and tape up, as well as fabricate. They also take much less time to produce. The problem is that they often require the extensive use of **jumpers:** insulated wires used to substitute for traces on the board. If the jumper situation gets out of hand (20 or more, for example), it may be time to consider a double-sided board, where jumpers are practically unheard of and greater circuit density is achieved.

Basically, the decision to go with a single- or double-sided PC board design is based on circuit complexity and the availability of double-sided PC board production facilities. If the circuit isn't too complex, and the jumper situation is likely to remain within controllable limits, it's far better to use a single-sided board. Furthermore, for prototyping, where time is often a factor, the single-sided method carries an added advantage. In this book, all PC board design and construction are for single-sided boards.

Estimating the Size and Shape of the PC Board

The size and shape of the finished PC board is determined by circuit complexity and enclosure requirements. Necessarily, the primary factor is how much circuitry (components and traces) must be placed on the board.

Although there are fancy formulas for determining the square inches of board space needed for particular components, a simpler, and often quite adequate, method of estimating board size is to lay out the actual components on a sheet of graph paper. Then, as you would manipulate paper cutouts of furniture patterns to determine their placement in a room, move the components around to get a rough idea of minimum space requirements. The concern here is not so much with where the components are to go but simply if there is room for all of them within a given space.

It is not enough, however, to consider only component size. Circuit traces must also be accounted for. The rule of thumb is to add about one-third the component space for traces. For example, if the components take up 6 in.^{2}, add at least 2 in.^{2} more for traces. You would want a board with a minimum of 8 in.^{2}.

Choosing the Scale to Work In

The photographic negative, placed over the photoresist-coated PC board, is naturally full scale, or 1×. The trace and pad patterns on the negative are the same size as those that will appear on the board, but the artwork (which is produced by tracing over the design layout) need not be done at full scale. It is often produced 2× or 4×, double and quadruple the final size, respectively (Figure 7–9). The artwork produced at these scales is then photographically reduced in size, just before the negative is make.

Of course, if the artwork is to be larger than the final PC board layout, the design layout, from which it is traced, must also be larger, or of the same scale. Why would anyone want to work at these larger scales? Because when artwork is reduced in size, any layout imperfections and inaccuracies are also reduced. In many cases, they are all but eliminated. Still, working at 4× is not practical in most cases. Double-up, or 2×, is the preferred scale in the PC board industry, yet this scale still requires you to reduce the original artwork by half. Because we wish to avoid the expense and time involved in making such reductions, especially for prototype projects, we work at full scale, or 1×, with all PC layouts in this book.

A View from the Component or the Trace Side?

The final decision to be made before beginning the actual design layout is to determine how you wish to view the board during the layout process. You have a choice of seeing the design from the **component** or the **trace** side of the board. With the component view, you look down on the layout as though it were the top of the PC board, with the components on it. The pads and traces are seen in "X-ray vision" through the board (Figure 7–10a). If you work from the trace side, it's as if you turned the board over and laid the traces out on the back (Figure 7–10b).

The preferred method is to view the layout from the component side, that is, looking down through the board. With this method, which is used

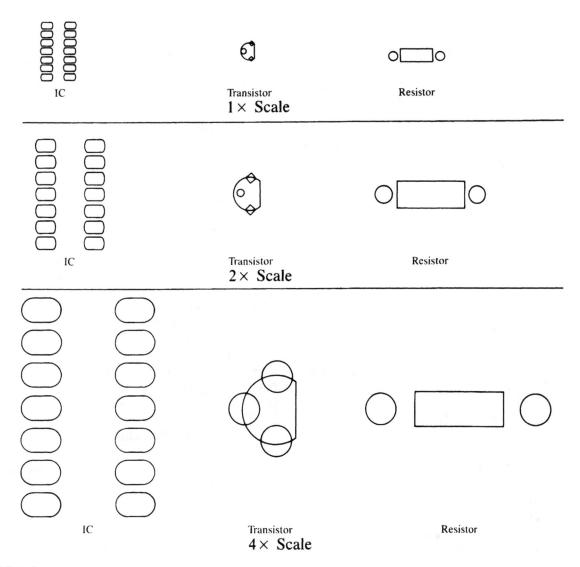

FIGURE 7–9
Drawing scale

FIGURE 7–10
View during layout process

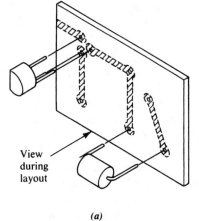

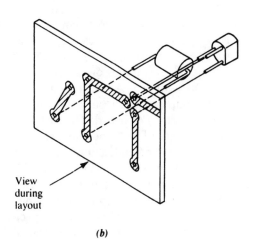

(a)
Component side

(b)
Trace side

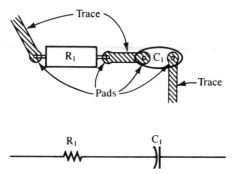

FIGURE 7–11
Component leads and PC pads

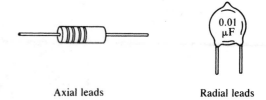

Axial leads   Radial leads

FIGURE 7–12
Component leads: axial and radial

throughout the book, it is much easier to keep track of the layout and to understand how the completed circuit will look.

The Design Layout: Basic Techniques

Now that the preliminaries are out of the way, we can examine the basic procedures for laying out a PC board. We will look at component leads and pads, component size criteria, optimum component placement, the use of connectors, trace routing techniques, and trace-width and spacing issues.

Component Leads and Pads

Every component lead must have its own pad on the PC board (Figure 7–11). Resistor R_1 has two leads; therefore, it has two pads. Because capacitor C_1 also has two leads, it, too, must have two pads. The resistor and capacitor cannot share a pad. Note that the two components are connected together with a short trace, rather than a common pad.

All pads are positioned on the intersections of a $0.100'' \times 0.100''$ grid (the graph paper used in the layout). Because almost all electronic components have lead spacing in $0.10''$ increments, locating pads on such a grid helps ensure a more precise layout.

Electronic components, as discussed in Chapter 3, are passive or active. Passive components have either axial or radial leads (Figure 7–12). **Axial** leads come out the ends, or along the axes, of a component. These leads must be bent to pass through the PC board. **Radial** leads, on the other hand, extend out the side of a component, rather than from the end. Such leads do not have to be bent to pass through the PC board.

Active components (transistors, SCRs, triacs, diacs, ICs, and the like) come in a wide variety of component packaging styles. You will probably need to refer to Appendix D as you proceed with the printed circuit design layout.

Component Size Criterion

Determining the component size is fundamental to the layout process. Not only must the component's body dimensions be known but also the lead spacing that is associated with it. Knowing body size and shape helps you keep components off one another's backs and out of one another's way. Establishing standard lead spacing lets you set down pad locations quickly and accurately.

Fortunately, you don't have to measure every component every time you lay one out on the drawing. This is taken care of by the PC board design **template** (Figure 7–13). On such a template, body size and pad locations for all standard components, with both axial and radial leads, are provided. To use the template, just position the correct component cutout over the desired location (keeping the pads, or holes, on grid intersections) and draw the outline. It's really that simple.

Figure 7–14 shows the use of the PC board design template in the layout of three small circuits. Note how neat and uniform the components are.

Component Placement

The best possible component placement, or layout, is achieved by addressing three relevant issues. If the layout results in (1) wasted board space, (2) the insertion of components in the wrong direction, or (3) difficulty in troubleshooting, it is not an optimum design.

Space is saved by alternating large and small components and by keeping them parallel to a board edge. Notice how, in Figure 7–15, valuable space is conserved simply by interchanging large and small components. Furthermore, keeping components in a vertical or horizontal plane, which usually means parallel to a board surface, results in a tighter layout.

Whenever possible, groups of polarized components (electrolytic capacitors, diodes, ICs, transistors) should be placed facing the same direction. This will aid assemblers in their board-stuffing

FIGURE 7–13
PC design template

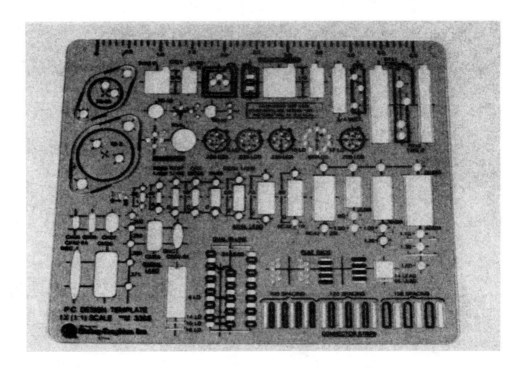

FIGURE 7–14
Design layouts using PC
board design template

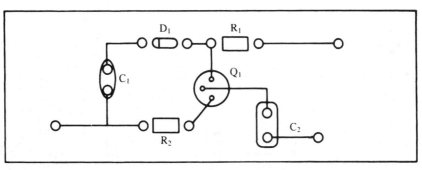

(a)

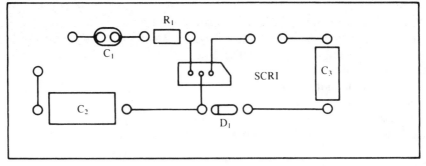

(b)

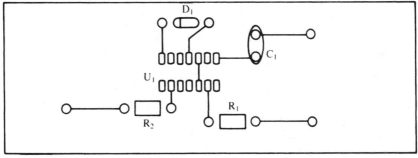

(c)

FIGURE 7–15
Alternating large and small
components

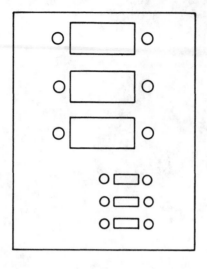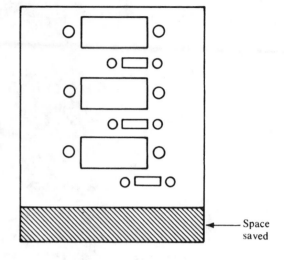

Space saved

FIGURE 7–16
Polarized components

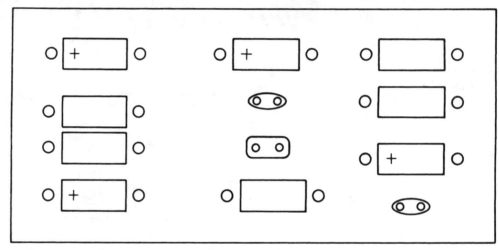

tasks (Figure 7–16). For example, if the cathode ends of most diodes face to the right but one or two face to the left, it will be easy to forget this fact and reverse the left-facing diodes.

Also, the component layout should take into consideration troubleshooting techniques. Easy-to-find components, predictably arranged, go a long way in making the repair technician's job that much easier.

Connecting to the Board

Figure 7–17 illustrates two ways to get off or on the PC board with connectors and terminals. The use of such connectors must be "designed into" the board at an early stage. It is important not to let them become an afterthought.

The etched finger connector (Figure 7–17a) is the most popular method used to make contact with the off-board "world." As shown, the male copper fingers mate with a female connector containing a ribbon of wire. The terminal post (Figure 7–17b) is also widely used when only one wire connection is required.

Trace Routing

Once the component patterns are drawn with the PC board design template, lines (representing copper traces) connecting the various pads are penciled in (Figure 7–18a). Remember, however, that on a single-sided circuit board, it is not possible for an etched trace to cross another without making a connection. One way to avoid the problem is to reroute the trace so that it crosses between the leads of a nearby component (Figure 7–18b). If this is not practical, a jumper may be the only solution. A jumper wire, remember, lifts one conductor off the etched surface so that it can cross the other conductor (Figure 7–18c).

Keep in mind that these trace lines, drawn with colored pencil to distinguish them from the component configuration layouts, are only a guide for the final artwork to come. Nevertheless, an accurate line layout at this point will save much time later.

You have a choice of using curved lines or straight lines with angled corners (Figure 7–19); however, your computer software package will most

FIGURE 7–17
Connectors

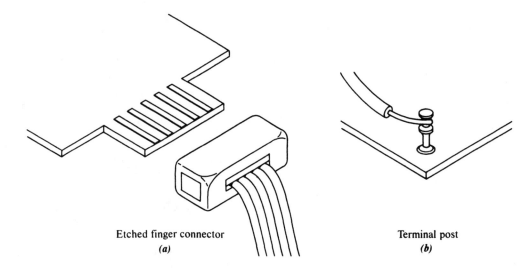

Etched finger connector
(a)

Terminal post
(b)

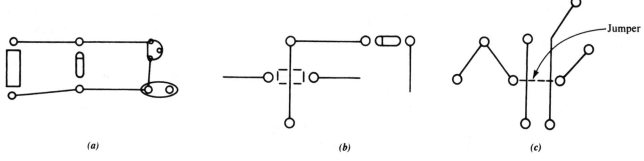

(a)

(b)

Jumper

(c)

FIGURE 7–18
Trace routing

FIGURE 7–19
Layout: curved- and
straight-line approaches

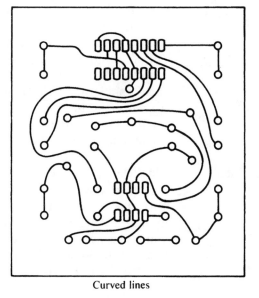

Curved lines

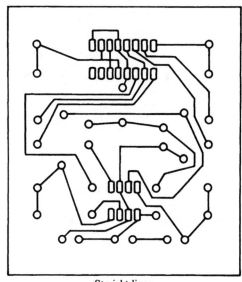

Straight lines

Table 7–1 Maximum Recommended Current-Carrying Capacity for Various Conductor Widths

Conductor Width (in.)	Current (A) 1-oz foil
0.005	0.50
0.010	0.80
0.020	1.40
0.030	1.90
0.050	2.50
0.070	3.50
0.100	4.00
0.150	5.50
0.200	6.00

likely work with straight lines only. Therefore, today the straight-line approach is the preferred choice.

Trace Width and Spacing

In designing your PC board, you must pay attention to final trace width and spacing: the former to account for maximum current levels, the latter to handle maximum voltages.

Table 7–1 shows the maximum recommended current-carrying capacity for various conductor widths, assuming a 1-ounce (oz) copper foil. Table 7–2 shows the recommended minimum conductor spacing for a given voltage range.

To illustrate the use of Tables 7–1 and 7–2, let's look at an example. Assume the following specifications:

Maximum current: 1 A

Maximum voltage: 60 V

Table 7–1 shows that a conductor width of 0.020″ will safely handle 1 A. Table 7–2 shows that the minimum spacing for a 0- to 150-V range is 0.025″. Thus we have:

Conductor width: 0.020″

Conductor spacing: 0.025″

Be sure to keep current and voltage considerations in mind when proceeding with your design layout.

Table 7–2 Recommended Minimum Conductor Spacing

Voltage between Conductors (V)	Minimum Spacing (in.)
0–150	0.025
151–300	0.050
301–500	0.100
> 500	0.0002 (in./V)

Doing the Design Layout: Two Steps

For relatively simple circuit boards, a two-step layout process is recommended. First, a **freehand trial layout sketch** is drawn, then the **final design layout** is produced, with all components, connectors and terminals, and traces neatly drawn with template and straightedge. Let's examine both steps.

The Freehand Trial Layout

The purpose of the freehand layout is to get a feel for where the components and traces will go. Components are usually represented by schematic symbols and are oriented in the direction that they will finally be placed (Figure 7–20). Accurate scaling is unimportant at this point. The main idea is to see if the layout will work, that is, if all components can be accounted for and all trace connections made.

The general technique is to begin by placing the main components (ICs and transistors, for example) as they are physically drawn on the schematic. Work from left to right, sketching the schematic symbol for each component on the graph paper. Next, with a colored pencil, connect the components with trace lines. Don't forget to include a pad for every component lead, wire, terminal post, and connector. It's also a good idea at this point to identify the tentative location of any PC board mounting holes, those that will provide the screws or rivets to hold the board to the chassis.

The freehand trial layout sketch is an important first step in the layout process. Treat this step seriously, and don't be in a hurry to move on. More than one such sketch may be needed before you are ready to proceed to the final layout. Errors caught here will be much less costly than those discovered later on.

To aid in producing such a layout, thin cardboard "cutouts," or "dolls," may be used. Simply cut out a bunch of component outlines for the most popular components. Be sure the outline takes into consideration the lead placement as well. Now all you have to do is arrange and rearrange them on the layout sheet until you have the optimum placement.

Final Design Layout

The final design layout is drawn on graph paper with a PC design template, straightedge, and colored pencils. Place the freehand sketch in front of you and copy (not trace) as you produce the final design layout to scale (Figure 7–21). First, draw the printed circuit board outline and locate all mounting holes. Then, with the PC design template and a black lead pencil, draw all component outlines (and

FIGURE 7–20
Freehand PC trial layout

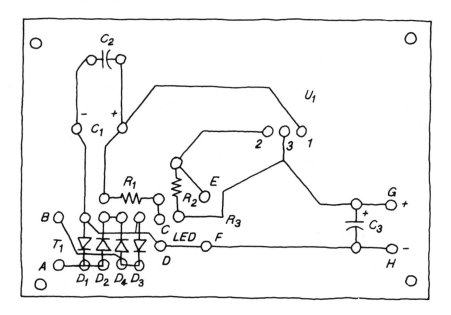

FIGURE 7–21
Final PC design layout

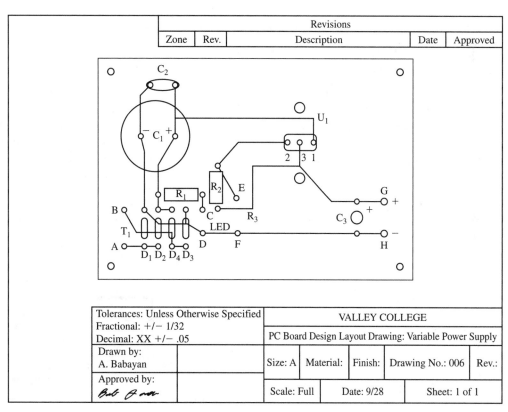

any terminal patterns). It is important that the holes for lead pads be located at grid intersections and that every component be oriented in the correct direction.

The Printed Circuit Board Artwork

Given that you will most likely be producing your final PC artwork on a computer, let's look at what factors to consider when selecting appropriate soft- ware, the issue of "backward" lettering, and the role of the acetate sheet in producing artwork ready for making a negative.

Computer-Generated Artwork

As we just said, it is assumed that you will be producing the final PC board artwork with the aid of a computer-generated CAD software package (see Chapter 11 for a detailed discussion on CAD, the project design process, and the availability of software for schematic drawing, PC board design, and artwork generation). Which one you choose, is, of course, up to you. You will need to consider a number of factors.

Although we have chosen in this text to use AutoCAD to produce the artwork for the Power Supply and 3-Color Organ projects, you may not have AutoCAD available. The program is indeed powerful, and the results are of professional quality. Nonetheless, AutoCAD is expensive. Alternatives for student use are available.

Whichever package you choose, be sure that it can, at a minimum, do the following:

- ☐ Draw lines
- ☐ Draw rectangles
- ☐ Draw circles
- ☐ Draw donuts
- ☐ Draw conductor patterns

In addition, it is beneficial if the software can

- ☐ do scaling and labeling, and make modifications;
- ☐ do reverse lettering.

If you have used Multisim 7.0, from Electronics Workbench, to run a circuit simulation of your project (see Chapter 3, Section 3.4), or a similar software design tool, you can now most likely "capture" your schematic and import it into a PC design package,

like Ultiboard, also from Electronics Workbench. If not, relatively inexpensive (and, in some cases, free) schematic drawing and linked PC board design software is available that will let you "build" your artwork, one component at a time. See Chapter 11 for suggestions.

Backward Lettering

Because you are producing artwork traces, pads, and the like, as seen "through" the PC board, from the component side, you will turn the final, acetate sheet over when pressing the artwork to the blank PC board prior to fabrication. Only then will your PC board print correctly, on the trace side.

Thus, any nomenclature must "read right," that is, so that you can read it. Consequently, when producing the computer-generated PC board artwork, you must reverse the lettering; that is, it must appear backward. The backward lettering will also ensure that the correct view is used when the PC board is processed (see Figure 7–22).

On to the Acetate

Once your PC board artwork is generated and printed out, most likely on a laser printer, you will have to produce a positive on clear acetate (an overhead projector sheet). Run the clear acetate sheet through a copy machine (or, if more convenient, a thermofax machine), and copy the laser-printed PC

FIGURE 7–22

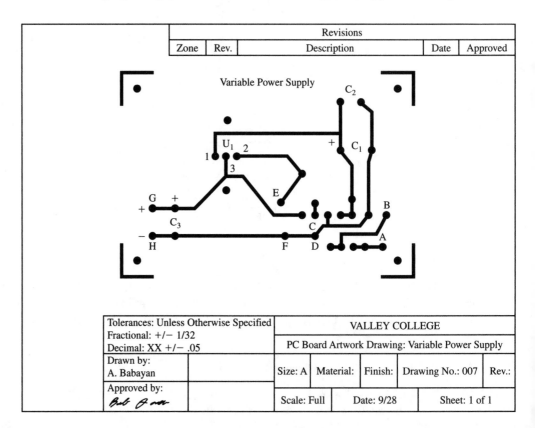

board artwork image onto it. The result is a positive artwork image that is both opaque and transparent; that is, images of pads, traces, and nomenclature are black, and the artwork is transparent everywhere else. It is from this acetate positive that we will make a negative to use in fabricating our PC board (see Chapter 8).

Something to Think About

Becoming knowledgeable about the exact size and shape of electronic components is critical for anyone doing PC board design layout. Why not gather as vast an array of components as you can, then examine each one with a magnifying glass. Did you discover any unusual physical characteristics?

7.3 PROJECT PRINTED CIRCUIT BOARD DESIGN

Let's turn to the Sample and Exercise Projects and analyze their PC board designs. As in previous chapters, the discussion of the Variable Power Supply will be more extensive than that of the 3-Channel Color Organ.

Printed Circuit Board Design for the Variable Power Supply

We begin by exploring the on-board, off-board component selection and the trial and final PC design

layouts. Then, we'll look at the final, computer-generated artwork.

Designing the PC Board

Working from the schematic drawing, parts list, and packaging plan, we can determine which components go on the PC board, which ones go on the cabinet (chassis), and which mount on the box's outer surfaces (Figure 7–23). Note that only two components, the transformer and the terminal lug, will attach to the bottom surface of the box. The line cord will connect to switch S_1, and the primary lead of transformer T_1 through a strain relief on the rear of the chassis.

With the components selected for the PC board, the board's rough size and shape can be estimated and the trial design layout sketch begun. Figure 7–24a illustrates the schematic drawing for the Variable Power Supply, showing only the components that will be mounted on the PC board. Note the points marked A–H. These points will be pads on the PC board that will accept wires connected to the off-board components.

Figure 7–24b shows the freehand trial layout sketch for the PC board. It is rough, with no attempt made to draw components to exact scale. It may take one or two steps to get even this far, or a lot of erasing. But the important thing is that the layout is workable and electrically correct. It may not be the way you would lay out each component; ask 10 people to do the layout, and you will get 10 different layouts. Yet what we have is suitable, and a good start.

FIGURE 7–23
On-board and off-board components for Variable Power Supply

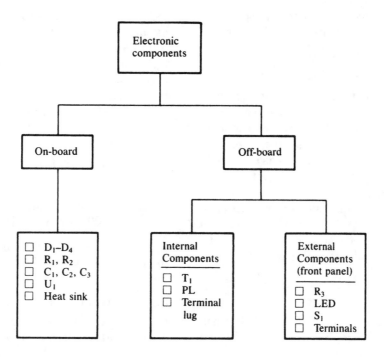

FIGURE 7–24
Freehand trial layout of PC
board for Variable Power
Supply

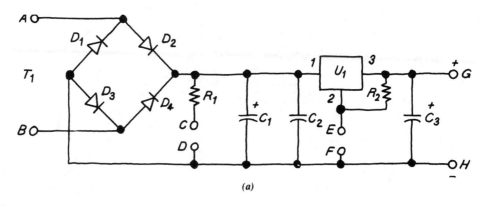

(a)

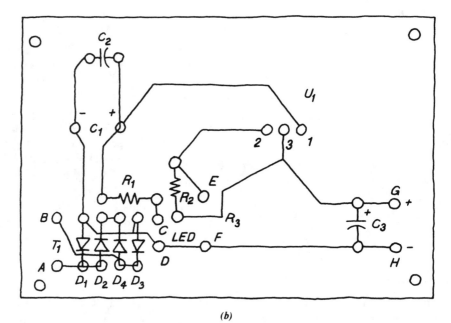

(b)

The final layout, which becomes the **printed circuit board design layout drawing,** is shown in Figure 7–25. All components are drawn to scale using a 1× PC design template. Also, all traces are drawn with a straightedge, in a color other than black. Note, furthermore, that the leads on capacitor C_3 and voltage regulator U_1 have been spread apart a bit. Both components have a "natural" lead spacing in multiples of 0.010″. To install these components on the board as is would require placing pads extremely close together. Although this can be done, it's best to leave as wide an air gap between pads as possible.

The board size and dimensions are now determined, and mounting holes located. All components are labeled, as are the off-board connecting pads. Let's move on to the final, computer-generated artwork stage.

Artwork Generation

In Figure 7–26 we see the final PC board artwork for the Variable Power Supply. Note the backward lettering. Again, once the artwork is transferred to an acetate sheet, that sheet will be turned over to produce the negative. The negative will read right, and the resulting traces, pads, and nomenclature will be correct for producing the trace side of the PC board.

The PC artwork for the Variable Power Supply is now complete. The next step, to be examined in Chapter 8, is to produce a photographic negative from this artwork.

Printed Circuit Board Design for the 3-Channel Color Organ

This section gives a few tips to consider when working on the 3-Channel Color Organ PC design.

Design Layout Tips

Figure 7–27 shows a chart suggesting a possible categorization of on- and off-board components.

FIGURE 7–25
PC board design layout drawing: Variable Power Supply

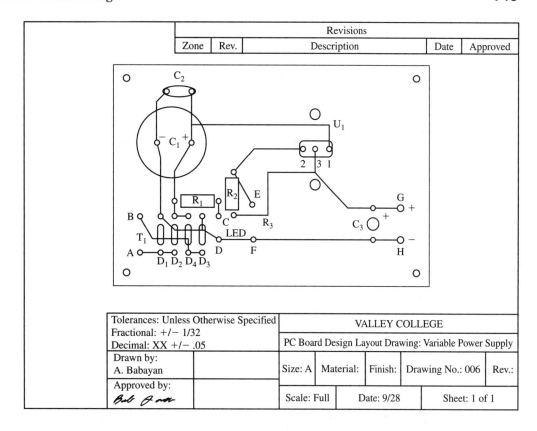

FIGURE 7–26
PC board artwork drawing: Variable Power Supply

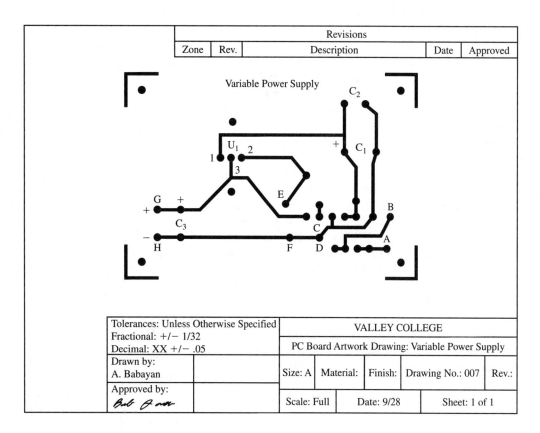

FIGURE 7–27
Possible on-board and off-
board components for 3-
Channel Color Organ

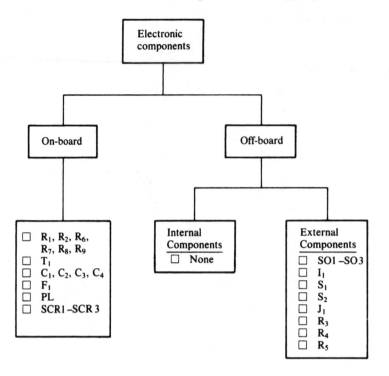

Note that in the case of the 3-Channel Color Organ Project no components are actually placed on the chassis. Everything goes on the PC board or is attached to the enclosure's outer surfaces, where you must touch it, look at it, or plug something into it.

With regard to the PC board layout itself, keep in mind that the circuit will be operated from the 120-V ac line. Although the circuit should rarely draw more than 1 A, just to be on the safe side, plan for trace widths of at least 0.060″. You may also want to spread the component leads of the three SCRs or offset the leads in a triangular pattern. They come spaced only 0.010″ apart, and that may result in too tight a spacing for the PC board pads. Also, don't forget to include a pad for every off-board component connection. Figure 7–28 shows the schematic for the 3-Channel Color Organ redrawn to include only the components to be mounted on the PC board. Note that 23 pads are needed for off-board connections.

You might begin the layout by following the schematic from left to right. When doing so, take advantage of the three bus lines that exist: one each at the top, middle, and bottom of the drawing.

Computer-Generated Artwork

As we stated in the previous section, with regard to computer-generated artwork for the Variable Power Supply Project, it is assumed that here, too, you are using a software artwork-generation tool, such as AutoCAD or something similar. If you are building one of the elective projects or a project of your own design, now is the time to produce the design layout and computer-generated artwork for the project.

> ### Something to Think About
>
> *Although commercial color organs are hard to find, the same is not true for power supplies. Why not gather as many power supplies as you can and examine closely their PC boards? What is the widest trace you notice? Are ground planes used?*

SUMMARY

In this chapter we looked at the advantages of printed circuit board use and gave an overview of the PC board design and fabrication process. We then delved into the printed circuit board design layout by first examining such preliminary considerations as the way to choose on- and off-board

FIGURE 7–28
Schematic for 3-Channel
Color Organ showing
possible on-board
components

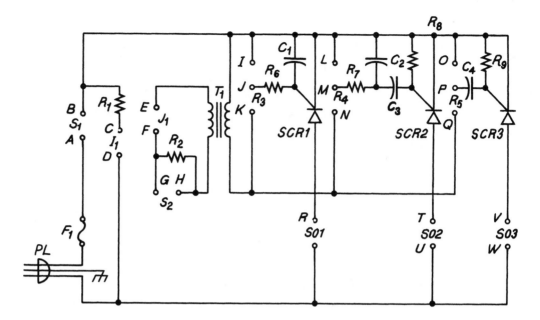

components, scale, and viewing methods. Finally, we studied design layout techniques and analyzed the two-step procedure that produces a freehand trial sketch and a final design layout.

We examined the procedures necessary to produce the printed circuit board design and artwork drawings for the Variable Power Supply Project,

along with some tips and suggestions for doing the same with the 3-Channel Color Organ Project.

In Chapter 8 we get "down and dirty" by discovering how to fabricate a printed circuit board, first by producing the photographic negative, then by etching and drilling the actual PC board. Let the fabrication begin.

QUESTIONS

1. List the three types of printed circuit boards in wide use throughout the electronics industry: _____, _____, _____.

2. List five advantages to using printed circuit boards over any other circuit construction method:

   a.

   b.

   c.

   d.

   e.

3. In the PC board design process, which comes first: the design layout drawing or the artwork drawing? _____

4. If a component is quite bulky or heavy, or if it is likely to generate excessive heat while in operation, it should _____ be mounted on the PC board.

5. Single-sided PC boards often require the use of _____, insulated wires used to substitute for traces on the board.

6. When laying out a PC board, the designer has a choice of viewing the layout from the _____ or _____ side.

7. In producing the design layout, two drawings are usually required: the _____ trial layout sketch and the _____ design layout.

8. All good PC board design software should be able to:

   a.

   b.

   c.

   d.

   e.

9. Relatively _____ PC board design software is available that will let you "build" your artwork, one component at a time.

10. When producing computer-generated PC board artwork, you must _____ any nomenclature.

REVIEW EXERCISES

1. Identify the circuit components on the PC board design layouts shown. Mark polarized components with a + and a − sign.

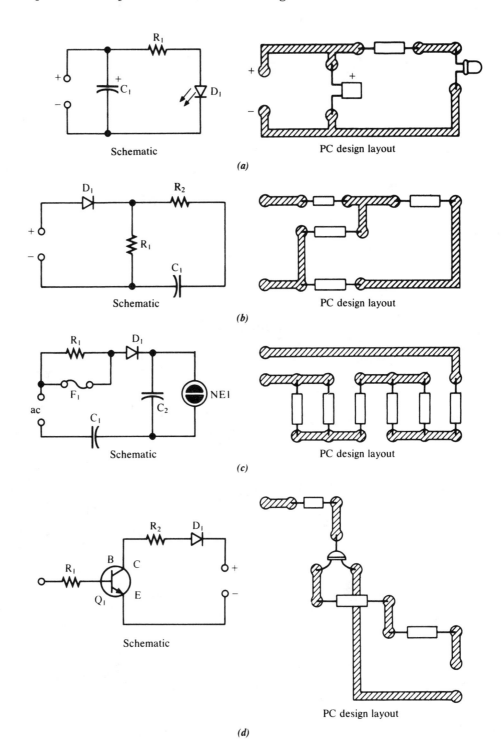

2. Draw each component in the correct location and, where applicable, in the correct direction on the PC board layouts shown. Label each component.

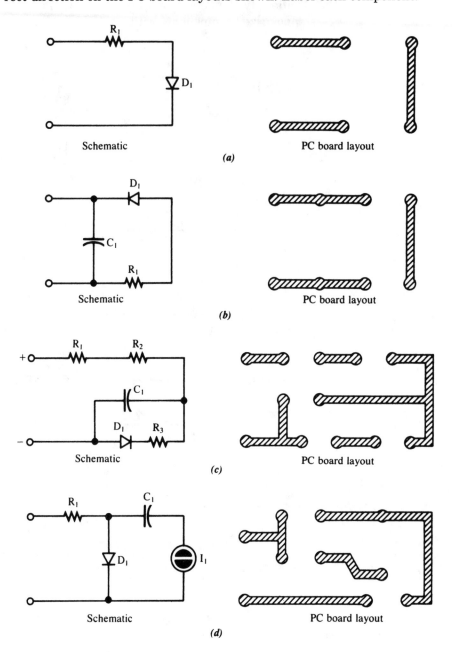

Schematic   PC board layout

(a)

Schematic   PC board layout

(b)

Schematic   PC board layout

(c)

Schematic   PC board layout

(d)

3. From the schematic shown, produce a 1× freehand trial layout sketch, viewed from the component side. Draw all trace lines with a red pencil.

Light detector

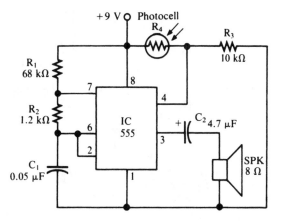

4. From the freehand trial layout sketch produced in Exercise 3, draw the final design layout. Use a PC board design template, straightedge, and colored pencil.

5. From the schematic shown, produce a 1× freehand trial layout sketch, viewed from the component side. Draw all trace lines with a colored pencil.

Unusual-sound-effects synthesizer

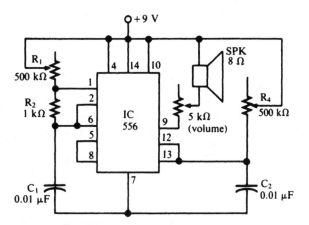

6. From the freehand trial layout sketch produced in Exercise 5, draw the final design layout. Use a PC board design template, straightedge, and colored pencil.

7. From the schematic shown, produce a 1× freehand trial layout sketch, viewed from the component side. Draw all trace lines with a colored pencil.

Audible alarm

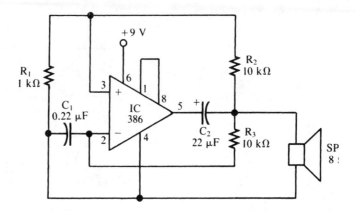

8. From the freehand trial layout sketch produced in Exercise 7, draw the final design layout. Use a PC board design template, straightedge, and colored pencil.

9. From the schematic shown, produce a 1× freehand trial layout sketch, viewed from the component side. Draw all trace lines with a colored pencil.

Door-ajar alarm

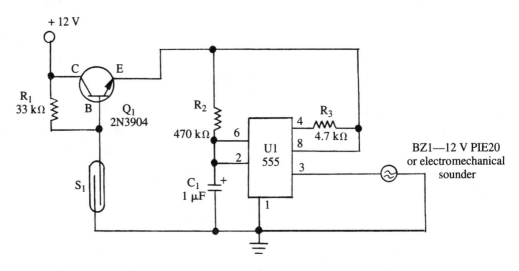

10. From the freehand trial layout sketch produced in Exercise 9, draw the final design layout. Use a PC board design template, straightedge, and colored pencil.

8 Prototyping—Printed Circuit Board Fabrication

OBJECTIVES

In this chapter you will learn

- [ ] How to expose film using the contact printing method.
- [ ] How to process a 1:1 negative from the positive PC board artwork.
- [ ] How to produce the PC board fabrication drawing.
- [ ] How to create a photoresist pattern on the PC board.
- [ ] How to etch a printed circuit board.
- [ ] How to drill a printed circuit board.
- [ ] How to repair an under- or overetched printed circuit board.
- [ ] How printed circuit boards are manufactured in industry.
- [ ] How to fabricate the printed circuit boards for the Sample and Exercise Projects.

The printed circuit board is designed, and it's now time to make it. In doing so, the working environment may come to resemble more a chemistry than an electronics lab. You must set aside the multimeter and soldering iron and gather the chemicals and related materials necessary to produce a film negative and an etched and drilled PC board. The process of fabricating such a board is not difficult. Little in the way of equipment and supplies is required, and if you follow correct procedures and observe simple safety precautions, you can obtain excellent results with little fuss, right in your home or school electronics laboratory.

In this chapter we find out how the photographic negative is made. We will examine the printed circuit board fabrication drawing, and see how to apply a resist pattern to the blank PC board. Next, we will etch, drill, and finish the board, getting it to the point where it is ready to be stuffed with electronic components. Then we'll investigate the industrial printed circuit board fabrication process. Finally, we'll look at how the PC boards for the Sample and Exercise Projects are to be made. So roll up your sleeves and put on your apron; this is definitely a hands-on experience.

8.1 THE PHOTOGRAPHIC NEGATIVE AND PRINTED CIRCUIT BOARD FABRICATION DRAWING

In this section we look at how the photographic negative is made and examine the PC board fabrication drawing. With regard to the former, we explore two methods of producing such a negative: the traditional liquid "photographic" approach and two types of "negativeless" negative procedures.

Creating the Photographic Negative: An Overview

You don't have to be a professional photographer to make a photographic negative (referred to from now on simply as a negative) from the PC artwork. The exposure and processing procedures are not complex and are tolerant of human error. All the necessary equipment and materials can be purchased for well under $75, and much of it is reusable.

Because you will be using the contact printing process, for which *no camera* is required, making a negative is easy and straightforward. (Of course, if you have access to a print shop where a vertical camera is available, your negative-making problems are solved.) But for those who want to do it themselves (or have no choice), let's see just what a negative is, examine the darkroom environment, and run through the list of needed supplies.

From Positive to Negative: Reversing the Artwork

A 1:1 negative of the PC artwork must be created. A negative is the reverse of the artwork, which is called the *positive*. This means that what is black on the artwork acetate (sheets) will be transparent on the negative, and what is transparent on the artwork will be black on the negative (Figure 8–1).

Why produce such a negative? Because the photoresist used to form the image of traces and pads on the PC board is negative-acting, which

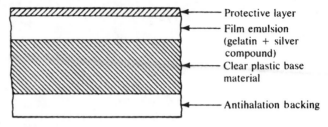

FIGURE 8–2
Cross section of sheet of photographic film

means it requires a negative, not a positive, to create the image. True, there are positive-acting photoresists that will work directly from the positive PC artwork, but although such a method does offer the advantage of not having to produce a negative, it also has several disadvantages that make this procedure undesirable. Therefore, we will go through the seemingly extra step of creating a negative from the PC artwork.

A negative is made from a sheet of film. As shown in Figure 8–2, the film consists of an emulsion (gelatin plus silver compound) behind a protective transparent layer. The emulsion adheres to a clear plastic base material that in turn is coated with an antihalation backing (an antihalation backing diffuses light). We will be exposing and processing this film. Because we are interested only in producing a full-scale negative that is pure black-and-white (actually the white area will eventually become transparent), as opposed to one containing shades of gray, the procedure is considerably simplified.

When film is *exposed,* a latent (invisible) image is produced on the film emulsion. The film is exposed by placing the PC artwork over it and then quickly turning on and off a nearby incandescent light—a process equivalent to opening and closing the shutter of a camera. The clear areas on the artwork acetate admit light to the film, exposing that portion of it. The opaque areas on the artwork (the traces) block light, and the film beneath these surfaces is not exposed. The result is the latent image.

In film *processing,* that latent image is developed, or "brought forth." First, the film is placed in a developer solution, which causes the exposed portions of the film (the nontrace areas) to turn black. The unexposed portions of the film (the trace areas) turn white. The result is a black-and-white negative. Next, the film is placed in a stop bath solution that, as its name implies, stops the developing action. From there, the film goes into a fixer solution, which dissolves the emulsion of the unexposed areas (the white traces) and leaves a transparent pattern. The result is a black-and-transparent (clear) negative. Finally, the film is rinsed and hung to dry. The nega-

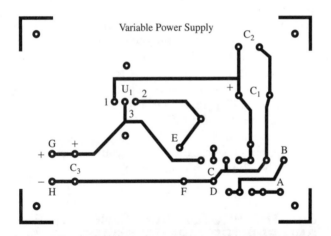

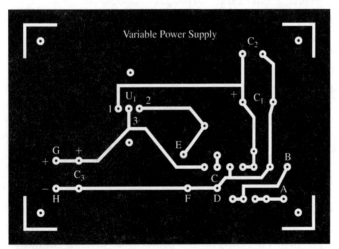

FIGURE 8–1
Printed circuit board artwork and negative

Equipment	Quantity	Equipment	Quantity
☐ Plastic rocker trays (also known as developing trays) 8″ × 10″ × 2″ Different colors	4	☐ Squeegee Plastic with sponge surface	1
☐ Plastic tongs	4	☐ Timer (mechanical or electronic) 0–15 minutes (alarm clock will do)	1
☐ Weighing scale (1-lb capacity)	1		
☐ Plastic funnel Intake diameter 6″–8″; spout diameter ½″	1	☐ Incandescent light bulb 150-W bulb; base	1
☐ Graduates—Large—32-oz capacity; graduated in ounces, milliliters, liters—Small—2-oz capacity; graduated in milliliters	2	☐ Beakers 3-qt size; 1-gal size Dark brown	4

Supplies	Quantity

Equipment	Quantity	Supplies	Quantity
☐ Thermometer 0°–120°F (0°–50°C) Glass; minimum 5″ in length	1	☐ Film Litho film, Type A, polyester base; 50 4″ × 5″ sheets	1 pkg
☐ Safelight Red 15-W bulb with lamp base	1	☐ Developer Type A and Type B (Kodak Cat. 1465152)	2 pkg
☐ Printing frame 8″ × 10″ with glass and cardboard backing	1	☐ Stop bath 1 U.S. pint (Kodak Cat. 1464247)	1
☐ Film clips Metal "tooth" type	2	☐ Fixer 1 U.S. quart (Kodak Cat. 1464080)	1

FIGURE 8–3
Equipment and supplies for making a negative

tive is complete and ready to be used in producing the PC board resist pattern.

Keeping It in the Dark: The Right Environment

Much of the process of producing a negative requires that you remain in the dark—literally. A darkroom environment is needed. Essentially, there are two choices. You can seal a work area from all outside light or simply do all film processing at night. Working at night restricts your time, so it's best to find a way of making your work space light-tight.

Cover all windows in the darkroom with a flat black paint. If that is not possible, cover the windows with a dark shade and use black electrical tape to seal the edges. The door, too, should be taped around the perimeter, and a towel or similar cloth placed beneath the door to prevent light from seeping through.

Make sure, also, that the darkroom has a source of hot and cold running water. Now you have it, a place to expose and process your negative.

Equipment and Supplies for Making a Negative

The equipment and supplies discussed here allow you to make good contact negatives from any

FIGURE 8–4
Photographic materials

reasonable size PC artwork. Everything can be purchased at any photo supply store.

A list of the materials necessary to produce the photographic negative is given in Figure 8–3; the actual equipment and supplies are shown in Figure 8–4. A brief explanation of each item follows.

Equipment Needed

☐ *Rocker Trays* (also called developing trays). Rocker trays are plastic trays used to hold the developer, stop bath, fixer, and water solutions during film processing. Four different colors should be selected. This is so that the same tray is used for the same chemical each time, thus cutting down on contamination.

☐ *Plastic Tongs.* Plastic tongs are used to carry the film from one tray to another.

☐ *Weighing Scale.* The scale is used to weigh the developer chemicals that come in powder form. A 1-lb postal scale for weighing letters is the best type.

☐ *Funnel.* Used to pour the chemicals, the funnel should be made of plastic and have an intake diameter of 6″ to 8″ and a spout diameter of $\frac{1}{2}$″.

☐ *Graduates.* Used to portion out the various chemicals, two plastic graduates, with well-identified index marks, are needed. The larger graduate should have a 32-oz [1-liter (L)] capacity and be graduated in milliliters (mL) and liters. The smaller graduate should have a capacity of 2 oz (60 mL) and be graduated only in milliliters.

☐ *Thermometer.* Used to measure the temperature of the various chemicals, the glass thermometer should range from 0° to 120°F (0° to 50°C).

☐ *Safelight.* To avoid working in complete darkness, a red safelight can be used. It should be a 7- to 15-W bulb in any convenient lamp base.

☐ *Printing Frame.* Used to clamp the artwork firmly to the negative, this printing frame is available commercially, or it can be made from a simple picture frame. It should contain a piece of window glass, cardboard backing, and pressure clips to hold the individual pieces together.

☐ *Film Clips.* Similar to "Bulldog" paper clips, film clips are used to hold the negative when it is hung to dry. They are available commercially. Or simply use "Bulldog" paper clips.

☐ *Squeegee.* Squeegees quickly remove excess water from the negative prior to hanging it to dry. They are inexpensive and readily available.

☐ *Timer.* Many of the procedures in film exposure and processing must be timed. Commercial timers are available, or you can make your own; an ordinary wristwatch or alarm clock will do. Regardless of the type of timer, make sure it does not have a lighted dial, which can cause problems during film exposure and developing.

☐ *Incandescent Light Bulb.* Used for exposing the film, the light bulb should be a 120-V, 150-W lamp, placed in a lamp socket with a switch.

☐ *Assorted Beakers.* Various assorted beakers, used to store the mixed chemicals, are required. Use dark brown beakers, to prevent light from penetrating, and ones with sealable caps.

Supplies Needed

☐ *Film.* The widely used Litho Film, Type A, with a polyester base is what you will need. A box of 50 sheets, each 4″ × 5″, will get you started.

☐ *Developer.* Developer comes in powder or liquid form. The powder lasts longer and thus is the recommended choice. Two packages of developer, Part A and Part B, are required.

☐ *Stop Bath.* One U.S. pint is enough.

☐ *Fixer.* One U.S. quart will do it.

Producing the Negative: Setup and Procedure

It's time to get set up and go through the procedure of making a negative.

Setup: Getting Everything in Order

To expose the negative, you will need the 150-W incandescent lamp and the printing frame. Set things up so that the printing frame is directly in front of the lamp, approximately 5 ft away.

Film processing involves chemicals. Fortunately, the chemical mixing required can take place under normal lighting conditions. Follow all directions and safety precautions printed on the chemical containers and proceed as follows:

1. Set out the four rocker trays, four tongs, a squeegee, and a film clip (Figure 8–5).
2. Place the red safelight nearby.
3. Mix the developer solutions. Part A developer is mixed (diluted) at a ratio of 1 part developer to 5.1 parts warm water. The exact amount should be 2.2 oz of powder to 11.6 oz of water. Warm the water to approximately 100°F. Let hot tap water run

FIGURE 8–5
Setup for making a negative

in a graduate until the submerged thermometer reads 100°F. Pour out all but 11.6 oz of water and add 2.2 oz of powder. Mix thoroughly. When powder is dissolved (in about 2 minutes), pour the entire contents of the graduate into an empty sealable container marked "Part A." Set it aside.

Part B developer is mixed at a ratio of 1 part developer to 5.4 parts warm water. In this case, the exact amount is 2.3 oz of powder to 12.4 oz of water. Mix as you did with Part A developer and pour contents into a separate sealable container marked "Part B." Set it aside.

4. Prepare the stop bath solution. The stop bath is highly concentrated and should be mixed at a ratio of 1 part stop bath to 63 parts water. Mix 8 mL of stop bath with 500 mL (half a liter) of slightly warmed water. Pour the dissolved contents into an empty sealable container marked "Stop bath." Set it aside.

5. Prepare the fixer solution. The fixer is mixed at a ratio of 1 part fixer to 3 parts water. Use 8 oz of fixer with 24 oz of water that has been warmed to approximately 70°F. When the fixer is dissolved, place it in a separate sealable container marked "Fixer." Set it aside.

6. Fill the fourth rocker tray with cool tap water.

Figure 8–6 summarizes the chemical mixing ratios and amounts.

Pour the contents of developer A and developer B containers into the first rocker tray, then pour the contents of the stop bath into the second rocker tray. Finally, pour the fixer solution into the third rocker tray. Let all chemicals cool for awhile (5 minutes will do) before immersing the film. If the chemicals are too hot (especially the developer), the film emulsion will soften. You are now ready to expose and process a negative.

Chemical	Ratio	Amount
Developer Part A	1:5.1	2.2 oz developer 11.6 oz water
Developer Part B	1:5.4	2.3 oz developer 12.4 oz water
Stop bath	1:63	8 mL (0.27 oz) stop bath 500 mL (16.9 oz) water
Fixer	1:3	8 oz fixer 24 oz water

FIGURE 8–6
Photographic chemical mixing ratios and amounts

Procedure: Producing the Negative

To *expose the film,* place the positive PC artwork against the inside of the glass in the printing frame, so that the artwork is right-reading (the information can be read) from the outside (Figure 8–7). Then, in complete darkness or under safelight conditions, remove a sheet of film from its package. Immediately reseal the package and set it aside. Place the sheet of film against the artwork, with the dull side of the film down, or toward the artwork. (The dull, or emulsion, side feels rougher or less glossy.) Put the cardboard backing in place, and lock the frame to secure all materials. Now, with the frame placed 4 to 5 ft from the 150-W incandescent bulb, expose the film by turning the lamp on and off quickly. About half a second exposure time is all you need. This done, the film is exposed and you are ready to process it.

Before following the film processing steps, keep three simple cautions in mind:

1. Use separate plastic tongs in each tray. Do not mix the tongs, for by doing so you might inadvertently transfer chemicals from one tray to the next.
2. Handle the film only on the edges or corners. Rubbing the tongs over the film surface may scratch the film.
3. When removing the film from a tray, let it drain into the tray before transferring it to the next tray.

Now let's get started.

To begin **film processing**, under safelight conditions place the exposed film in the *developer tray.* Rock the tray gently. In approximately 30 seconds the exposed areas will start to turn black, and the unexposed areas (traces and pads) will turn white. How long the film should be left in the developer depends on such factors as the length of exposure, type of film, freshness of chemicals, and temperature of film. Two minutes should be enough time.

You *can,* however, under- or overdevelop the film. When the white areas are clear and sharp, developing is complete. If you underdevelop, the black areas will be too gray and transparent. On the other hand, if you overdevelop, the white areas will become fuzzy around the edges. You will have to experiment a few times to get it just right.

When developing is complete, place the film in the **stop bath** tray. Rock the tray gently for about 30 to 40 seconds. Time is not critical. No change in the negative's appearance will be seen during this stage.

Next, place the negative in the **fixer solution** and rock the tray for $2\frac{1}{2}$ to 3 minutes. Again, time is not crucial. During this stage, the white areas of the film will become transparent. If you are using a tray

FIGURE 8–7
Film exposure

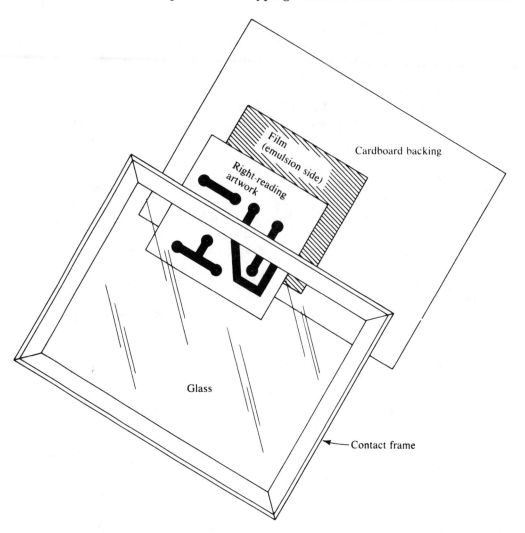

of a color other than black, you should see the color of the tray "come through" the traces and pads of the negative.

After removing the negative from the fixer tray, place it in the *water bath* for 1 or 2 minutes. This, and subsequent operations, can be done under normal lighting conditions.

Remove the negative and *rinse* it under cold running water for 2 to 3 minutes. Next, with the squeegee remove excess water from the film and, using a film clip, hang the negative to dry.

Figure 8–8 summarizes film processing times and reactions.

What do you do with the chemicals? After the day's negatives have been made, the developer cannot be reused and it must be discarded. (All processing chemicals can safely be poured down the drain. Before doing so, however, it is best to dilute them with running tap water and keep the water running awhile afterward to clear the drain tap.) The stop bath and fixer solutions may be saved for another day, but probably not another week. They

Step	Time	Reaction
1. Developer	2 minutes (approximately)	Exposed areas turn black in about 30 seconds
2. Stop bath	30–40 seconds	No change in appearance of negative
3. Fixer solution	$2\frac{1}{2}$–3 minutes	White areas of film become transparent
4. Water bath	1–2 minutes	No change in appearance of negative
5. Rinse	2–3 minutes	No change in appearance of negative

FIGURE 8–8
Film processing times and reactions

both deteriorate with age, and by the fifth or sixth day they may be too weak to be of much use.

Are you through? Perhaps. After the negative has dried, examine it carefully. Are there any pinholes? **Pinholes** are tiny, clear (transparent) areas caused by dust trapped between the film and the PC artwork during exposure. If there are pinholes, they will need to be filled in, or made opaque. You can do this with a special red pen designed for that purpose, or simply use white correction fluid, the kind used for removing typing mistakes. With this touch-up process complete, the negative should be ready for use.

The "Negativeless" Negative Approach

How would you like to avoid this whole negative-making process and yet wind up with a "negative" of equal quality? No chemicals, no darkroom, no mess, no cleanup. There are two ways to go about this, and we'll look at both. One uses 3M *Reverse Image Film*. The other simply employs the negative image feature found on most photocopiers. Let's take a look.

A "Negativeless" Negative with 3M Reverse Image Film

You're going to need two things: a transparency maker (Thermofax machine) and a quantity of 3M IR1511 Reverse Image Film.

The transparency maker (3M 4550 Desktop Transparency Maker or similar machine) is a piece of equipment that exists in virtually every school. It has been used for decades to make overhead projector transparencies on acetate of printed material. The machine is a breeze to use, making a transparency (in our case a "negative") in seconds. A transparency maker is so simple to operate, there isn't even an on-off switch—you just adjust a timer knob, insert your sheet, and a few seconds later out comes a "negative."

3M IR1511 Reverse Image Film is available in $8\frac{1}{2}'' \times 11''$ sheets in a package of 100. The cost amounts to about 25¢ a sheet. The film requires no special care: it is not light-sensitive, and you can handle it like you would any thin piece of paper. Contact your 3M distributor for availability and prices. Be sure to ask for the Reverse Image film.

To make a "negativeless" negative with 3M Reverse Image Film, we assume that you have a transparency maker, a few sheets of the film, and your original PC board artwork. To make a "negative," follow these six simple steps:

1. Make a good photocopy of your PC artwork on white paper.

2. Taking a sheet of the 3M IR1511 in your hand, carefully peel off the film (black sheet) from its backing. If you don't need the whole $8\frac{1}{2}'' \times 11''$ sheet, simply cut off what you do need before peeling off the film.

3. Place the film, shiny side up, over the photocopy of your artwork. You will notice that the film is quite thin, so handle it with care.

4. Adjust the timer setting on the transparency maker to the recommended setting. If you can't determine what that is, you will have to experiment a bit to get the image just right.

5. With the film on top and the photocopy of your artwork on the bottom (face up), run the pair through the transparency maker.

6. When the sheets emerge from the machine (it takes only a few seconds), gently peel off the film from the photocopy. Presto, you have a "negative."

Now, carefully inspect your "negative" by placing it on a light table or holding it up to a light bulb. Are the black areas completely opaque and the traces and pads transparent? Are there any grainy black particles filling in the transparent areas? If the negative is not as clear as you would like, experiment with the timer setting a bit more to get the "exposure" correct.

To improve negative quality, you can also run the original photocopy with a new sheet of 3M IR1511 Film through the transparency machine a second time. You will notice that after the photocopy of your artwork went through the machine the first time, a thin black film was deposited on the opaque traces and pads. That's good, because your photocopy is now even more opaque than it was when you made it. This added opaqueness will improve your "negative" when the 3M IR1511 and photocopy is run through the transparency maker a second time.

A "Negativeless" Negative Using a Photocopier

Most photocopiers, particularly those found at local copy centers, have the ability to reverse an image: turn black into white and white into black, and do so, literally, at the press of a button. Thus your positive, with its black traces and pads, becomes an image with white traces and pads, and black everywhere else.

Of course, that's not exactly what you want. You need a negative, on which the traces and pads are transparent against a black surrounding area. No problem! With your copy machine set to reverse the positive artwork, just place an overhead transparency in the paper tray and presto—you have your negative. You might have to mess with the copier's contrast setting a bit to get the black areas to come out as opaque as possible, but that should be no problem. Good luck.

The PC Board Fabrication Drawing

Before placing a negative on the sensitized PC board and producing the photoresist pattern, you must know specifically how the board will be fabricated. To do that, you need a **PC board fabrication drawing**. Let's explore the purpose of this drawing and examine a few of its main characteristics.

Purpose of Drawing

A PC board fabrication drawing is like any other construction drawing, in that it is used to describe graphically how something is to be made. The mechanical drawing of a flange (Figure 8–9) states the material it is to be made of, gives its dimensional configurations (size and shape), and identifies any particular machining operations and metal finishing required. Similar information is necessary to describe how a PC board is to be fabricated. Specifically, we want to know what the board consists of, its dimensional configurations, hole sizes and locations, and any finishing process called for. Even for a one-of-a-kind prototype PC board (the

type we are going to make), this is the minimum information required.

A Typical Drawing

The PC fabrication drawing is made to the same scale as the positive, or artwork, with dimensions and notes added (Figure 8–10). Note the following points with regard to this drawing:

☐ *Material.* Here you identify the material the board is made of, its thickness, copper-clad characteristics, and color. The drawing note states that the board is to be made of laminated glass epoxy, 0.062″ thick, have a 1-oz copper clad on one side, and be green in color.

☐ *Dimensions.* The board's shape and size are referred to with drawing dimensions. All dimensions have a tolerance, and for the typical PC board it is ±0.010″.

☐ *Hole Size and Location.* Hole size and location information is usually presented in a hole data chart included in the drawing. A letter symbol identifies each hole called out, hole size and tolerance are given, and the quantity of each hole size is stated.

☐ *Finish.* In many cases, especially with prototype PC boards, no finish is required. However,

FIGURE 8–9
Mechanical drawing

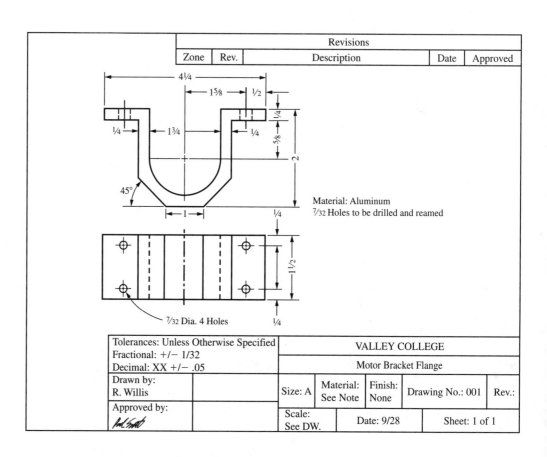

FIGURE 8–10
PC board fabrication
drawing

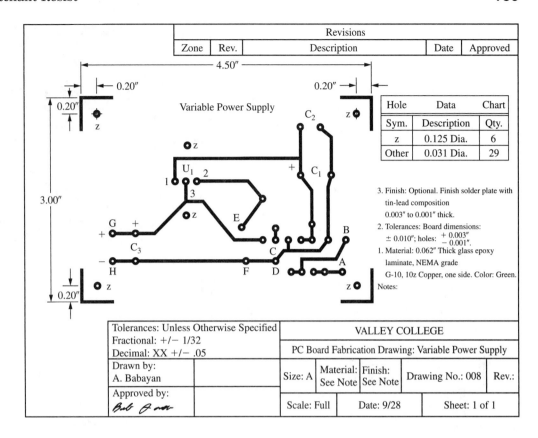

Revisions				
Zone	Rev.	Description	Date	Approved

Variable Power Supply

Hole	Data	Chart
Sym.	Description	Qty.
z	0.125 Dia.	6
Other	0.031 Dia.	29

4.50″   0.20″   0.20″   0.20″   3.00″   0.20″

C_2 C_1 U_1 1 2 3 E G C_3 H F D C B A

3. Finish: Optional. Finish solder plate with
   tin-lead composition
   0.003″ to 0.001″ thick.
2. Tolerances: Board dimensions:
   $\pm$ 0.010″; holes: $^{+\,0.003″}_{-\,0.001″}$.
1. Material: 0.062″ Thick glass epoxy
   laminate, NEMA grade
   G-10, 10z Copper, one side. Color: Green.
   Notes:

Tolerances: Unless Otherwise Specified Fractional: +/− 1/32 Decimal: XX +/− .05		VALLEY COLLEGE		
		PC Board Fabrication Drawing: Variable Power Supply		
Drawn by: A. Babayan		Size: A	Material: See Note	Finish: See Note	Drawing No.: 008	Rev.:
Approved by:		Scale: Full	Date: 9/28	Sheet: 1 of 1

almost all production boards have at least a solder coat applied to the traces and pads. (Solder
coating is discussed in more detail in the section
on PC board manufacturing in industry.) A typical solder coat call-out is shown on the drawing
(as an option). It states that the solder coat must
be a mixture of 40% lead and 60% tin and must
be 0.003″ to 0.001″ thick.

There you have it. The PC board fabrication
drawing isn't complicated; it merely tells graphically, and with a few words, what the finished board
is to look like.

Something to Think About

*Although creating a photographic negative
does not require that you be a photographer,
knowing something about the photo developing
process is always a plus. Why not get together
with a photographer, amateur or professional,
and have him or her show you around the
darkroom? Doing so should throw light on the
overall negative-making process.*

8.2   THE PC BOARD ETCHANT RESIST

Acid resist is material that will resist the copper
etching caused by acids such as ferric chloride and
ammonium persulfate. Basically, there are two
types, photoresists and the rest. Here we look at
two photoresist methods: wet and dry; and then at
a direct iron-on method that eliminates the need for
a negative and all the chemicals we have discussed
so far.

Creating the Photoresist Pattern

Creating a photoresist pattern involves cleaning the
PC board, sensitizing it, exposing it to ultraviolet
light, developing the pattern, and rinsing and drying the board. Before going through the process, let's
examine the equipment, supplies, and working environment required.

Photoresist Materials

If the equipment and supplies (Figures 8–11 and
8–12, respectively) seem familiar, it is because creating the photoresist, like producing a negative, is a
photographic process. Keep this in mind as we discuss the materials.

Equipment	Quantity
☐ Ultraviolet light source 15-W, 16″ GE black light (F15T8-BLB) and fixture (or sun)	1
☐ Printing frame 8″ × 10″ with glass and cardboard backing	1
☐ Heat gun (hair dryer) 1000- to 1200-W handheld type	1
☐ Plastic funnel Intake diameter, 6″–8″; spout diameter, ½″	
☐ Metal tray 8″ × 10″ × 2″	1

Supplies	Quantity
☐ Photoresist: choose one of these	
1. Kodak KPR (preferred choice), Cat. 1892074	1 qt

2. Oakak Corp., photoresist
     liquid spray pump, Cat.
     ER-71
	1–4 oz jar

☐ Developer—must be matched to photoresist. Buy from same company from which you purchase photoresist. Recommended: Kodak photoresist developer, negative type, Cat. 1763572	1 gal
☐ Paintbrushes, 1″ wide	6
☐ Steel wool, #00 or #0000	1 pkg

FIGURE 8–11
Equipment and supplies for creating a photoresist pattern

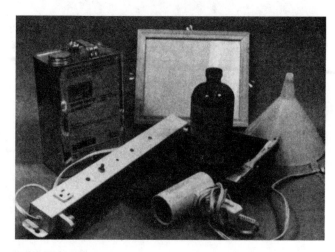

FIGURE 8–12
Photoresist materials

Equipment Needed

☐ *Ultraviolet Light Source.* The best source of **ultraviolet light** is the sun. But if you want to work entirely indoors, you will need a GE **black light** rich in ultraviolet energy. A 15-W, 16″ bulb in a fluorescent fixture of the appropriate size is ideal. Be prepared to pay $15 to $20 for the bulb alone.

☐ *Printing Frame.* You can get by with the same frame used in producing the negative. If you wish to keep the equipment in both areas separate—a good idea—purchase (or make) a second frame of the same size.

☐ *Heat Gun or Hair Dryer.* Used to dry the sensitized PC board, a 1000- to 1200-W handheld model with warm and hot adjustments is a good choice.

☐ *Funnel.* Purchase a funnel similar to the one used in processing the negative.

☐ *Tray.* Be sure the tray is metal or glass, not plastic. Photoresist developer poured into a plastic tray will eat through it.

Supplies Needed

☐ *Photoresist.* **Photoresist** is available in two forms: liquid and air pump. The Kodak KPR liquid is strongly recommended, as it consistently produces the best results. Regardless of which method you use, be sure to purchase only **negative-acting resist,** not the positive-acting type.

☐ *Developer.* The **developer** must be matched closely with the photoresist material. The two materials are usually purchased together. Again, be sure to get only negative-acting developer.

☐ *Paintbrushes.* You will need a half-dozen 1″-wide paintbrushes for cleaning the PC board of steel wool particles and applying the liquid photoresist. Don't buy expensive ones; you will be going through them quickly.

☐ *Steel Wool.* Purchase a package of #00 or #0000 non-oil-based steel wool pads for cleaning the PC board. Alternatively, you may want to use 3M Scotch-Brite 96 pads.

☐ *Printed Circuit Board Material.* Of course, you are going to need printed circuit boards. You can buy the copper-clad board at most electronics stores, or you can scrounge around and get surplus pieces free. Try to pick up single-sided board (of almost any thickness), as that's the only kind you will need; however, you may use double-sided blank board (which, interestingly, is more readily available). It just means that you will use additional acid to etch away all the copper on the nontrace side.

Working Environment

☐ The work area must be dark, although pitch-black is not required. The area should have a source of cool running water, preferably with a sink. Also, it must be well ventilated. Vapors from the developer solution are hazardous when inhaled, making good air circulation crucial.

Photoresist Process

Figure 8–13 shows a flowchart of the six steps required to produce a photoresist pattern and get the PC board ready for etching. Follow each step carefully. Remember, the two chemicals you will be using, the photoresist and developer, can be injurious to your health if handled improperly. Read and fol-

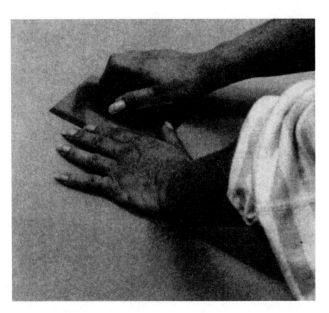

FIGURE 8–14
Cleaning the PC board

low all safety instructions on the jar or can before using these chemicals.

Step 1

Clean the Blank PC Board. This first step is probably the most important. Give it your full effort and attention. Select a copper-clad board large enough to allow at least a 1″ border around the negative. Then, with a #0000 steel wool pad, or a 3M Scotch-Brite 96 pad, rub the PC board vigorously in one direction (Figure 8–14). The idea is to shine the board, cleaning it of all dirt and oily contaminants. Once it is clean, brush off any excess steel wool particles with a paint brush reserved for that purpose only. Do not touch the copper surface after it has been cleaned. Your fingers have body oil, which will stain the copper.

Step 2

Apply the Photoresist. Although liquid photoresist can be applied by spraying, dipping, speed whirling, and roller coating, these methods are used primarily for mass-production in industry. Here, for prototyping, we will stick with the time-honored "paint-brush" method.

When the liquid photoresist is first applied, it really is not yet a resist, but a sensitizer. To take on the properties of a resist that will protect copper underneath, it must be dried and exposed to ultraviolet light. It is important to remember this as we discuss the process.

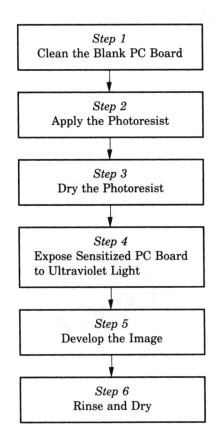

```
┌─────────────────────────────┐
│         Step 1              │
│   Clean the Blank PC Board  │
└─────────────────────────────┘
              │
              ▼
┌─────────────────────────────┐
│         Step 2              │
│    Apply the Photoresist    │
└─────────────────────────────┘
              │
              ▼
┌─────────────────────────────┐
│         Step 3              │
│     Dry the Photoresist     │
└─────────────────────────────┘
              │
              ▼
┌─────────────────────────────┐
│         Step 4              │
│ Expose Sensitized PC Board  │
│    to Ultraviolet Light     │
└─────────────────────────────┘
              │
              ▼
┌─────────────────────────────┐
│         Step 5              │
│     Develop the Image       │
└─────────────────────────────┘
              │
              ▼
┌─────────────────────────────┐
│         Step 6              │
│       Rinse and Dry         │
└─────────────────────────────┘
```

FIGURE 8–13
A half-dozen steps for producing a photoresist pattern

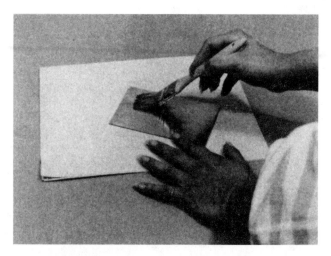

FIGURE 8–15
Applying the liquid photoresist

In a darkened room or under red safelight conditions, paint the PC board with photoresist while it lies flat on a table (Figure 8–15). Wiggle the brush a bit as you apply the resist in one direction. Only two or three passes are needed to work the photoresist in. Don't apply too much liquid. One dip of the 1″ brush should cover a 3″ × 3″ board.

Step 3

Dry the Photoresist. You can leave the board in a dark cabinet overnight to let it dry. Better yet, dry it with a heat gun or handheld hair dryer. (Make sure the light from the gun's heating coil does not shine directly on the sensitized PC board. Such light could affect the production process.) Set the dryer to warm, not hot, and, holding it 8″ to 10″ from the board, move it in a circular pattern (Figure 8–16). In

3 to 4 minutes the board should be dry. Test by touching the very corner with your finger. If it is still tacky, it needs more drying time.

At this point your board is sensitive to ultraviolet light. Keep it away from sunlight and even fluorescent bulbs.

Step 4

Expose the Sensitized PC Board to Ultraviolet Light. If the board was now placed under an ultraviolet light, the entire sensitized surface would turn to a resist, and no copper could be etched away. Obviously, this is not what you want to do. By positioning the negative over the board to act as a *mask,* before exposing it to ultraviolet light, you will produce a resist only in the pattern of the PC traces and pads. To do so, place the negative on the inside of the glass of the printing frame so that it is right-reading from the outside. Put the sensitized PC board against the negative with the sensitized surface facing toward the glass. Now, seal the unit.

If you are using sunlight, expose the board for 6 to 8 minutes. If you have an ultraviolet black light, set the printing frame so that it is directly facing the light, approximately 1 ft away. Be sure the surface is exposed to ultraviolet light and that the lamp holder does not cast shadows (Figure 8–17). Expose for a minimum of 10 minutes (it can be as long as 20 minutes). Avoid looking directly at the ultraviolet source; ultraviolet light can damage your eyes.

When the exposure is complete, remove the PC board from the frame. The sensitized areas exposed to ultraviolet light are now hard and have formed

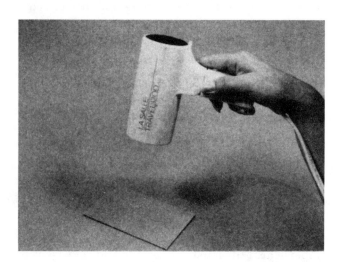

FIGURE 8–16
Drying the photoresist

FIGURE 8–17
Exposing the sensitized PC board

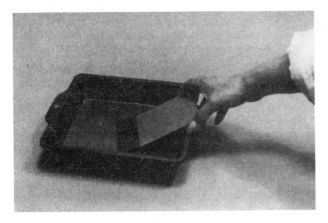

FIGURE 8–18
Developing the exposed PC board

into a resist. Do not, however, touch the board surface with your hands. The resist coating can still be easily removed with only slight rubbing.

Step 5

Develop the Image (Pattern). The purpose of the developing step is to remove the unwanted remaining sensitized material (where there was no ultraviolet exposure) and bring out the latent image of the traces and pads. While still under safelight conditions, place the board in a metal or glass tray filled to a depth of $\frac{1}{2}''$ with developer. Cover the tray to avoid breathing harmful fumes, and rock the tray gently for about $2\frac{1}{2}$ minutes. Then, remove the board and let it drain (Figure 8–18). You may at this time return to normal lighting conditions. An image of the traces and pads should now be standing out in clear relief from the surrounding copper.

Step 6

Rinse and Dry. Under cool, slow-running water, rinse the board for 2 to 3 minutes. Do not let the water hit the board directly, since it could smear the resist pattern. Break up the stream of water with your fingers.

Next, examine the board thoroughly under a bright lamp. The trace patterns should be very clear and distinct. If the pattern is broken up, fuzzy, or washed away in many spots, you will have to begin anew. Start with the very first step, cleaning the board with steel wool.

Dry the board by shaking off any excess water. Do not dry the PC board with a paper towel; the resist pattern is still soft enough to be smudged.

If all is OK, you are ready to begin etching; but if you are having difficulty, take a look at the troubleshooting guidelines in Figure 8–19. They should help you overcome some of the more common prob-

Etching the PC Board

Problem	Probable cause
The nontrace areas will not etch away	☐ The etchant is exhausted
	☐ A positive photoresist was used by mistake
	☐ A positive developer was used by mistake
The pattern is attacked by the etchant	☐ Exposure time was inadequate
	☐ Sensitizer was applied too thinly

Creating the Photoresist

Problem	Probable cause
Pattern appears on PC board but is washed away by developer	☐ Board was not well cleaned with steel wool
	☐ Sensitizer was applied too thinly
	☐ Sensitizer was not completely dry
No image appears when board is developed	☐ Exposure time was insufficient
	☐ The developer is exhausted and needs to be replaced
The pattern blisters and flakes away	☐ Sensitizer was applied too thickly
	☐ Board was not well cleaned with steel wool
	☐ Board was not completely dried

FIGURE 8–19
Photoresist and etching guidelines

lems. Remember, this stage, like the one that produces the negative, requires a little practice, a little patience, and usually more than one run-through to get it just right.

Dry-Film Photoresist

There is a way to avoid the purchase, application, and disposal of liquid photoresist—use a dry-film photoresist. The latter consists of a thin (0.5 to 2.5 mils) sheet of light-sensitive photopolymer, which is applied to a copper-clad board at the factory. A very thin (1.0 mil thick) sheet of plastic covers the photoresist to protect it from smudges and scratches. The completed board comes packaged in an opaque black plastic envelope to ensure a light-tight seal. You can purchase board sizes from a few square inches to a square foot or more.

To use, simply remove the board from the opaque envelope in subdued light. Next, peel off the protective plastic. You can now proceed as in steps 4–6 outlined in the previous section. Dry-film photoresist is simple and easy to use—but it is also expensive. Its purchase is perhaps best left for the fabrication of double-sided printed circuit boards.

Press-n-Peel PC Board Transfer Film

Photoresists aren't the only types of resist—far from it. Laser and photocopier toners will work as resist materials, too. With that in mind, Techniks, Inc., of Ringoes, New Jersey, came up with an interesting product called *Press-n-Peel PC Board Transfer Film,* with which the PC artwork is ironed onto the blank PC board; no negative or chemicals are involved. Let's take a look at what Press-n-Peel is, what you'll need to do to effect the transfer, and tips on how to make it work—every time.

Press-n-Peel: Back to the Ironing Board

Press-n-Peel is a printed circuit board artwork transfer method whereby a PC board artwork image, generated by you and printed with a laser printer, or artwork from a published source, is pressed onto a blank, copper-clad PC board. To accomplish this feat, you must first transfer the PC board artwork to a *transfer film,* a blue-colored sheet with emulsion applied to one side. Then, placing the film onto the black PC board, image-side down, you apply heat with an ordinary household iron to press the artwork image onto the board.

The method works by transferring ink-cartridge toner from a laser printer or photocopier onto the emulsion sheet. Because the image transferred is reversed when it is applied to the PC board, to get a right-side pattern, you must be sure that the artwork is itself reversed. If you are printing from a laser printer, your PC board artwork-generation software may be able to do the reversal for you. If not, you can perform the reversal as you would with an image from, say, a magazine.

The best way to accomplish either of the preceding reversals is simply to transfer your original PC board artwork image to an *overhead transparency.* In other words, you make a photocopy of your PC board artwork image onto a transparency. Then, to reverse the image for transfer onto the

blue-sheet transfer film, you simply turn the photocopied image over when placing it in the copy machine.

Here is what you will need to accomplish your task:

- [ ] An image of your PC board artwork transferred onto an overhead transparency
- [ ] Blue-sheet transfer film from Techniks, Inc.
- [ ] Blank PC board
- [ ] Household iron
- [ ] Access to a photocopier

Making the Transfer

Here are the basic steps for placing a PC board artwork image, to act as an acid resist, directly onto a blank PC board:

1. Obtain a good, black-on-white PC board artwork image, either from a laser printer or from a published source, such as a magazine or a book.
2. Using a photocopier, place an overhead transparency in the paper tray.
3. Placing the artwork image in the copy window, make a photocopy of the image onto the overhead transparency.
4. Place a sheet of the Press-n-Peel transfer film into the photocopier's paper tray, such that the artwork image will be transferred to its dull (emulsion) side.
5. Place the overhead artwork image, reverse-reading, in the copy window and copy the image to the transfer film.

You should now have a black (toner) image of your PC board artwork that you cannot read, on the blue emulsion sheet. Check to see that every pad, every trace, and all nomenclature are intact. Proceed as follows:

1. Clean your blank PC board with #0000 steel wool. Brush off all leftover particles with a paintbrush. Be sure to scrape any burrs that appear on the edge of the board that may have resulted from the cutting/shearing process.
2. Cut your sheet of Press-n-Peel, leaving a $\frac{1}{4}''$ border around the circuit image.
3. Place the Press-n-Peel image facedown onto the clean copper board.
4. Plug your iron in and set the temperature to between 275° and 325°F. Iron settings generally are between "acrylic" and "polyester."
5. Place a sheet of paper over the Press-n-Peel image.
6. Press the iron over the sheet of paper, and slowly move the iron back and forth for approximately 2 minutes.

7. Quench the board/film combination under cold running water. Slowly peel off the film.

8. With the transfer complete, your PC board is now ready for etching.

Full Court Press

At least that's the theory. To make this process work, every time, you will need to practice and follow a few simple hints.

Remember this! Your purpose is not just to transfer artwork pads, traces, and nomenclature to the PC board; it is to transfer the image in its entirety. Although 98 or 99% may be more than good enough for an A on your next test, it is not good enough here. No matter what you do, you will transfer some part of your artwork image, guaranteed; but your goal is to achieve a 100% transfer: every trace, every pad, every number and name. If not, you will have to touch things up, adding resist here and there. You don't want to do that if at all possible.

To get a great transfer, you must press the image onto the PC board at every point. Slowly, but with a fair amount of pressure, move your iron about, over the entire image, pressing down firmly as you go. Wiggle the iron a bit, "working in" the toner image. You will have to experiment, no doubt about it. Getting it completely right the first time is most unlikely; but with practice, adjustment of the iron heat setting, and attention to detail, it will work. And, think of it—no chemicals to deal with. What a break!

8.3 PC BOARD FABRICATION

In this section we look at the etching and drilling of the PC board, repair of PC boards, PC board manufacturing in industry, and the PC board fabrication of the Sample and Exercise Projects.

Etching and Drilling the PC Board

Let's see how to etch and drill a printed circuit board, and how to dispose of harmful chemicals.

Etching Materials

Figure 8–20 is a list of equipment and supplies necessary to etch and drill the printed circuit board. Figure 8–21 shows a picture of these materials. Only a few points about each need be made.

Equipment Needed

☐ *Etching Container.* The acid used for etching, ferric chloride, does its job best when heated from 100° to 120°F. A good container, ideal for this purpose, is a commercially available Crock-Pot™. A 1-gal size with warm and hot temperature

Equipment	Quantity
Etching container	1
Crock-Pot, 1 gal, or glass or plastic tray at least 2" deep	
Plastic tongs	1
Avoid metal type	

Supplies	Quantity
Etchant (acid)	1 qt or
Ferric chloride (liquid or dry concentrate)	1 gal
Safety supplies	
☐　　Eye goggles	1
☐　　Rubber gloves	1
☐　　Apron	1

FIGURE 8–20
List of etching materials

FIGURE 8–21
Etching materials

adjustments is what you want. Be sure it is made of glass; the acid will eat through metal. You can get by with an ordinary glass or plastic tray, but the Crock-Pot is the preferred choice.

☐ *Plastic Tongs.* Because acid reacts with metal, plastic tongs are the only type to use; however, the tongs will not remain in the acid very long, so in a pinch you can use a pair of ordinary slip-joint pliers. Just be sure to wash them off in water from time to time.

Supplies Needed

☐ *Etchant (Acid).* Two popular copper etchants are on the market: ammonium persulfate and ferric chloride. Ferric chloride is relatively safe (compared with ammonium persulfate), readily available at moderate costs, and produces excellent results and thus is the better choice. Ferric chloride

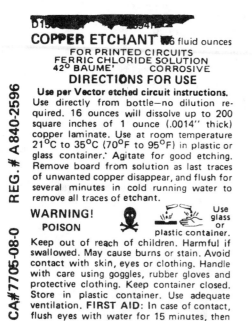

FIGURE 8–22
Etchant warning label

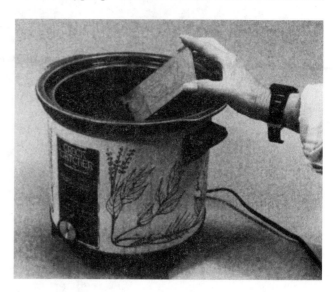

FIGURE 8–23
Placing the board in the acid

comes in liquid or dry concentrate. Either will do the job.

☐ *Safety Supplies.* Because working with ferric chloride may at the least stain your skin and clothes and at the worst cause serious injury, you must protect yourself. Wear rubber gloves, an apron, and eye goggles when working with the acid. And follow all safety precautions printed on the etchant container. Figure 8–22 shows such a warning label.

Etching Process

Of all the processes involved in making a PC board, the etching procedure is probably the easiest. Fill the etching container with acid to a depth that will cover the submerged PC board. Ideally, the board should be standing on edge, but if it must lie horizontally, be sure the copper side is facing up. Don't plunge the board into the acid; place it in the container (Figure 8–23).

Three factors determine how long it takes a PC board to etch: (1) the *strength of the acid;* (2) whether the acid is *agitated;* and (3) how *warm* the acid is. Let's take a moment to examine each factor.

Ferric chloride can, and should, be used over and over again, but each time a board is etched and copper is removed, the acid is weakened, or diluted,

a little. Obviously, the fresher the ferric chloride, the better job of etching it will do. How do you know when the acid has "had it"? If it takes more than 1 hour to etch a small board (20 to 30 minutes is a good average etching time), you probably need a new batch of ferric chloride.

If the acid is agitated, that is, made to swirl around the submerged PC board, the etching action will be speeded up considerably. There are many ways to move either the acid past the PC board or the PC board through the acid. Figure 8–24 shows a

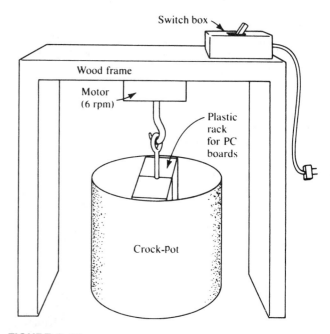

FIGURE 8–24
Agitating the acid

setup that rotates a submerged plastic rack used to carry a half-dozen or so PC boards. Rotational speed is set at 6 rpm. You don't, however, need to get this fancy. Just stirring the acid bath from time to time is enough.

As mentioned, warming the ferric chloride from 100° to 120°F will also speed up the etching process. If the etching container has a built-in temperature control, set it to "warm." Another way to heat the acid is simply to suspend a 250-W heat lamp a few inches above the etching container. Raise or lower the bulb to adjust the heat level.

Now, back to the etching. Check from time to time to see how the board is doing. When the last bit of unwanted copper is removed, withdraw the PC board and thoroughly rinse it under cold running water to stop the chemical action of the etchant. You can now handle the board with your bare hands.

Next, dry the PC board thoroughly with a paper towel. At this point, remove the photoresist layer by lightly rubbing the board with steel wool. The copper traces should be sharp (well formed) and shiny.

If you are having trouble with the etching process, check the troubleshooting guidelines in Figure 8–19.

Drilling the PC Board

If the board looks good—all traces are complete and there is no excess copper—the holes are ready to be drilled.

A number of drills on the market will help you do a professional job of drilling holes in the PC board. Probably the best, least expensive, and most readily available is the Dremel drill (described in Chapter 1). Choose one with a variable speed, and purchase a drill press stand to go with it. Add a dozen or so #60 drill bits, and you are ready to drill.

Place the PC board on a flat surface (preferably the stand of a drill press) with the copper trace side up. Using the appropriate bit size, drill the holes with a slow drill feed and a fast drill speed (Figure 8–25). Let the bit "find" the tiny pilot hole created in the etching process (Figure 8–26). Be prepared to go through a few drill bits. The glass epoxy base material of the PC board is particularly hard on these tiny bits. Larger mounting holes can be drilled with a $\frac{1}{4}''$ or $\frac{3}{8}''$ electric hand drill. When doing so, be sure to secure the PC board with a clamp or vise so that it doesn't break loose and spin out of control.

Disposing of Hazardous Chemicals

Disposing of toxic and environmentally harmful chemicals is a big issue today. Eventually your

FIGURE 8–25
PC board being drilled

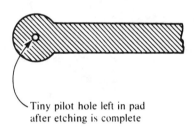

Tiny pilot hole left in pad
after etching is complete

11FIGURE 8–26
Pilot hole

etchant and developer will be weakened to the point where they won't do their job. You'll then want to dispose of them safely and effectively.

Your best bet is to take such chemicals to an authorized chemical disposal center, often listed in the phone book. Another solution is to "hand them off" to your school's chemistry department. They know exactly what to do. Finally, you can contact a local printed circuit board manufacturer and ask if it will dispose of your chemicals along with its own. I've found such firms most accommodating, and, incidentally, an excellent source of scrap printed circuit board.

Repairing the Printed Circuit Board

Problems with an etched printed circuit board tend to be of two types: the finished board either has copper where there shouldn't be any, or there is no copper where there should be some. Both problems result from a bad etch. Too much copper is due to underetching; too little, because of overetching. If the situation isn't too bad—copper left only here and there, a few traces broken up—the board can easily be repaired. Let's see what it takes to remove excess copper or add copper where it is needed.

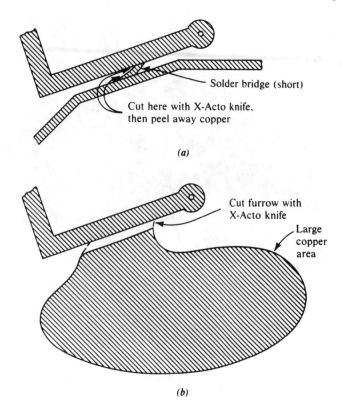

(a)

Solder bridge (short)

Cut here with X-Acto knife, then peel away copper

Cut furrow with X-Acto knife

Large copper area

(b)

FIGURE 8–27
Removing excess copper

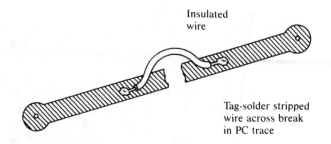

Insulated wire

Tag-solder stripped wire across break in PC trace

FIGURE 8–28
Replacing copper

Removing Excess Copper

The problem is to remove excess copper that has been left (bridged) between traces. If it is not cut out, an electrical short will result. The simplest way to proceed is to use an X-Acto knife to cut and peel the unwanted copper away. The copper clad is thin, only 0.0014″ thick, and can be peeled off once a slit is made around its perimeter (Figure 8–27a). Be careful not to let the knife slip and cut across nearby traces.

If large areas of copper remain on the board, resist the temptation to grind or gouge them off. All you need to do is to isolate such areas from any conductive traces or pads. Use the X-Acto knife to cut a reasonably wide furrow around the copper you wish to segregate (Figure 8–27b).

Replacing Copper

In this case we are referring to a copper trace that has etched away at some point along its length. You need to bridge the gap with a jumper wire made of copper. To do so, use a piece of wire no smaller than 24 gauge. If the wire is too thin, not enough current will flow, resulting in all kinds of problems for the circuit, such as false triggering, audio clipping, or spurious oscillations.

Use only insulated wire. Uninsulated wire could "wander" over to adjacent traces and short them out. Strip the two ends of the wire and tag-solder them

across the trace gap (Figure 8–28). Tag-soldering is best done by first melting a dab of solder on both sides of the trace, then placing the wire over the solder and reheating the connection. The wire should now be soldered in place.

It is also a good idea, if the jumper wire is quite long, to glue it down on the PC board at staggered spots. Clear nail polish works great.

Printed Circuit Board Manufacturing in Industry

If you were to compare your finished printed circuit board with one that has been mass-produced industrially, you would see great similarities—and most likely, a number of clearly observable differences. This is because the industrial boards usually go through three additional steps we have not yet talked about. These boards often have a solder coat, a solder mask, and a marking mask. It is important to become familiar with these added production processes so that you can feel just as much at home with the industrial PC board as you do with your own.

The Solder Coat

Have you ever noticed that many production PC boards have no copper visible on the surface? The copper is certainly there; it's just covered with a thin (0.00002″ to 0.0005″) tin–lead alloy (solder) layer. This **solder coat** provides a protective and highly conductive layer on the trace side of the PC board. The coat protects the etched surfaces from oxidizing and allows for easier soldering by providing a "pretinned" surface that gets solder flowing as soon as the hot soldering iron contacts the board. The result is a board that will better withstand adverse environmental conditions and be easier to solder.

In industry, the solder coat is applied by the electroplating method. An etched PC board is placed in a tank of 60% tin and 40% lead plating solution. Electrodes are submerged in the tank, and a high electric current is passed between them. The result

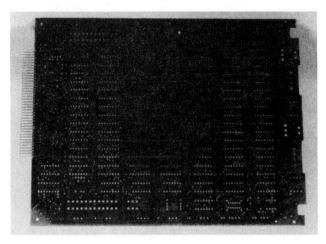

FIGURE 8–29
Solder masking of PC board

FIGURE 8–30
PC board with marking mask

is a solder-coated PC board with a smooth fine-grained deposit, dull gray in color.

Electroplating is not the only way to solder-coat a PC board. A reasonably good job can be done right at home or in the school lab with a tin plating solution available from GC Electronics Company. A 6-oz jar sells for just $5 and it is quick and easy to use. Mix the contents of a small vial with 6 oz of solution and pour the result into a glass tray. Place the etched PC board in it, agitate for 5 minutes, remove, rinse—and you are done. At the very least, your coated PC board will provide good oxidation protection.

The Solder Mask

Solder masking is a process in which a polymer coating (usually epoxy resin) is screened onto the entire surface of the PC board—*except* pads that require soldering (Figure 8–29). Thus, only the pads (and, of course, any component-mounting configurations) have their solder-coated surfaces exposed. Why is this done? Again, protection is the main purpose—in this case, to prevent solder bridging of closely spaced traces during the automatic wave-soldering process. The solder mask also provides for good surface protection and insulation, in addition to causing less solder to be applied while the board moves over the wave solder bath.

The Marking Mask

The **marking mask** is used to aid in component assembly. Silk-screened onto the component side of the PC board, it provides such information as component outlines, positioning, and sequential reference designations (Figure 8–30).

It is important to remember that the information provided by the marking mask is different from the type etched onto the trace side of the PC board. That information, you will recall, is used to aid the technician in troubleshooting the circuit. What is provided in this step helps the assembler do a good job.

Figure 8–31 summarizes, in graphic form, the various layers and their names. Figure 8–31a shows a cutaway for a single-sided board; Figure 8–31b, for a double-sided board.

Printed Circuit Board Fabrication of the Sample and Exercise Projects

We have, for the most part, completed the printed circuit board fabrication process for the Variable Power Supply in the previous sections, so little more need be said here. The same is true for the 3-Channel Color Organ Project.

TOP OR COMPONENT SIDE
BOARD LAYER
BOTTOM OR SOLDER SIDE

SINGLE-SIDED
(a)

TOP MARKING MASK
TOP SOLDER MASK
TOP SOLDER COAT
BOARD LAYER
BOTTOM SOLDER COAT
BOTTOM SOLDER MASK
BOTTOM MARKING MASK

DOUBLE-SIDED
(b)

FIGURE 8–31
PC board layers

PC Board Fabrication for the Variable Power Supply Project

The negative of the Variable Power Supply artwork is shown in Figure 8–1. Note that pinholes have been covered with an opaqueing material.

The PC board fabrication drawing for the Variable Power Supply is shown in Figure 8–10. Note, again, the optional note pertaining to the solder coat finishing process.

PC Board Fabrication for the 3-Channel Color Organ Project

In producing the negative, PC board fabrication drawing, and etched and drilled PC board for the 3-Channel Color Organ, proceed in the manner outlined. You may wish to choose the option of solder-coating your board, as was done with the Variable Power Supply.

If you are building one of the elective projects or a project of your own design, now is the time to make the negative, PC board fabrication drawing, and etched and drilled PC board for that project.

Something to Think About

Touring a commercial printed circuit board fabrication facility can be an eye-opener. Why not contact such a plant in your area and simply ask for a tour, explaining that you are a student in electronics. Tell the company you want to see how it is "really" done. Most manufacturers will be happy to show off and explain what they do.

SUMMARY

In this chapter we looked at how the photographic negative is produced. After assembling the necessary equipment and supplies, we exposed and processed a 1:1 negative from the positive PC board artwork.

Next, we examined the PC board fabrication drawing. We studied the four elements of every fabrication drawing: materials, board dimensions, hole size and location, and finishing requirements.

The actual PC board fabrication process began with the creation of the photoresist pattern. We reviewed the materials and process necessary to create such a pattern. We then saw how the PC board was etched (using ferric chloride) and drilled. We also examined ways to repair a printed circuit board that has been under- or overetched.

Before turning to the PC board fabrication of the Sample and Exercise Projects, we investigated PC board manufacturing in industry. We also explored such additional industrial steps as the application of a solder coat, solder mask, and marking mask.

In Chapter 9 we begin the final assembly process. We look at how to design and fabricate a cabinet, and how to install and to wire up everything. It's all coming together.

QUESTIONS

1. Producing a negative involves _____ and _____ a sheet of film.

2. Four steps are involved in film processing. The film is first _____, then it is placed in a _____ bath, from which it goes into a _____ solution, and is then finally _____ and dried.

3. A PC board _____ drawing is necessary to show how the printed circuit board is to be made.

4. On the PC board fabrication drawing, hole sizes and location are usually presented in a _____ _____ placed right on the drawing.

5. Kodak KPR is a liquid _____.

6. List the six steps in producing a photoresist pattern on a printed circuit board:

    _____,   _____,   _____,   _____,

    _____,   _____.

7. The most popular etching acid is _____ _____.

8. The etching process can be speeded up in three ways: _____, _____, _____.

9. Problems with an etched printed circuit board tend to be of two types. The finished board either has copper where there _____ be any, or there is no _____ where there should be some.

10. Industrial printed circuit boards usually have _____ coat, solder _____, and a _____ mask applied.

REVIEW EXERCISES

1. With PC artwork developed from previous chapter exercises, practice making a photographic negative.

2. Using nail polish, enamel paint, contact paper, or similar materials as a resist, apply each to a blank PC board and etch the board. Observe how these various materials work as a resist.

3. Obtain etched, but not drilled, surplus PC boards from industry. Practice drilling holes with a #60 drill.

4. With surplus PC boards, cut breaks in various traces and practice jumping the gaps with stripped pieces of insulated wire.

5. Obtain from industry, or any appropriate source, samples of single- and double-sided PC boards. Also get samples of PC boards with a solder coat, solder mask, and marking mask.

9 Prototyping—Final Assembly and Project Packaging

OBJECTIVES

In this chapter you will learn

- [ ] How to produce a printed circuit assembly drawing.
- [ ] How to assemble a printed circuit.
- [ ] The criteria for a good sheet metal cabinet design.
- [ ] How to produce a sheet metal drawing.
- [ ] How to cut, drill, and punch the sheet metal pattern.
- [ ] How to form sheet metal using a bending brake.
- [ ] How to finish the sheet metal cabinet by spray painting or applying a self-adhesive vinyl coating.
- [ ] How to label a cabinet with dry-transfer lettering.
- [ ] How to produce a wiring diagram.
- [ ] How to wire up a prototype project.
- [ ] How to produce the final packaging drawing.
- [ ] How to assemble the final prototype project.

In this chapter we put everything together to finish the prototype project construction process. Essentially, we need to do four more tasks. We must assemble the printed circuit (stuff the completed PC board with electronic components). We must design and fabricate a cabinet, the box to house the project. We must complete the final wiring; we must connect the off-board components to one another and, where appropriate, to the PC board. Finally, we must assemble all the elements—circuit board, chassis and enclosure components, hardware items, and cabinet—into a completed prototype project.

To achieve these objectives we require the help of four drawings. A **printed circuit board assembly drawing** shows what components are to go where on the PC board. A **sheet metal drawing** tells how to fabricate the cabinet. The **wiring diagram** explains how to connect all components not mounted on the PC board. A final **packaging drawing** illustrates how everything is fastened together to form the completed project, ready for presentation. These drawings, and the topics related to them, are discussed as this chapter unfolds.

9.1 PRINTED CIRCUIT ASSEMBLY

To assemble the printed circuit—to fill the PC board with electronic components—we need a drawing to tell us where things are to go and to instruct us in assembly techniques. Let's look at the drawing first and then at how components are to be attached to the PC board.

Printed Circuit Assembly Drawing

The printed circuit assembly drawing is like any other assembly drawing in that it shows how various parts fit together. Specifically, with a PC assembly drawing we want to see what components go where and how they are oriented. Let's examine what such a drawing should consist of and how it is produced.

PC Assembly Drawing Contents

All PC assembly drawings should contain at least three items: a view of the components showing their layout; notes describing various assembly processes; and a parts list. Furthermore, the choice of a drawing scale and how we represent components and their orientation, wire jumpers, and noncomponent parts have to be considered. Let's take each item in turn while referring to Figure 9–1.

☐ *Scale.* The PC assembly drawing is usually drawn to the same scale as the artwork. In the project example, that means full scale, or 1×.
☐ *View.* The assembly drawing is viewed from the component side—that is, we look down on the board with the components coming up at us. A side view of the assembled board that identifies the highest point above the PC board is also shown.
☐ *Component Identification.* Electronic components are shown in outline form and with standard reference designations, as used on the schematic drawing.
☐ *Component Orientation.* Many electronic components, such as diodes, electrolytic capacitors, transistors, ICs, and SCRs, must be oriented in the correct direction during assembly. The component orientation is indicated with + and – signs, references to component leads (such as E, B, C), and with pin numbers for integrated circuits.
☐ *Jumper Wires.* Jumper wires, when required, are shown on the drawing as straight lines and identified as J_1, J_2, and so on.

☐ *Noncomponent Parts.* Fasteners, spacers, heat sinks, and other hardware items are shown with ballooned item numbers on the drawing.
☐ *Parts List.* Often produced on a separate sheet, the parts list shows item numbers, component reference designations, parts descriptions, and quantity required. (See Figure 9–47 for the Variable Power Supply assembly drawing parts list.)
☐ *Notes.* Notes that reference the schematic drawing, clarify soldering procedures, and describe special assembly techniques are included.

Producing the Assembly Drawing

The easiest method for producing a PC assembly drawing is to draw the component outlines in their proper location within a blank board outline (Figure 9–1). Make sure your drawing contains, at a minimum:

☐ Borders and title block.
☐ A side view to show maximum height above the PC board surface. Only a dotted line is necessary to illustrate height; you need not draw in the components.
☐ All components and parts.
☐ Any notes.
☐ A parts list.

FIGURE 9–1
PC board assembly drawing: Variable Power Supply

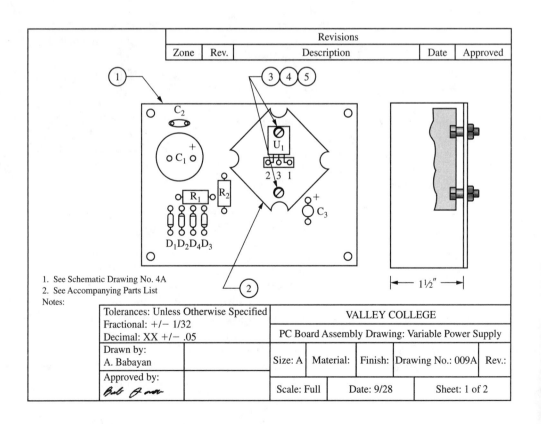

That's all there is to it. The PC assembly drawing is relatively simple; few dimensions are required. The PC assembly plan is now finished.

PC Assembly Techniques

Once a drawing is complete, the actual printed circuit assembly may begin. In examining this process, we'll look at component lead preparation, along with component insertion and soldering techniques.

Component Lead Preparation

Most components have to be prepped prior to insertion into the PC board. Axial lead components need to have their leads bent at a 90° angle with a needle nose pliers (Figure 9–2a). Do not make the bend at

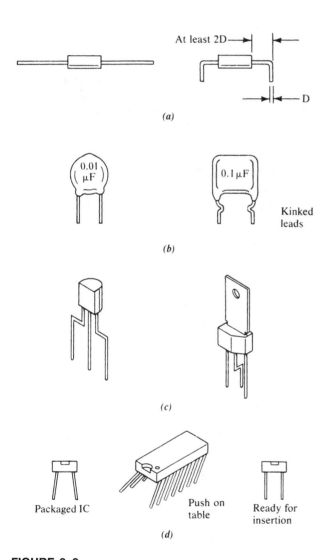

(a)

(b)

(c)

(d)

FIGURE 9–2
Component lead preparation: (a) axial lead components; (b) radial lead components; (c) solid-state discrete components; (d) prepping IC leads

a point too close to the component body. Leave at least twice the lead diameter straight before making the bend.

Radial lead components, of course, drop straight into the board. Some of these components, however, have a kinked pair of leads that allow the component body to stand slightly above the board (Figure 9–2b). Raising the component in this manner increases the cooling effect and reduces the risk of body chipping and cracking.

You may need to bend and reposition the leads of transistors, SCRs, triacs, and other solid-state discrete components (Figure 9–2c). When doing so, as with axial lead components, do not bend the lead too close to the component body, and treat each lead as though it may break at any moment. Solid-state components are delicate items, and they must be handled as such.

Finally, integrated circuits are often packaged with each row of pins spread a bit farther apart than the designed spacing (Figure 9–2d). One way to pull the pins in is to use chain-nose pliers. Another approach to bring them together is to press one row of pins in at a time. Work each row in slightly, as shown, until the correct spacing allows the chip to fit easily into the board.

As a general rule, all components should be mounted flush to the PC board (unless they have kinked leads). The exception is components that may generate excessive heat, such as many types of transistors. In that case, the component body should be raised above the board about $\frac{1}{8}''$. A good spacer to use in positioning the component is an ordinary Popsicle stick (Figure 9–3).

Component Insertion and Soldering

Before any components can be inserted into the PC board, the board must be cleaned thoroughly. Use fine, non-oil-based steel wool and a nonabrasive cleanser to scour the trace side until it is shiny. Remove particles of steel with a paint brush. Be sure oil from your hands does not contaminate the solder pads. Failure to take these simple steps will prevent good solder joints from forming, and unreliable connections will be the result.

After the board has been cleaned and dried, insert components from the noncomponent side. Be sure to watch component orientation: Those with polarity and designated lead configurations must be inserted in the correct direction. When the component lead is pressed through the hole, bend it at about a 30° angle. This configuration is known as a **service bend** because it allows the component to be readily serviced, that is, easily removed (Figure 9–4a). If a full clinch is made, with the component

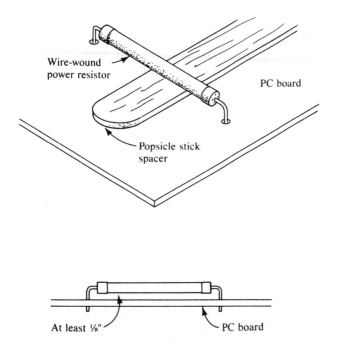

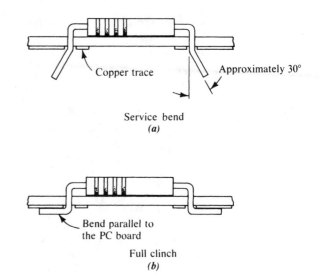

FIGURE 9–4
Component insertion

FIGURE 9–3
Components not mounted flush to the board

lead bent flat against the board, replacing the component becomes very difficult (Figure 9–4b). On the other hand, if the lead isn't bent at all, the component is likely to fall out before it can be soldered in place. To prevent ICs from doing the same thing, bend over the pins on only two opposite corners. Bending over all the pins is unnecessary and makes removing the integrated circuit a real problem.

PC board soldering techniques are discussed in Chapter 6. It might be wise to review that discussion before soldering any components in place. Remember, heat both the component lead (or wire)

and the copper pad, but don't apply too much heat: Not only might the copper pad rise up off the board, but the component also could be damaged. When soldering transistors and small diodes, it is best to heat-sink each lead. As shown in Figure 9–5, a heat sink can be improvised from a needle nose pliers with a rubber band around its handle to provide tension if the special tool designed for the purpose is not available. The metal heat sink intercepts heat from the soldering joint before it can reach the component body and cause any damage.

When soldering is complete, clip the component leads as close to the PC board as possible. Any excess lead length can easily bend over and bridge, or short, to adjacent PC traces and components.

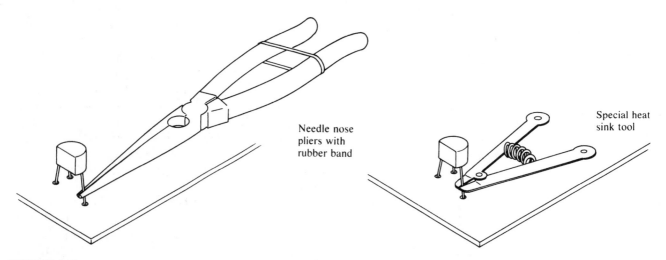

FIGURE 9–5
Heat-sinking component leads

9.2 CABINETS: THE DESIGN PHASE

We have the assembled printed circuit, we know which components are to go on the chassis and which ones will attach to the enclosure, and we have a preliminary packaging sketch developed with our early design drawings. We are now ready to design, fabricate, and finish a project cabinet.

Elements of Design

Why design and fabricate a cabinet? What about buying one? We'll examine that issue first. Then, regardless of the buy/make outcome, we will explore some of the important packaging design considerations.

The Buy/Make Decision

Numerous companies sell many types of ready-made chassis, enclosures, boxes, cabinets, and panels constructed of metal, plastic, or wood (Figure 9–6). The advantages of purchasing a completed box are obvious: A great deal of construction time is saved, and the product is usually of high quality. Yet even with a purchased cabinet, some fabrication on your part will

most likely be required. Drilling holes and punching or cutting slots are often necessary. Performing these operations "in the fold"—that is, with the box already bent and formed—is not easy. This factor needs to be considered when deciding to buy or make.

Although buying a formed cabinet for the prototype project has its advantages, if you want complete control over the design, fabrication, and finish of your project housing, you will need to make it yourself. Furthermore, the learning experience and skills acquired in producing a box from scratch will be of great value to any prospective technician. And when your future boss says, "You even built the cabinet, too," you'll know that you made the right buy/make decision.

Design Considerations

The overriding design consideration has to do with purpose. Specifically, what is the cabinet to be used for? A power supply? A color organ? A stereo amplifier? All subsequent design criteria stem from this one factor. With purpose or intent uppermost in mind, we consider such elements as materials, size and shape, ease of fabrication, safety of use, appearance, and cost. As we examine each item in turn, note how interrelated the factors are, and how planning for one often requires thinking about all the others.

Materials. Most cabinets used to house prototype projects are made of metal, either cold-rolled steel or, more commonly, aluminum. Plastic is used when electrical safety is of primary concern, and wood when appearance is of chief importance. Both plastic and wood, however, are fairly difficult to work

FIGURE 9–6
Commercial cabinets

FIGURE 9–7
Common cabinet styles

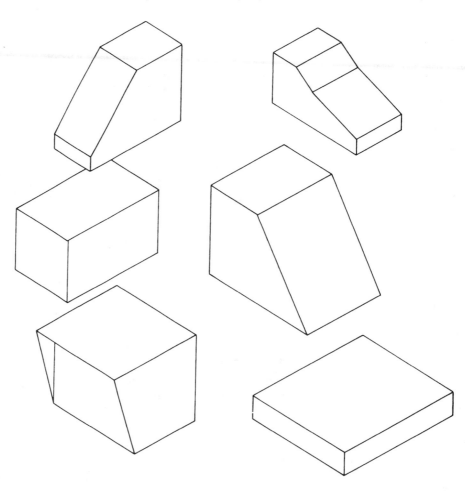

with. Unless other considerations dominate, aluminum sheet metal serves most purposes quite well.

Size and Shape. Size, of course, is determined primarily by the bulk of the electronic circuitry. The cabinet should be no larger than necessary to house and mount the electronics, yet consideration must also be given to such factors as access to component parts, heat dissipation, and project handling.

In choosing the cabinet shape, thought must be given not only to appearance but also to function and ease of repair and fabrication. Figure 9–7 shows some of the more common cabinet styles one can either buy or make. It is important not to overdesign by creating a fancy-looking box. As with most design considerations, the simpler the better is the rule to follow.

Ease of Fabrication. The "keep it simple" philosophy is especially relevant to cabinet fabrication. In reference to sheet metal, the various cuts, holes, and bends can get out of hand if you're not careful. In fact, too many bends may simply be impossible to achieve. The rule, again, is do only what is necessary to accomplish the task.

Safety of Use. Two considerations are important here. You want the cabinet to protect the project and to protect you from the project. The box should be designed and fabricated so that it can support the various circuit components and provide ventilation if necessary. In turn, if dangerous currents are present, you want the cabinet to prevent you, at least under normal circumstances, from reaching them.

Appearance. You want the cabinet to look good—be visually appealing. In addition to size and shape, the outer finish contributes most to that end. What type of paint or vinyl coating is applied will be a factor in the aesthetics of the project. Also, consideration must be given to the type of labeling (lettering, numbers, lines, etc.) to be applied. Such labeling must be durable as well as attractive.

Cost. Cost is normally a secondary consideration with regard to the prototype cabinet but of primary concern with the mass-produced product. Because the prototype is to reflect as closely as possible the final mass-produced item, cost should also be given some consideration at this point. You do not want to design and produce a cabinet that will cost a great deal to fabricate in large quantities just because it is appealing for the prototype. Remember, the prototype is a model of what, it is hoped, is to come—full-scale production.

All these design considerations must be kept in mind when creating a prototype cabinet. In the sections to follow, the cabinet is built out of aluminum sheet metal. Such a material will meet the cabinet needs for all the projects presented in this book. It is also the most widely used material for packaging electronic prototype projects.

Sheet Metal Design Layouts

In this section we look at the sheet metal drawing, bending allowances, and the sheet metal template.

The Sheet Metal Design Drawing

The design and fabrication of a sheet metal cabinet takes place in four stages. First, a design drawing is made. Next, the sheet metal pattern is cut out and holes are drilled or punched in it. Third, the cabinet is formed into shape. Finally, a surface finish, either

paint or a vinyl coating, is applied and the cabinet assembled, usually with sheet metal screws.

Figure 9–8 illustrates many of the common sheet metal patterns used to create functional and attractive cabinets for electronic prototype projects. (Pattern layouts for these cabinets are shown in more detail in Appendix B.) Before you select one for your project, it would be a good idea to redraw the pattern on a sheet of stiff construction paper. Then, cut out the pattern and assemble it into a box using transparent tape. In that way you will know exactly what you are getting, before you start slicing up a costly sheet of aluminum.

The choice of a sheet metal pattern, and naturally the resulting sheet metal cabinet, must take into account the design factors. The key element is to determine how much space will be needed for the printed circuit board, the components that will be mounted on the chassis and will not need to be exposed, and components such as lights and switches that must be seen or touched.

After you have chosen the appropriate sheet metal pattern, or designed one of your own, it is time

FIGURE 9–8
Sheet metal cabinet patterns

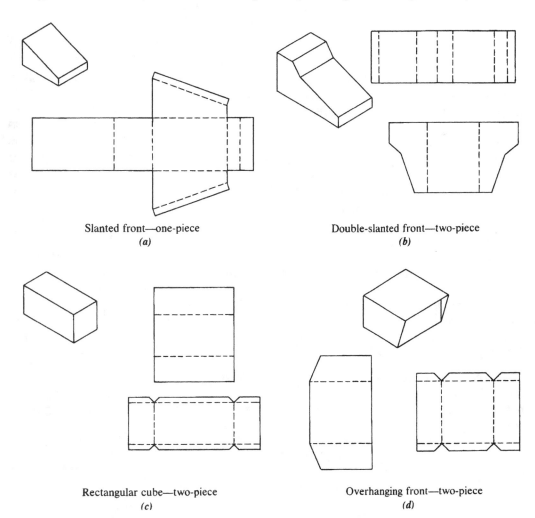

Slanted front—one-piece
(a)

Double-slanted front—two-piece
(b)

Rectangular cube—two-piece
(c)

Overhanging front—two-piece
(d)

FIGURE 9–9
Sheet metal drawing:
Variable Power Supply
(chassis)

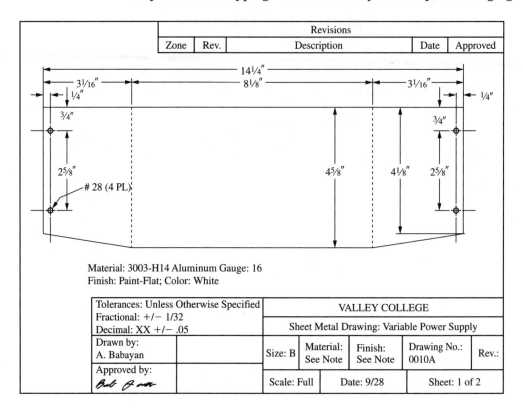

		Revisions		
Zone	Rev.	Description	Date	Approved

Material: 3003-H14 Aluminum Gauge: 16
Finish: Paint-Flat; Color: White

28 (4 PL)

Tolerances: Unless Otherwise Specified Fractional: +/− 1/32 Decimal: XX +/− .05		VALLEY COLLEGE				
		Sheet Metal Drawing: Variable Power Supply				
Drawn by: A. Babayan		Size: B	Material: See Note	Finish: See Note	Drawing No.: 0010A	Rev.:
Approved by: *Bob ___*		Scale: Full	Date: 9/28		Sheet: 1 of 2	

to begin the sheet metal drawing (Figure 9–9). Note the following with regard to this type of drawing:

1. The drawing is made full scale. Because the sheet metal template (to be discussed shortly) will be a photocopy of the sheet metal drawing, a full-scale drawing is the obvious choice.
2. The view is from the inside or bottom of the cabinet. The box will thus fold up, toward the drafter.
3. Fold lines are indicated with dotted lines.
4. Relief holes, $\frac{1}{8}''$ in diameter, are provided at fold intersections. Such holes prevent bulging or fracturing of the metal at fold points of high stress.
5. Hole sizes are given in diameter, or when rectangular, by a width × length dimension.
6. Holes are usually located on center from a sheet metal edge or fold line.
7. Material type and size are specified.
8. Cabinet finish is stated.

A completed sheet metal drawing gives enough information to fabricate and surface finish the cabinet. From this drawing a **sheet metal template** is created, to be taped to the sheet metal and used as a guide for cutting, drilling and punching, and folding the box. But before discussing the template, let's examine the "bending allowance" problem, to make sure that separately folded sheets of metal fit together correctly.

Bending Allowances

When sheet metal is bent, the thickness of the metal must be taken into account when arriving at the new overall folded dimension. As shown in Figure 9–10, the 5″ width of the bottom base becomes $5\frac{1}{8}''$

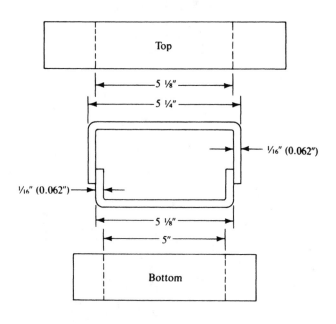

(drawing not to scale)

FIGURE 9–10
Sheet metal bend allowances

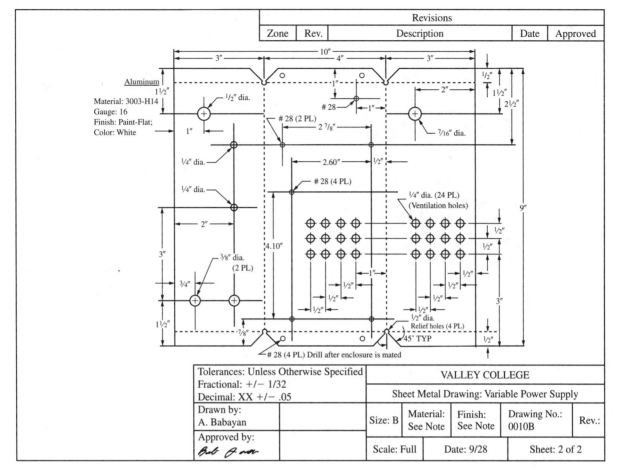

FIGURE 9–11
Sheet metal template: Variable Power Supply

when the sides are bent at right angles (the base width of 5″, plus the two metal thicknesses of $\frac{1}{16}$″). In designing the top piece to fit over the bottom, its base width must be $5\frac{1}{8}$″ before the bend takes place. The final overall width of the project will be $5\frac{1}{4}$″.

Figuring sheet metal bend allowances is not difficult; just take into account the metal thickness. Failure to do so will result in an improper fit, and you will have to fabricate at least one of the sheets again.

Sheet Metal Layout Template

The sheet metal layout template is a photocopy of the sheet metal drawing taped down onto the metal to be cut, drilled, and punched (Figure 9–11). Using a template in such a manner eliminates the need to transfer dimensions from the sheet metal drawing to the sheet metal. Not only is time saved but errors associated with redimensioning are eliminated.

After making a photocopy of the sheet metal drawing, cut it approximately $\frac{1}{4}$″ larger around all borders. Don't worry if dimensions and related information are cut away in the process. You need only the

drawing pattern itself, which is presented full scale. Next, lay down several strips of double-sided adhesive tape onto the sheet metal surface (Figure 9–12). Carefully place the template over the tape and smooth it down. Be particularly careful to position the template accurately; once it contacts the tape, it is all but impossible to reposition. Actually, it is a good idea to make two or three copies of the sheet metal drawing, just in case a second or third template will be needed.

An alternative to double-sided tape is rubber cement. Rubber cement is easy to use and to clean up.

Something to Think About

Professionally rendered sheet metal drawings used to depict the fabrication and assembly of ductwork can be quite instructive. Why not try to obtain a few? When you do, examine the bends and fittings closely. The drawings are a classic study in descriptive geometry techniques.

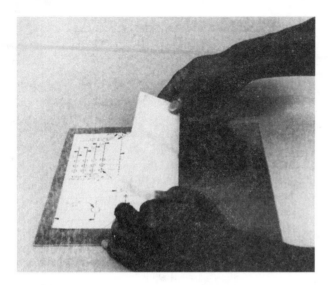

FIGURE 9–12
Laying down the sheet metal template

9.3 CABINETS: THE FABRICATION PROCESS

It is time to go into the shop and build a cabinet. First, we will see how the sheet metal is cut and how holes are drilled, punched, or notched. Then we'll find out how the sheet metal is folded into place.

Working in the Flat: Cutting the Blank and Making Holes

In this section we examine sheet metal cutting techniques, see how round holes are drilled and punched, find out how to nibble any shape hole, and learn how to smooth things out by filing.

Cutting the Blank Metal

For general cabinet work, an aluminum alloy, either 5052K32 or 3003-H14, should be used. It is readily available, and it provides enough strength as well as malleability for our purposes. A good thickness is 16 gauge (0.062″ or $\frac{1}{16}$″).

Once the metal is selected, the sheet metal layout template is taped down, as described in the previous section. The sheet metal blank should extend over the template on all sides for about an inch.

Two tools are used for cutting or shearing the sheet metal: a foot-operated squaring shear and a handheld tin snip (Figure 9–13). The shear consists of two cutting blades: a ridged bottom blade and a movable top blade. Pressure on the foot bar causes the top blade to come down, shearing the sheet metal placed between the two blades (Figure 9–14). The result is an even, clean cut of the metal.

The tin snip (Figure 9–15) can be used instead of a squaring shear, but the results are usually not as satisfactory. The main purpose of the tin snip is to cut notches and corners that would otherwise

FIGURE 9–13
Cutting, or shearing,
sheet metal

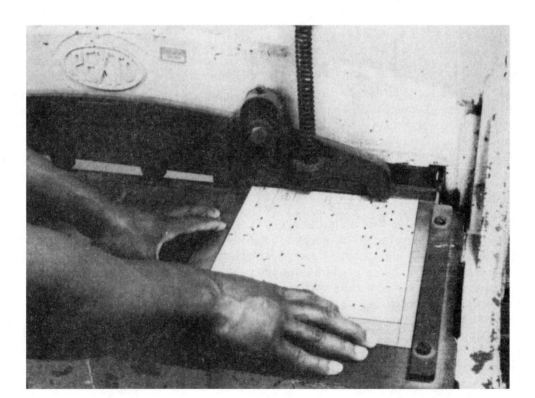

FIGURE 9–14
Shearing operation

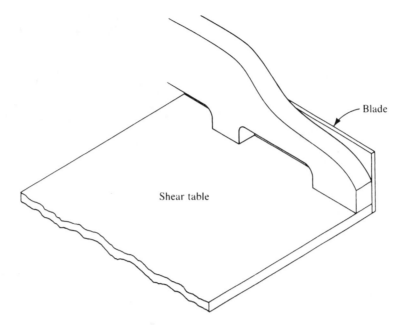

Blade

Shear table

require an expensive hand-operated notching machine. Compound shears (aircraft type) of the
straight, right curve, and left curve type are also useful for this purpose, and should be used if available.

Making Round Holes: Drilling

We can make round holes in the sheet metal by
drilling or punching them. Generally, holes up to $\frac{1}{4}''$
in diameter are drilled; anything larger is punched.

The procedure for drilling a small hole in sheet
metal is as follows:

1. Center punch all holes that require drilling
   (Figure 9–16). With the use of a center punch and
   hammer, make a small indentation at the center

of the hole to be drilled. One light tap of the hammer is all that is needed to make the mark. The
purpose of this indentation is to provide a "home"
for the drill bit—to prevent it from "walking"
(moving off center) at the initial moment of drill
and metal contact.

2. Back up the sheet metal with a piece of wood at
   least $\frac{3}{8}''$ thick (Figure 9–17). Not only does this
   provide a solid surface for the drill bit to enter after it has penetrated the sheet metal but it also
   reduces the severity of any burrs that may form.

3. Fasten the sheet metal and wood backing securely to the drill press table, preferably with a
   C-clamp.

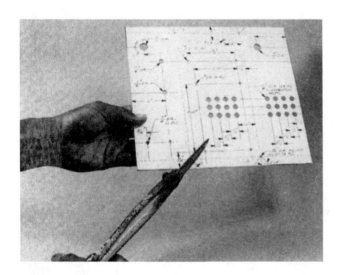

FIGURE 9–15
Tin snip used for notching

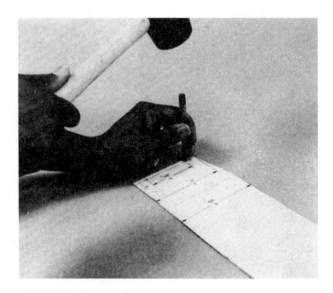

FIGURE 9–16
Center punching a hole

FIGURE 9–17
Drilling holes in sheet metal

4. Drill the hole using appropriate drill speed and feed. Generally, the larger the hole, the slower the drill speed. Drill feed should always be at a slow, steady pace and extend at least $\frac{1}{4}''$ into the wood backing. If holes need to be enlarged slightly, a tapered hand reamer can be used for that purpose (Figure 9–18).

All holes, when drilled, will form burrs on one or both sides of the sheet metal. These sharp, jagged protrusions must be removed (Figure 9–19). To remove them, use a tool designed specifically for this purpose, or a large drill bit wrapped with electrical tape to prevent cutting your hand. A Dremel Moto-Tool with a

FIGURE 9–18
Using the tapered hand reamer

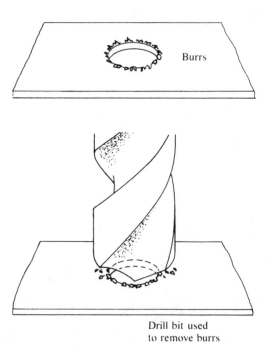

Burrs

Drill bit used
to remove burrs

FIGURE 9–19
Removing burrs

FIGURE 9–20
Using the chassis punch

small grinding wheel also does an excellent job of deburring. If burrs still remain, they may need to be cleaned up with a round-nose file.

Making Round Holes: Punching

For round holes larger than $\frac{1}{4}''$, good, clean (burr-free) holes can be cut with either a chassis punch or a multistation turret punch. Let's examine each tool.

The **chassis punch** (Figure 9–20) comes in many sizes, from $\frac{3}{8}''$ to more than $2''$ in diameter. The chassis punch consists of three parts: a punch, die, and draw screw. Explaining how the chassis punch is used will also describe how it works.

1. A hole is drilled in the sheet metal large enough to accept the draw screw.
2. The top end of the chassis punch (known simply as the punch) is secured in a vise. It has two parallel flats that make it easy to grip between the jaws of the vise.
3. The sheet metal is positioned over the punch at the point where the hole is to be formed.
4. The die is positioned over the sheet metal in line with the punch.
5. The draw screw is inserted through the die, the sheet metal, and into the punch.
6. The draw screw is tightened down with a wrench.

As the screw is turned it draws the die and punch together. When the sheet metal pops through, the screw is loosened and the die and punch are separated. The sheet metal is then lifted off, and the blank punched piece of metal is extracted from the

punch. The cutting action that results from this type of punching operation ordinarily produces a clean and burr-free hole.

The **turret punch** (Figure 9–21) contains many different-sized die and punch combinations that can easily be swung into place for instant cutting. In addition to the advantage of having so many punches available at the swing of a turret, with this machine no predrilling of a draw screw hole is required; however, all holes produced with the turret punch should still be center-punched to allow the die to align correctly.

FIGURE 9–21
Turret punch

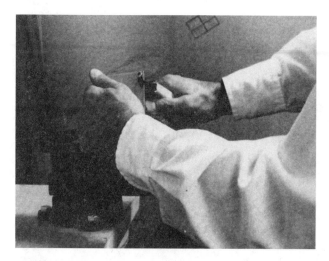

FIGURE 9–22
Nibbler

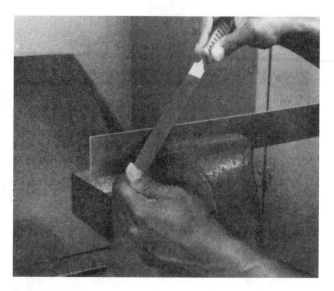

FIGURE 9–23
Filing

Note also that some large-diameter hole saws (for use in drill presses) are on the market for use on aluminum.

Making Rectangular Holes and Slots

Rectangular holes, slots, and notches are best made with a nifty handheld minishear known as a **nibbler** (Figure 9–22). With each squeeze of the handles, the tool cuts away a small piece of metal in front of it. Its operation is quite simple:

1. A hole is drilled or punched (when cutting internal shapes) large enough to accept the tool's cutting head. A $\frac{3}{8}''$ hole is usually adequate.
2. The tool is inserted up through the hole until the metal to be cut rests on a platform directly below the cutting blade.
3. The handles are squeezed as the tool is guided along, nibbling the metal to the size and shape desired.

Although the nibbler is easy to use, it is best to practice on a sheet of scrap metal before taking a bite out of the cabinet you will be fabricating.

Finishing Up

Before the sheet metal can be formed, or bent, it should be smoothed along its external and internal edges with a file. Use a single- or double-cut file, with a wooden or plastic handle, of medium or bastard grade. Figure 9–23 illustrates the correct way to file sheet metal edges. In this method, known as *draw filing,* the file is placed at an angle against the sheet metal edge. As the file is pushed and pulled

across the metal, uniform pressure is applied. To prevent metal particles from clogging the file, be sure to clean the file often with a file card.

Forming the Sheet Metal Cabinet

A special tool known as a **bending brake** is needed to form a sheet of metal. Let's see what it is and how it is used.

The Bending Brake

The bending brake (Figure 9–24) consists of a worktable with metal fingers on top. The sheet metal to be bent is placed between the fingers and worktable. A movable bending wing is rotated against a portion of the metal, causing it to bend at the desired angle.

A five-step bending sequence is shown in Figure 9–25.

1. The fingers are raised up (via the clamping handle) and the sheet metal is inserted between the fingers and the worktable.
2. With the clamping handle moved forward, the fingers are pressed firmly and evenly down across the metal, holding it in place.
3. The handle on the bending wing is slowly raised, causing the sheet metal above it to bend.
4. The bending wing is raised to the desired angle of bend.
5. The clamping handle is moved to release the fingers, freeing the sheet metal.

The bending brake contains a number of fingers of various sizes that are positioned to cover the width of the sheet metal to be bent. Choose the

FIGURE 9–24
Bending brake

FIGURE 9–25
Sheet metal bending sequence

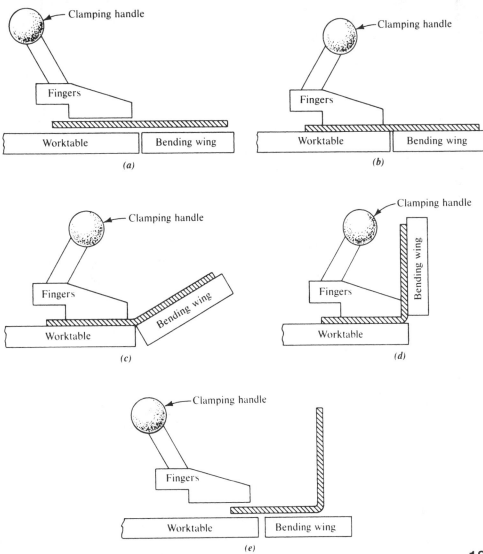

FIGURE 9–26
Bending brake fingers

fewest number of fingers that will cover the sheet metal without leaving excessively wide gaps in between. If possible, the fingers should extend beyond the sheet metal width slightly (Figure 9–26).

It is important that the correct distance be set between the finger edges and the edge of the worktable (Figure 9–27). For an optimum bending radius, this distance should be the thickness of the sheet metal to be bent. If the distance is less than the metal thickness, the metal will be gouged. If the distance is set greater than the thickness of the metal, too wide a radius will be formed.

Bending Sequence and Limitations

In forming any sheet metal cabinet, the correct bending sequence must be followed. If you don't follow a planned succession of steps, it may become impossible to complete all the necessary bends. Although the bend sequences for standard sheet metal patterns were worked out long ago (Figure 9–28), a good approach to forming any cabinet is to plan out the progression of steps in your mind ahead of time. An even better method is to cut out a copy of the sheet metal layout pattern and carefully fold it in the sequence to be followed. As you do so, note if the particular order will work—if the bending brake can accommodate each bend in succession.

All bending brakes place limits on the size of a sheet of metal that can be formed. The width, height, and depth of the folded sheet metal are limited, in turn, by the width, height, and depth of various elements that make up the machine. Figure 9–29 shows how the brake's overall width, crosshead height, and crosshead depth must be considered when planning the size and shape of the folded cabinet.

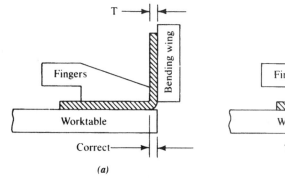
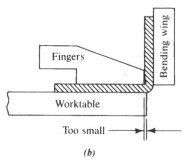
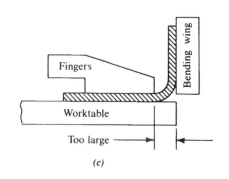

(a)   *(b)*   *(c)*

FIGURE 9–27
Finger settings

9.4 CABINETS: THE FINISHING PROCESS

With the sheet metal cut, drilled and punched, and folded into shape, we are ready to apply the proper surface finish. We need to lay on a protective and attractive coating to the aluminum sheet, and we must label the appropriate areas with numbers, letters, and graphic indicators.

Covering It Up

Let's examine how readily available and easy-to-use spray paint and self-adhesive vinyl coatings can be used to cover the prototype project cabinet.

Spray Painting the Cabinet

The key to success in any spray painting operation is the continual application of a strong dose of— patience. The willingness to apply two or three coats of primer and two or three coats of paint, and find something else to do during each drying time— that's what is meant by patience. Taking the time to do it right, rather than trying to paint the entire project in one operation, will result in a professional job, every time. Let's see how.

With the use of aerosol spray paints, available at any hardware or paint store, painting an aluminum surface is foolproof, if you obey the

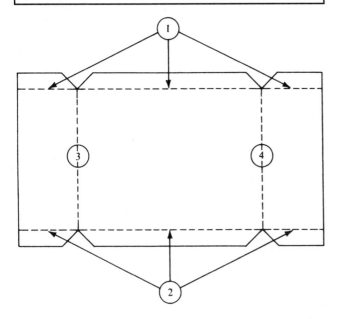

FIGURE 9–28
Bending sequence

FIGURE 9–29
Bending brake limits

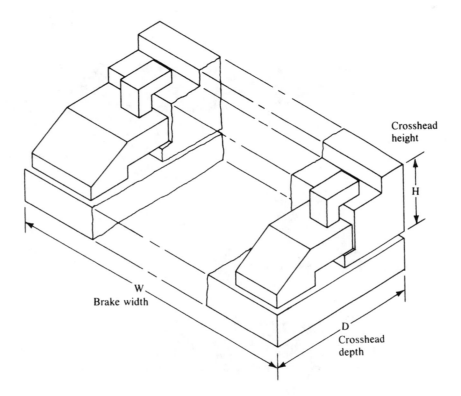

directions on the can and follow these three easy steps:

☐ *Clean the Metal Surface.* Using a sponge and ordinary kitchen cleanser, scour the cabinet thoroughly, removing all dirt, grime, and grease. These contaminants must be eliminated; there is no way paint primer will adhere to filthy surfaces. When the cabinet is completely clean, dry it with a paper towel. In some cases, use a hand-held dryer to remove all traces of moisture in hard-to-reach areas.

☐ *Apply a Primer.* The primer, available in an aerosol can, is used to seal and fill the metal prior to the application of paint. Choose a primer with a zinc chromate base. Construct a "quick-and-dirty" spray booth from a cardboard box (Figure 9–30). After shaking the can for a minute or two, hold it 10″ to 12″ from the surface to be sprayed and, in a back-and-forth motion, apply a thin, even coat, working from the top to the bottom (Figure 9–30). Let the primer thoroughly dry, and then apply two or three additional coats as necessary.

☐ *Apply the Spray Paint.* Enamel paint is also available in an aerosol can. Choose flat or semigloss. Apply it as you did the primer, in two or three thin coats. Allow complete drying between each spraying operation.

Do not be concerned if, after drying, the paint seems a bit "flat," or dull. A protective acrylic coating, to be applied after the labeling is done, will restore luster to the painted surfaces.

Self-Adhesive Vinyl

An attractive alternative to paint is self-adhesive vinyl. This contact paper comes in a variety of printed patterns, with wood grain being the most popular. When applied correctly, it provides a durable surface and professional look to any cabinet.

Self-adhesive vinyl is easy to use. After the cabinet is cleaned and dried, the vinyl sheet is cut slightly larger than the size necessary to cover the cabinet surface. A small section of protective backing is then peeled away, and the vinyl is positioned in place and pressed down (Figure 9–31). Then, slowly, as the protective sheet is pulled away, the vinyl coating is smoothed down over the cabinet surface. It is important not to remove the protective backing all at once, for once the vinyl sheet touches the metal surface, it is very difficult to reposition.

Once the vinyl is pressed in place, trim it with a safety razor, or better yet, a sharp X-Acto knife (Figure 9–32). For internal slots, carefully cut around the inner edges with the knife.

Surface Labeling

With the cabinet covered with a protective and attractive coating, all that remains is to make it literate: supply it with the letters, numbers, and symbols necessary to explain the project's function and operation. Let's examine two approaches to surface labeling: the more traditional dry-transfer lettering and the Dynamark Imaging System.

FIGURE 9–30
Spray painting

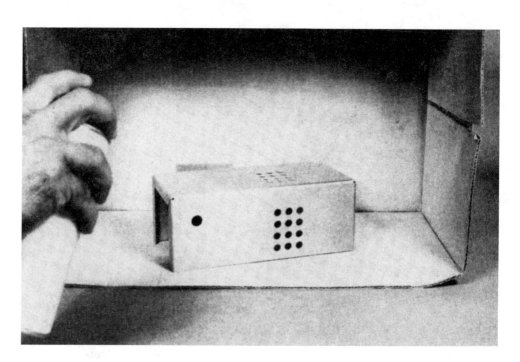

FIGURE 9–31
Self-adhesive vinyl

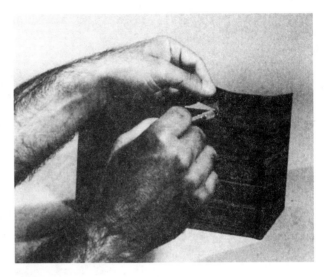

FIGURE 9–32
Trimming around the edges

The Dry-Transfer Lettering System

Dry-transfer lettering that consists not only of letters but also numbers and graphic symbols is available in a wide variety of sizes and styles at electronics, stationery, and art supply stores. Inexpensive sheets that contain letters, words, phrases, numbers, and index marks specifically designed for electronic prototype projects are easy to obtain and use (Figure 9–33).

Only a few simple items are necessary to get a good transfer of lettering to the cabinet surface. In addition to the lettering sheets, you will need a burnishing tool (a dull pencil will do), masking tape, graph paper, and a can of clear acrylic protective spray. With these materials in hand, follow this six-step process for applying dry-transfer lettering:

1. *Lay Out Design on Paper.* Using graph paper, lay out the overall design for the panel. Get everything in order on paper first. Know exactly where each symbol is to go before you start the rub-on process.
2. *Use Masking Tape as a Guide.* To line up a row of letters or numbers, place a strip of masking tape on the painted surface a fraction of an inch below where the lettering is to go. The tape will act as a guide to help keep your letters straight.
3. *Position Transfer Sheet and Rub on with Burnishing Tool.* With the symbols positioned

FIGURE 9–33
Dry-transfer lettering

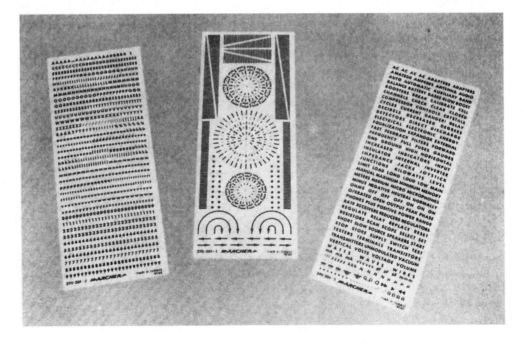

FIGURE 9–34
Applying dry-transfer
lettering

over the desired spot, rub over the surface with a burnishing tool (Figure 9–34). Start with light pressure, then gradually increase the force. As the transfer takes place, the lettering on the transfer sheet will go from black to dull gray (assuming you're using black lettering). Check to see that all lettering has been transferred by holding the sheet in place with one hand and lifting slightly with the other. If transfer is not complete, put the sheet back and simply rub a bit more.

If you haven't applied dry-transfer lettering before, it is recommended that you practice a bit first on a scrap piece of painted metal.

4. *Remove Unwanted Lettering with Plastic Tape.* If you make a mistake, simply use a piece of plastic tape to pull up the undesired symbol.
5. *Apply Index Marks Carefully.* When applying index marks for rotary switches, meters, and the like, be sure of the spacing ahead of time. Carefully plan the placement of each mark.
6. *Apply a Clear Protective Coating.* In order to protect the dry transfers from chipping and rubbing off in normal handling, apply a clear acrylic protective spray coating. Available in an aerosol can, it is sprayed on in the same manner as the primer and enamel paint. Again, it is best to apply two or three thin coats, rather than trying to do it all at once with one thick coating.

If you follow these steps, there is no reason why you shouldn't have a top-quality, professional-looking cabinet for the prototype project.

Dynamark Imaging System

The Dynamark Imaging System, a product of 3M, is designed to produce high-quality metal or plastic labels using a photosensitive process similar to that used in making photosensitive PC boards. With this system, attractive nameplates, dial faces, labels, and control panels can be made quickly and easily and at a reasonable cost (Figure 9–35).

Figure 9–36 outlines the process for making either a negative (the reverse of the original artwork) or positive image (identical with the original artwork). The original artwork can be composed of opaque copy on a transparent carrier or even a half-tone photograph. The results, as you can see in Figure 9–35, give your prototype project a truly

---

Something to Think About

Practice, practice, practice; patience, patience, patience—that's what it's all about. Using self-adhesive vinyl, why not practice by applying it to surplus cabinets or those you want to "refinish"? You might find equipment right in the lab that could use a new look.

Letraset®
Imaging Systems

Dynamark™
Self-Adhesive Sign and Label

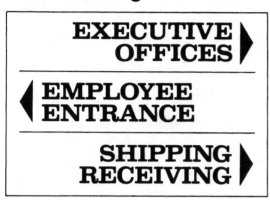

Dynamark™ Self-Adhesive Signs and Labels can be made yourself or ordered through Color Imaging Centers. Call 800-526-9073 to find the location of the Letraset Dealer or Color Imaging Center nearest you.

Dynamark is a registered trademark of 3M.

FIGURE 9–35
Dynamark Imaging System

professional look. For further information regarding the Dynamark Imaging System, contact Letraset USA, 40 Eisenhower Drive, Paramus, NJ 07653.

9.5 WIRING UP AND FINISHING UP

The printed circuit board and cabinet are finished. It is time to wire everything together and assemble the final parts into a finished prototype project. Let's look at wiring methods and wiring plans and then at the final project packaging.

Wiring Methods

We are ready to connect the final electronic components together. At this stage we treat the printed circuit board (and all the components on it) as one large component with many terminals. The off-board components, those that mount on the chassis and on the enclosure, will connect to the PC board and to one another, as necessary. How they all come together is illustrated with the use of a wiring diagram. However, before we discuss the wiring diagram, let's see exactly what wire and wiring are all about.

Wire

Hookup wire, which is almost always made of copper, is of two types: single strand and multistrand

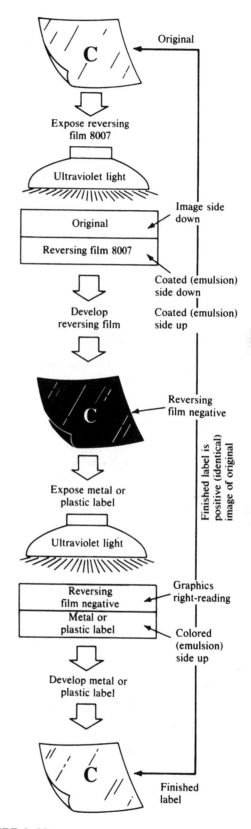

FIGURE 9–36
The Dynamark process
Courtesy Letraset USA

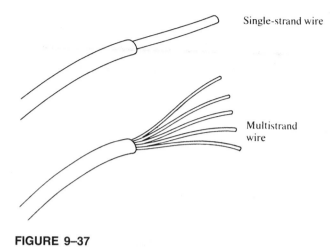

FIGURE 9–37
Wire

(Figure 9–37). Single strand is easier to work with, but multistrand is much more flexible and thus preferred.

Wire size is specified in diameter, circular mills, or gauge. The most common sizes for electronic assembly are #22 and #24 gauge (the higher the number, the smaller the diameter). A multistrand #22 gauge wire can handle almost 2 A of current, more than enough for most prototype projects.

Most wire comes with an insulated plasticlike coating that protects the wire from adverse environmental conditions and keeps the electrical current "in its place." You must strip the insulation off the wire edges before crimping and soldering.

If you wish to cover uninsulated wire or bare component leads, a hollow rubber tubing, known as "spaghetti," is available in different diameters (Figure 9–38). Some forms of spaghetti, known as heat-shrink tubing, will shrink tightly around the lead or wire they protect with the application of heat from a match, soldering iron, or heat gun. The result is a clean, effectively insulated connection (Figure 9–39).

Wire Splicing

On occasion you will need to splice wires, that is, connect two or more wires without a stationary terminating point. There are three common types of splices: *rat-tail, tee,* and *Western Union.* Let's briefly examine the characteristics and methods of forming each one.

With the rat-tail splice (Figure 9–40a), begin by stripping the insulation back about 1″ from the two wires to be joined. Next, twist the pair tightly with your fingers, forming approximately four turns. Finally, use a pair of needle nose pliers to

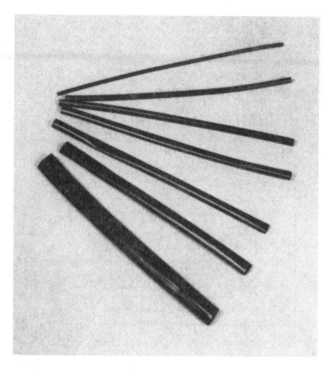

FIGURE 9–38
Spaghetti

twist the wires into an even firmer grip. Cut away any excess wire with a diagonal cutters.

Begin the tee splice by cutting insulation away (using wire strippers or diagonal cutters) from the base wire and the same amount from the wire to be joined (Figure 9–40b). With your fingers, twist the stripped end of the latter wire around the base wire, forming five or six turns. Cut off any excess wire with diagonal cutters.

Begin the Western Union splice by stripping approximately 4″ of insulation from the ends of two wires (Figure 9–40c). Then, tightly wrap three or

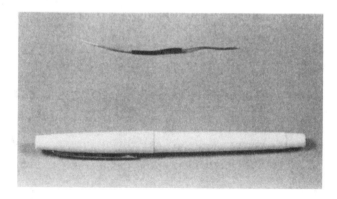

FIGURE 9–39
Application of heat-shrink tubing

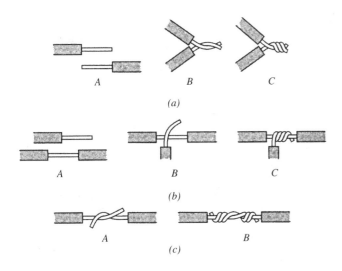

FIGURE 9–40
Wire splicing

four turns of one wire around the other wire, as shown in the figure. Cut off any excess wire using diagonal pliers.

Wiring Plans

Let's now turn to the subject of wiring plans. We'll look at the need for such plans and the elements of a wiring diagram.

The Need for a Plan

With the PC board considered one giant component, the wiring diagram shows how it and the various off-board components (potentiometers, lamps, switches, transformers, etc.) are wired together. Such a diagram should be easy to draw, yet clear and simple to read. There are two general types: the **pictorial diagram** and the **functional wiring diagram**. The pictorial, although certainly easy to understand, is a nightmare to draw (Figure 9–41). This drawing is primarily for those with little experience in proto-type project building. It is the type of drawing found in some electronic kit assembly instructions.

Because people building prototype projects are likely to be more experienced in electronic assembly than first-time kit builders, a drawing more functional and less pictorial is acceptable. Such a

FIGURE 9–41
Pictorial wiring diagram

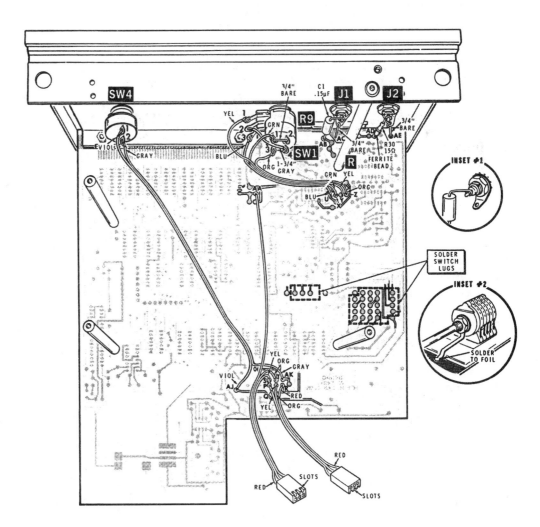

FIGURE 9–42
Functional wiring diagram

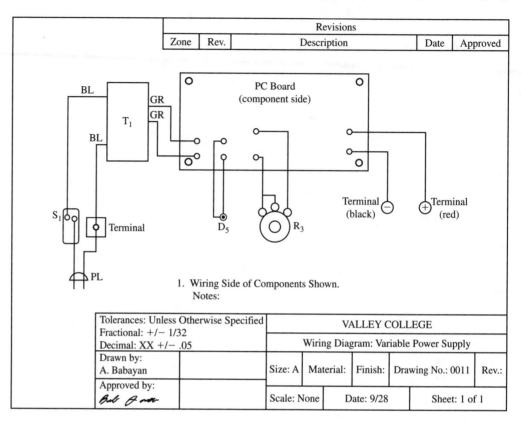

Elements of a Wiring Diagram

As mentioned, the functional wiring diagram (from now on referred to simply as a wiring diagram) contains three elements: electronic components, component terminals, and wiring. Let's examine how each element is represented on the wiring diagram.

Electronic components are represented by their basic physical shape. Round components (potentiometers, some lamps, switches, and so on) are usually shown as round (Figure 9–43a). Almost all other components are displayed as square or rectangular (Figure 9–43b). Each component is identified with the same letter–number designator used on the schematic drawing.

Component terminals are represented by a small circle (Figure 9–43). In some cases the circle is shaded in; in others it is open.

Wire or *wiring* is represented on the drawing with straight lines even in width and forming right angles when bends occur. Any hidden wiring is shown, as on all standard drawings, with dashed lines (Figure 9–44).

drawing (Figure 9–42) shows the basic component outlines, the wiring terminals for each component, and the interconnecting wires. It is the type of drawing we will use to help wire up the prototype project.

Pulling It All Together:
The Complete Wiring Diagram

Figure 9–42 shows the wiring diagram for the Variable Power Supply. In it, the electronic components are arranged in roughly the position they would be seen during the wiring-up process. Keep in mind that this is not a final assembly drawing

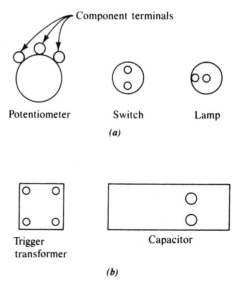

FIGURE 9–43
Electronic component representation

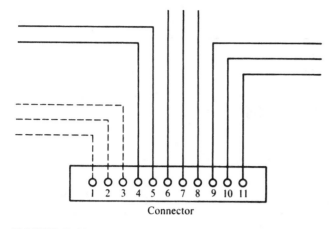

FIGURE 9–44
Wire representation

showing where the PC board and off-board components are to be mounted. That comes next. The wiring diagram is just that—it shows how components are wired together.

   The drawing is begun by placing a round, square, or rectangular representation of each component in the proper location. Next, small circles are drawn to indicate component terminals. Note that terminals not actually used as connection points are nevertheless shown. Doing so helps to establish a reference point for the terminals that do require connection. Once the components and terminals are in, lines depicting wires are drawn. Finally, component identifiers, the same ones used on the schematic drawing, are put in. Notes, title block, and borders complete the wiring diagram.

Final Packaging

It is time to assemble all the elements into a complete working prototype project. To do so, we need a final packaging drawing to show how everything fits together. It could be argued that this type of drawing should have preceded the wiring diagram, as we probably would want to mount any components before the final wiring takes place. In truth, both procedures, and thus both drawings, need to be "worked," or read, together. Depending on the particular component, it may actually be better to attach wires to it before fastening it to the cabinet; yet for a different component, the situation may be reversed: mount the component first, then wire to it. In prototype work, it is usually a matter of doing a little of one and then the other (wiring, component mounting, wiring, etc.) until the final assembly is complete.

Pictorial versus Functional Assembly Drawing

As with the wiring diagram, two types of packaging drawings exist. A **pictorial assembly drawing** (Figure 9–45) is easy to read but difficult and time-consuming to draw. To produce such an "exploded" view often requires the services of a skilled technical illustrator. The **functional assembly drawing** (Figure 9–46) is much easier to draw, yet to read such a drawing you must be reasonably experienced in interpreting orthographic projections. Because most technicians are experienced in this technique, we will be using the functional assembly drawing to illustrate the assembly of the prototype projects.

FIGURE 9–45
Pictorial assembly drawing

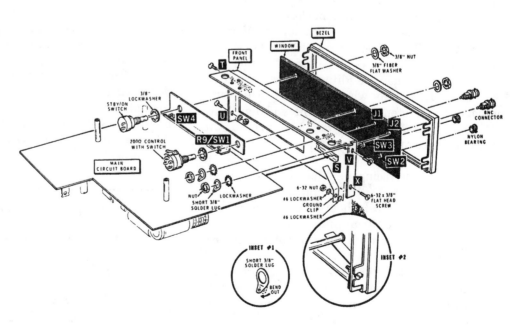

FIGURE 9–46
Functional assembly
(packaging) drawing

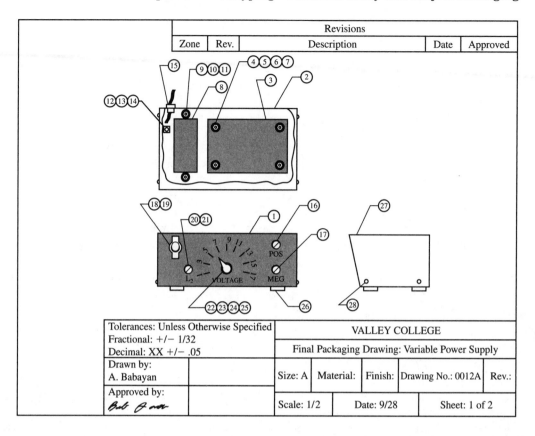

Final Packaging Drawing

The purpose of the final packaging drawing—that is, the functional assembly drawing—is to show what components go where and to identify all fasteners and hardware pieces. To accomplish this objective, we need to produce two items. We must make a drawing that shows the location of all elements and we must develop an assembly parts list based on what is known as the **indenture identification system**. Let's examine each item closely.

The final packaging drawing for the Variable Power Supply Project is shown in Figure 9–46. Top, front, and right side views are shown. We start with the top view and include additional views as needed. Each view shows only what is needed to depict a component or fastener. Component and hardware detail is kept to a minimum; the only purpose is to identify each component or part and place it in the right location.

Note that each component has a ballooned part number attached to it. Additional ballooned numbers may, in turn, attach themselves to a preceding number. These additional numbers identify the hardware items used to fasten the various components to the cabinet.

The parts list (Figure 9–47) includes item number, quantity, part number, indenture, part description, and comments. The indenture system (from the term *indent,* meaning "to set in from the margin") is just a way of indicating what smaller items are part of a larger item or subassembly. For example, item 1 is the completed Variable Power Supply. Its indenture position is all the way to the left. Everything else is a subassembly of it. The PC board assembly (item 3) is a subassembly of the Variable Power Supply chassis (column 2). Therefore, its indentured position is shown in the third column. The four hardware items used to mount the PC board assembly to the cabinet (items 4, 5, 6, and 7) are, in turn, indentured as part of the PC board assembly. Thus, they are in the fourth column from the left. And so it goes, as each item in the completed prototype project is accounted for.

With a functional assembly drawing and an indentured parts list, we can see easily where each item is to go and how all items fit together—all with a minimum of drawing detail. No exploded pictorial drawings are required to illustrate how the prototype project is put together or packaged.

Something to Think About

Does your lab have a wire splice board, one that displays the wire splices shown in Figure 9–40? If it does not, why not construct one so that everyone can see actual examples of each splice? That should tie things up nicely.

FIGURE 9–47
Assembly (packaging) drawing parts list

			Revisions				
			Zone	Rev.	Description	Date	Approved

PARTS LIST
Final Assembly

Item No.	Qty.	Part No.	Indenture				Description	Comments
1	1		X				Variable Power Supply	
2	1			X			Chassis (Bottom)	
3	1				X		PC Board Subassembly	
4	4					X	Screw, 4–40×$^{3}/_{4}''$	
5	4					X	Nut, 4–40	
6	4					X	Washer, $^{1}/_{4}''$	
7	4					X	Spacer, $^{1}/_{4}''$	
8	1				X		Transformer, T_1	
9	2					X	Screw, 4–40×$^{1}/_{2}''$	
10	2					X	Nut, 4–40	
11	2					X	Washer, $^{1}/_{4}''$	
12	1				X		Terminal Strip (2–LUG)	
13	1					X	Screw, 4–40×$^{1}/_{2}''$	
14	1					X	Nut, 4–40	
15	1				X		Strain Relief	
16	1				X		STD Terminal-Red	
17	1				X		STD Terminal-Black	
18	1				X		Switch, S_1	
19	1					X	ON/OFF Plate	
20	1				X		LED, D_5	
21	1					X	Grommet, $^{1}/_{4}''$	
22	1				X		Potentiometer, R_3	
23	1					X	Washer (Match POT)	
24	1					X	Nut (Match POT)	
25	1					X	Knob, Pointer	
26	4				X		Rubber Feet $^{3}/_{4}''×^{3}/_{4}''$	Self-adhesive
27	1			X			Enclosure (Top)	
28	4					X	Screw SM, $^{3}/_{8}''$	Self-tapping

Tolerances: Unless Otherwise Specified Fractional: +/– 1/32 Decimal: XX +/– .05		VALLEY COLLEGE		
		Final Packaging Drawing Parts List: Variable Power Supply		
Drawn by: A. Babayan		Size: A	Material:	Finish:	Drawing No.: 0012B	Rev.:
Approved by: *Bob Brown*		Scale: None	Date: 9/28	Sheet: 2 of 2

9.6 ASSEMBLING AND PACKAGING THE SAMPLE AND EXERCISE PROJECTS

Now, it's time to assemble and package the Sample and Exercise Projects. We have been developing the Variable Power Supply throughout this chapter. Many of the drawings and examples cited depict the assembly and packaging of the power supply. With the 3-Channel Color Organ, as all along, you are left to work things out with only a few hints and suggestions supplied.

The Variable Power Supply

Let's wrap up construction of the Variable Power Supply by examining the project's printed circuit assembly, sheet metal cabinet fabrication, and wiring and final packaging.

Printed Circuit Assembly

Figure 9–1 presents the PC assembly drawing for the Variable Power Supply. Re-create the drawing, following the procedures outlined. Remember to include a side view showing the height of the tallest component (probably capacitor C_1 or the metal heat sink) above the PC board.

Before you assemble, or stuff, the PC board, you need to prepare the components with axial leads. You must bend the leads of diodes D_1–D_4 and resistors R_1 and R_2 prior to inserting them into the PC board (Figure 9–2). Once all the components have been readied, insert them into the PC board from the component side. Watch the polarity of diodes D_1–D_4 and capacitors C_1 and C_3. Reposition the leads for the voltage regulator U_1 so that they fit into the in-line hole pattern. Bend the leads as close to the component body as possible, but be careful— the leads break easily. Also, make sure that the leads do not touch the metal heat sink when U_1 is installed. As each component is inserted, be sure to make a service bend of 30° with each lead (Figure 9–4).

With all components stuffed into the board, solder them in place, then trim each component lead as close to the board as possible with side cutters (or dikes). Except for installing wires to be connected from the board to off-board components (to be discussed shortly), the printed circuit for the Variable Power Supply Project is now complete.

Sheet Metal Cabinet

The sheet metal drawing and template derived from it for the Variable Power Supply chassis are shown in Figures 9–9 and 9–11. If you are building the power supply, first redo the sheet metal drawing, then make a photocopy (or two) to be used as the template.

Next, cut out the template and stick it down on the sheet metal surface with double-sided tape (Figure 9–12). Using a foot-operated squaring shear and tin snip, cut the sheet metal to size and shape (Figures 9–14 and 9–15). Now, with a center punch, drill bit, and chassis punch (or turret punch if available), make the necessary holes in the aluminum sheet (Figures 9–16 through 9–21). Holes less than $\frac{1}{4}''$ in diameter can be drilled; those $\frac{1}{4}''$ and larger should be punched. Ream and deburr all holes as necessary. Finally, smooth any rough sheet metal edges with a file.

Once the sheet metal is cut, drilled, and filed smooth along the edges, it is ready to be formed, or bent. Using the bending brake, follow the bending sequence presented in Figure 9–28. Be sure to set the bending fingers the correct distance back from the edge of the worktable. That distance should be the thickness of the metal, $\frac{1}{16}''$.

To finish the cabinet, use a primer and spray paint or a self-adhesive vinyl. Whichever method

you choose, follow the directions closely, as described earlier in the chapter.

Because the voltage scale for the Variable Power Supply will consist of dry-transfer patterns rubbed onto the front of the cabinet, you cannot apply those particular designations until the power supply is calibrated. (This is done in the next chapter.) Therefore, you might wish to wait until the project is fully assembled, tested, and calibrated before you rub on any dry-transfer patterns.

Wiring and Final Assembly

In completing the final wiring and assembly for the Variable Power Supply you will want to refer to three drawings: the PC assembly drawing (Figure 9–1), the functional wiring diagram (Figure 9–42), and the final packaging drawing (Figures 9–46 and 9–47). If you have your own versions of these drawings, use them.

Here is the step-by-step procedure for completing assembly work on the Variable Power Supply Project:

1. Insert the secondary leads of transformer T_1 into the PC board and solder them in place.
2. Cut, strip, and tin appropriate lengths of hookup wire to connect from the PC board to off-board components. Run the wire along the chassis edges, rather than letting it dangle in wide loops (Figure 9–48). Plan your lengths as carefully as possible, although it is better to have longer wire that can be cut down than to wind up with a piece that is too short.
3. Mount the PC board to the chassis. Use four spacers to elevate the board and appropriate hardware to hold the assembly in place.
4. Using screws, washers, and nuts, mount T_1 to the chassis.

FIGURE 9–48
Wiring: keeping it neat

5. Mount switch S_1 to the cabinet. If possible, use an on/off face plate that comes with most toggle switches. If one is not available, you will have to use dry-transfer lettering to create the on/off designation.

6. Mount the two-lug terminal post, using a screw, washer, and nut. (Only one lug is actually needed, but two-lug terminal posts are easier to find.)

7. Strip both ends of the line cord back about $\frac{1}{4}''$ and tin each wire. Then, crimp the strain relief connector around the line cord, approximately $6''$ from the end. Insert the strain relief into the chassis hole designed to receive it. You may need to file or ream the hole out a bit to secure a tight fit (Figure 9–49).

8. Connect one wire from the line cord and one from the primary lead of T_1 together at a terminal lug. Solder them in place.

9. Connect the remaining wire of the line cord to one terminal of S_1. Solder it in place.

10. Connect the remaining wire from the primary of T_1 to the other terminal of S_1. Solder it in place.

11. Using one of the screws and nuts making contact with the metal chassis and a wire nut (or teardrop), connect the green grounding wire of the line cord to the metal chassis.

12. Mount potentiometer R_3 to the cabinet. Connect wires from the PC board to R_3 and solder them in place.

13. Mount the red and black (+ and −) terminal posts to the cabinet. Be sure to include the solder lug (teardrop) with each terminal. Connect wires from the PC board to each terminal and solder them in place.

FIGURE 9–50
Applying heat-shrink tubing

14. Mount the rubber grommet to the cabinet and insert LED D_5. Place a short length (about $\frac{1}{2}''$) of $\frac{1}{8}''$-diameter heat-shrink tubing over each LED lead. Now, connect wires from the PC board to the LED and solder them in place. When the solder has cooled, pull the heat-shrink tubing over the connection and heat it with the tip of a soldering iron. Two good, well-protected connections should result (Figure 9–50).

15. Mount four self-sticking rubber feet to the bottom of the cabinet.

16. Place the pointer knob on the shaft of R_3 and secure it in place.

17. Seal the cabinet with four sheet metal screws.

Except for the application of dry-transfer lettering, to be rubbed on shortly, the Variable Power Supply is now ready for final testing, troubleshooting, and calibration.

The 3-Channel Color Organ

To finish up the 3-Channel Color Organ Project we will look at a few assembly and packaging hints and then at how best to construct a light box for the three strings of colored lamps.

Assembly and Packaging Hints

To complete the 3-Channel Color Organ, first read and study the material in this and the next section. Second, prepare the four project drawings:

FIGURE 9–49
Line cord strain relief

the PC assembly drawing, the sheet metal layout drawing, the wiring diagram, and the final packaging drawing. Third, finish constructing the project.

As regards printed circuit assembly, prep all axial lead components and reposition the leads for the three SCRs, then complete the PC assembly, except for the installation of off-board wiring.

The sheet metal cabinet design may be selected from one of the patterns in Appendix B, or you may wish to design your own. Regardless of which approach you take, you will have to use a nibbler to cut the rectangular holes for the three AC sockets. Once these and all other holes are cut or punched and the sheet metal is folded into shape, apply paint or self-adhesive vinyl finish. Finally, add dry-transfer lettering to identify the on/off switch S_1 (if no face plate is supplied), input signal bypass switch S_2, the three channel sensitivity controls (bass, midrange, and treble), and the three ac sockets (also bass, midrange, and treble).

Begin final wiring and assembly by cutting to length, stripping, and tinning the many wires needed to connect from the PC board to off-board components and, in turn, from one off-board component to another. Plan to run each wire along the chassis bottom or edge rather than directly through space to the component being wired. If necessary, use plastic cable ties to bunch parallel groups of wires together (Figure 9–51).

After all wires have been soldered to the PC board, mount the board to the chassis, using $\frac{1}{4}''$ spacers. Then, mount and wire off-board components in the most convenient sequence. Finally, close up the cabinet and set it aside. You're now ready for final

FIGURE 9–51
Plastic cable ties

testing and troubleshooting of the 3-Channel Color Organ Project.

The Light Box

In addition to the color organ electronics, you will need a box to house three strings of Christmas lights, one for each channel. The box, which is best made out of solid wood or particle board covered with a self-adhesive vinyl, may be any convenient width and height but no less than 6″ in depth. A sheet of light-diffusing plastic is placed over the front, and the back is usually covered with a thin sheet of Masonite. The bulbs are hung on stiff wires strung back and forth inside the box. Each string of lights will have an ac line cord plug at one end. These plugs are brought out of the back of the box, to be plugged into the 3-Channel Color Organ resting on top of the cabinet or nearby.

Plans for constructing a wooden light box are shown in Figure 9–52. Overall dimensions are omitted, so you can choose the size that best fits your needs.

SUMMARY

In this chapter we looked at the printed circuit assembly. We examined the PC assembly drawing contents and saw how a typical drawing is created. We studied various PC assembly techniques by investigating component lead preparation and component insertion and soldering procedures.

Something to Think About

Why not take time, now, to compare your completely assembled project with a similar piece of equipment professionally made. What are the differences? How are they similar? Are there ways you could have made your project more professional? Are you satisfied with how your project turned out?

Then it was on to the design and fabrication of the project cabinet. We explored the criteria employed in choosing a quality sheet metal design, and we saw how to produce the sheet metal design drawing and layout template. We went on to analyze the cutting, drilling, punching, and forming of the sheet metal cabinet using such tools as the foot-operated squaring shear, tin snip, chassis punch, turret punch, and bending brake. We finished the cabinet

FIGURE 9–52
Light box

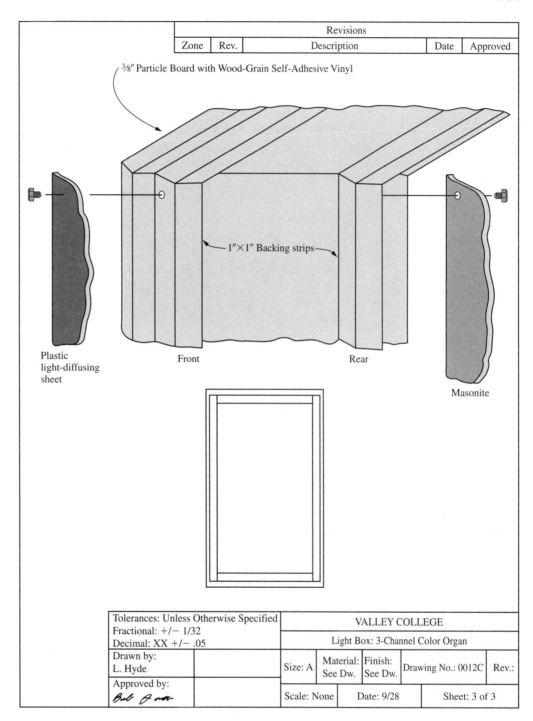

Revisions				
Zone	Rev.	Description	Date	Approved

⅜″ Particle Board with Wood-Grain Self-Adhesive Vinyl

1″×1″ Backing strips

Plastic
light-diffusing
sheet

Front

Rear

Masonite

Tolerances: Unless Otherwise Specified Fractional: +/− 1/32 Decimal: XX +/− .05		VALLEY COLLEGE				
		Light Box: 3-Channel Color Organ				
Drawn by: L. Hyde		Size: A	Material: See Dw.	Finish: See Dw.	Drawing No.: 0012C	Rev.:
Approved by: *Bob Jones*		Scale: None	Date: 9/28		Sheet: 3 of 3	

by spray painting or applying a self-adhesive vinyl coating.

After the printed circuit was assembled and the cabinet constructed, we examined wiring methods and final packaging. We looked at wire characteristics and the wiring diagram, then we saw how to pull everything together, with a final packaging plan.

We concluded the discussion of final assembly and project packaging with a look at how to complete the Variable Power Supply and 3-Channel Color Organ Projects.

We are not there, yet, however—but almost. In Chapter 10 we test, then, if necessary, troubleshoot, our project. Building a project is one thing, building a working prototype is quite another.

QUESTIONS

1. The PC assembly drawing shows what _____ go where and how they are oriented.

2. _____ lead components need to have their leads bent at a 90° angle prior to insertion into the PC board.

3. A _____ bend allows components to be easily removed from the PC board.

4. Most cabinets used for prototype projects are made of _____.

5. The sheet metal drawing must be drawn _____ scale because the sheet metal layout _____ will be produced from it.

6. Two tools for cutting sheet metal are the foot-operated _____ _____ and the handheld _____ _____.

7. Round holes are cut in sheet metal with a drill, _____ punch, or turret punch.

8. Rectangular holes, slots, and notches are cut in sheet metal with a handheld minishear known as a _____.

9. A bending _____ is used to form, or bend, sheet metal.

10. Before a cabinet is painted or covered with a self-adhesive vinyl coating, it must be thoroughly _____.

11. Before a cabinet is painted, a coat of _____ must always be applied.

12. Dry-transfer lettering is also known as _____ - _____ lettering.

13. _____ wire is much more flexible than single-strand wire.

14. The _____ wiring diagram is much easier to draw than the pictorial wiring diagram.

15. The final _____ drawing shows how all components and hardware come together to form a complete prototype project.

REVIEW EXERCISES

1. Obtain surplus drilled and etched PC boards. Stuff them with surplus axial and radial lead components. Solder them in place.

2. Using $8\frac{1}{2}'' \times 11''$ graph paper, sketch the sheet metal layout patterns for the cabinet depicted. A chassis and an enclosure are required.

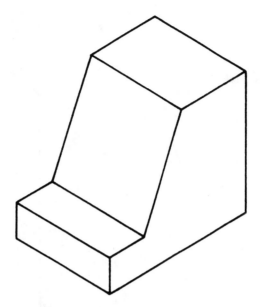

3. Using $8\frac{1}{2}'' \times 11''$ graph paper, sketch the sheet metal layout patterns for the cabinet depicted. A chassis and an enclosure are required.

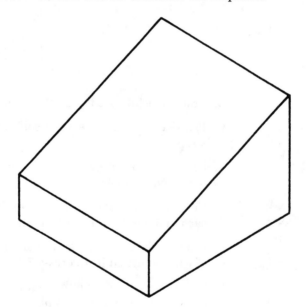

4. Label the correct sequence of bends for the sheet metal pattern shown.

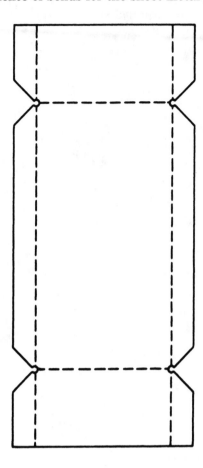

5. Practice bending surplus sheet aluminum at 90°, 60°, 45°, and 30° angles.

6. Practice applying sheets of self-adhesive vinyl to aluminum surfaces. Trim to size.

7. Using a sheet of graph paper, practice rubbing on dry-transfer letters, numbers, and symbols.

8. Cut and strip various lengths of multistrand hookup wire. Twist the strips of each end together and tin them.

9. Practice the cut-slash-and-hook method of wiring and soldering. Use the pieces of wire produced in Exercise 7.

10. Obtain various sizes of heat-shrink tubing, cut them to length, and place them over component leads. Apply heat from a soldering iron or other appropriate source to shrink the tubing.

10 Testing, Troubleshooting, and Final Documentation— The Moment of Truth

OBJECTIVES

In this chapter you will learn

- ☐ The four basic testing and troubleshooting steps.
- ☐ How to perform preliminary tests to prevent project burnout.
- ☐ How operational tests are carried out.
- ☐ How to use human senses in troubleshooting.
- ☐ How to read a troubleshooting flowchart.
- ☐ Why performance tests are necessary.
- ☐ How to write a Test Results Document.
- ☐ How to write a Summary and Recommendations Document.
- ☐ How to compile the final Project Report.

To finish the prototype project and Project Report, you need to test the project to see if it works. If it doesn't, you have to find out why and fix it. Also, two short written documents—the Test Results Document and the Summary and Recommendations Document—are needed. These and other written documents, as well as all graphic documents (drawings), are then brought together to form the completed Project Report.

10.1 PROTOTYPE TESTING AND TROUBLESHOOTING

The prototype project is now built. But does it work, and under all relevant conditions? Through an interweaving process of testing and troubleshooting, you will find out if the project performs as it should, and if it does not, how to go about making it do so.

Let's begin by examining how to test and troubleshoot a prototype project. Then we will explore the four procedures required to get any project up and running and doing what it was designed to do.

Testing and Troubleshooting Procedures

To test and, if necessary, troubleshoot a prototype project, four steps, or procedures, are required: (1) preliminary tests are performed, followed by (2) operational tests. If the prototype fails any of the operational tests, (3) troubleshooting begins. When the project is finally functioning, then (4) a series of performance tests is conducted. Only after all these efforts can the project be evaluated to see if it should proceed to the production stage.

Let's examine each test and troubleshooting procedure briefly and see how they relate to each other. Next, we will explore how the testing and troubleshooting of a prototype project differ from doing the same for a product that is already in production.

The Four Basic Testing and Troubleshooting Steps

In Figure 10–1 the four test and troubleshooting procedures are listed in flowchart form. The flowchart allows you to see how each step is related to the others.

Preliminary testing is done first. These tests are performed *before* power is applied to the prototype project. They are designed to catch errors that

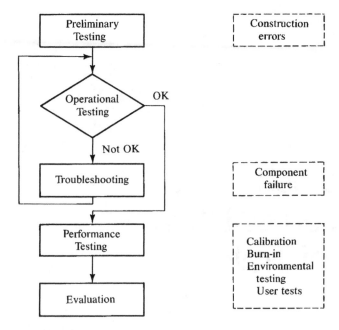

FIGURE 10-1
Testing and troubleshooting procedures

could cause serious problems should the wrong voltages and currents be allowed to reach critical components. Preliminary testing is preventive testing—an example of the "better safe than sorry" approach.

Once the project passes all preliminary tests, **operational testing** begins. Here, power is applied for the first time and basic project functioning is determined. If all appears well at this point, performance testing is done. If there is a problem, however—lights do not come on, sound is not heard, relays do not click in, displays aren't visible—the project must move to the troubleshooting stage.

Troubleshooting is done to determine what is wrong, why it's wrong, and what to do about it. In other words, we identify the cause of a problem and fix it.

Once you think the problem is solved, you must run the project through the operational tests again. If all is now fine, it's on to performance testing. If there is still a problem, more troubleshooting is required. As you can see in Figure 10–1, you remain in a loop of operational testing–troubleshooting–operational testing until all problems are solved.

Performance testing is designed to put the prototype project through its paces—"give it the treatment," so to speak. You want to see if it will work, not just in the lab but under all adverse environmental conditions. Will the project function in the real world?

After performance testing is completed, all results are evaluated.

Testing and Troubleshooting a Prototype Project

The procedures for testing and troubleshooting a prototype project differ, at least in one important respect, from those used to correct problems in production models. When the product is returned to the factory for repair, at least we know that at one time it did work. We assume it left the plant in functioning order and that it probably performed correctly, at least for a while. Therefore, we know the production model was wired up properly. However, if the prototype is not working as it should, the problem could be a wiring error, because the prototype has yet to perform as presently constructed. That is why construction errors are of primary concern when evaluating a prototype project.

True, the prototype did work at the breadboarding stage. That is important, because we know the design is correct and that the components used in the breadboarded version were operating at one time. Essentially, the problem now is to determine what, if anything, went wrong between the breadboarding stage and the final wired-up prototype. It's usually either a *construction error* or a *component failure* (Figure 10–1). During preliminary testing you probe for construction errors; during troubleshooting you go after component failure.

Testing—One, Two, Three

The procedures involved in preliminary and operational testing are examined in more detail in this section.

Preliminary Tests: Better Safe Than Sorry

Preliminary testing is designed to check for construction errors. That means looking for two things: *wiring mistakes* and *incorrect component placement*. Let's explore each in turn.

Wiring errors can be of four types: a wire is going where it shouldn't go; a wire is missing (it is not going where it should go); a wire is bad (open); or there is a faulty wire connection (poor solder joint).

First, look for wires or circuit traces that shouldn't be there, or ones that are missing. Check against the schematic drawing, using a colored pencil to mark your progress. If the project is fairly complex, instead of checking every wire or trace, do a random check. Pick a wire here and there, and see if it belongs where it is.

Next, look for bad or open wires. Although a broken wire may be visible to the eye, chances are the only way you will find an open connection is with an ohmmeter. Set it to test for continuity, and in a random manner pick various wires and PC traces to scrutinize, bearing in mind that there may be parallel paths.

Faulty solder connections (cold solder joints) are usually visible as dull gray balls of solder. If you suspect a faulty solder connection, check the connection with an ohmmeter.

Checking for component placement error means seeing if a component is in the wrong place, if the component is in backward, or if the component leads are inserted incorrectly.

Wrong component placement is a particular problem with color-coded resistors. It is easy to mistake a 100-ohm for a 100,000-ohm resistor, for example. The only difference is the third band color. Check all resistors carefully. It is also easy to install electrolytic capacitors, diodes, and integrated circuits backward. The only way to be sure is to check each one individually.

Even though the right component may be in its appropriate place and installed in the correct direction, there could still be a problem if the designated leads are not going to the proper points. This is particularly true with transistors, SCRs, voltage regulators, and the like. Again, your best check is a visual check. Examine each component meticulously.

Finally, although it is going beyond looking for wiring mistakes and incorrect component placement, it is a good idea to check for shorts between any input voltage pins and ground. These must be eliminated before power is applied.

Operational Testing: Ready or Not

There comes a time when you just have to go for it: when the moment of truth arrives. If you have been patient and taken the time to perform all relevant preliminary tests, then the operational tests shouldn't turn into smoke tests. If the supply voltage terminals are not shorted, the project will not blow up. So now is the time to apply power, preferably through an isolation transformer, to provide complete ac line isolation, and see what happens.

Depending on the type of project, lights may flash, sounds may rumble forth, motors may turn, relays may click in, or dc voltages may appear. Play around with the project a bit; let it warm up and settle into a normal operating mode. Your purpose at this point is not so much to perform an extensive evaluation of what is going right but rather to see what, if anything, may be wrong. If, after a time, all seems well, you are ready to perform any necessary project calibration and then move on to performance testing. But if the project fails to operate satisfactorily, troubleshooting is called for.

Troubleshooting the Prototype Project

It is difficult to generalize about component failure; each project is different, and the problems vary. All we can hope to do in this section is to explore, in nonspecific terms, component failure rates and component substitution procedures. By doing so, you will get some idea as to what components are likely to fail, how to check for failure, and how to correct it.

Component Failure Rates

If your car won't start on a cold morning, you're unlikely to suspect a transmission problem. Similarly, if the string of lights on your Color Organ Project turns on but does not flash to the sound of music, you won't check the line cord. You know the line cord is good because power is getting to the light bulbs. The point is, depending on the problem, some components are more likely to be the cause of trouble than others. An experienced technician doesn't waste time checking in the wrong places but will go right to the likely source of difficulty.

It is possible to list the failure rate of common electronic components in relative terms and by so doing to provide a general troubleshooting guide. From the list in Figure 10–2 it is clear that passive components are much less likely to fail than active components. At the bottom of the list are the resistors and disc capacitors. Suspect them last. Moving up the list, you encounter the solid-state components, such as ICs and transistors. Suspect them first.

Most likely to fail	☐ Transistors
	☐ Integrated circuits
	☐ Light-emitting diodes
	☐ Diodes (rectifiers)
	☐ Switches
	☐ Relays
	☐ Potentiometers
	☐ Lamps
	☐ Speakers
	☐ Transformers
	☐ Coils
	☐ Capacitors (electrolytic)
Least likely to fail	☐ Capacitors (nonpolarized)
	☐ Resistors

FIGURE 10–2
Component failure rates

Once you think a component may be bad, how do you check to be sure? Surprisingly, four of the five *human senses* can come in very handy at this time.

Use your power of *sight*. Look closely for trouble, and you just may find it. If components have received excess current, they often show it by being charred or split open. It is the first thing you should look for.

Sniff around; use your sense of *smell*. Transformers, coils, and other components with insulated linings often have a burnt smell when they have been given an overdose of current.

Listen up! The snap, crackle, and pop you *hear* could indicate loose wires, bridged connections, or frying components.

Be sensitive to the *touch*. As you gain experience in troubleshooting, you will begin to know which components are running too hot and which ones too cold, just by feeling them. Generally, if you can't keep your finger on it, it is probably "gone." Conversely, if solid-state components, such as transistors and ICs, are cold to the touch, they are unlikely to be receiving any voltage.

Using only your senses of sight, smell, hearing, and touch won't get you through the more troublesome problems. For those, you'll need test instruments: meters, oscilloscopes, and logic probes. They will act as your eyes and ears as they delve into the circuit itself.

Component Substitution

Although the component substitution approach has its limitations, it is a good first step to rapid circuit repair. In this method, a suspect component is identified and a good one is substituted. As long as this

is done one component at a time, the procedure works. Do not, however, replace a string of diodes, transistors, or the like all at once. Not only is this a waste of time, but you'll never know for sure which component was faulty to begin with.

In many cases, you need not completely remove the suspected component from the circuit; often, unsoldering one or two leads is all that is required. Then, clip the replacement component into place (Figure 10–3). If the circuit now functions as it should, solder the new component permanently in position.

Through a process of selected component substitution, you should eventually get the prototype project up and running. As you go through each troubleshooting step, remember to repeat the operational tests. Inevitably, you will break out of this operational testing–troubleshooting loop and proceed to performance testing.

Performance Testing

Performance testing has one main purpose: to determine if the prototype project will work whenever and wherever it is supposed to. With performance testing the project is exposed to the real world. It is taken out of the laboratory and given the "once over."

Playing the Critic

If you built (and maybe even designed) the prototype project, naturally you want it to work, to pass all tests. This can be a problem. Are those who built the project the ones who should test it? Will they be objective no matter what the results may be, or are

FIGURE 10–3
Clip-lead component replacement

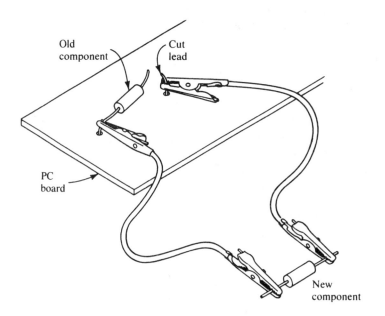

they likely to overlook specific faults? You have to be careful about this and make every effort to avoid letting wishful thinking interfere with thorough and precise testing and evaluation of the project.

To do so, play the critic. Constantly question your own test results. Never give your stamp of approval until you are absolutely sure that the project has "had it all"—seen all it is likely to see in real-world operation and performance.

Giving the Project Away

After you have run the project through some of the more obvious performance tests—leaving it on for extended periods of time, placing it in hostile environments (excess heat, cold, dirt), driving it beyond its rated capacity—it is time to give it away. By that, we mean handing the project over to a *potential user* and telling the person to use, even abuse, it to its limits.

This potential user should be someone who doesn't have a vested interest in making the project work or look good, someone who will use the prototype under the rigorous conditions it is likely to see every day. If the project satisfies the potential user, usually the most critical judge, you'll know it has probably been through the worst it's going to see. The project is now ready for an honest and complete written evaluation.

What specific performance tests the prototype project goes through naturally depends on the type of project it is. A variable power supply wouldn't be subjected to the same evaluation as a color organ. As a general rule, run the device for long periods of time in the roughest environment it is likely to encounter. Record all observations about its performance, then evaluate the results.

Something to Think About

Though the procedures used for troubleshooting a prototype project differ from those employed with a production model, much can be learned from working with the latter. Why not gain troubleshooting experience by taking on the repair of faulty equipment found in your lab, at home, or in your car? Any troubleshooting experience you gain will be worth the effort.

10.2 FINAL DOCUMENTATION

It's time to produce two final documents: the **Test Results Document** and the **Summary and Recommendations Document**. These, along with other written and graphics documents, are placed in

a folder to make a completed Project Report. (The report will also need a title page and table of contents.) Two such finished Project Reports are shown in Chapters 15 and 16. Look them over carefully and use them as a guide when putting together your own report.

Test Results Document

The Test Results Document records, or documents, the results of the test and troubleshooting procedures. Essentially, it is a checklist of all that took place during these tests. It is a record and summary of the data collected.

What Happened?

The Test Results Document is compiled during the testing and troubleshooting procedures. It tells what happened by following, in outline form, the flowchart shown in Figure 10–1. In addition to a place for the project name and the name of the person who performed the tests, there is a section containing preliminary test results (Figure 10–4). Simply answer the questions in this section. When the answer to each question is yes, initial it, supply the dates, and move on to the next section.

The operational testing section requires that you note any catastrophic failures (the project blew up), minor problems, and the results of project calibration, if required. Will troubleshooting be necessary, and, if so, what do you anticipate needs to be done? When the operational tests have yielded satisfactory results, sign off in the space provided.

If troubleshooting is called for, and it often is, you must define the problem and document the steps taken to solve it. Then, it's back to operational testing to see if everything works. If it does, it's on to calibration (if required). When everything is functioning correctly, sign off and move on to performance testing.

In the performance test results section, you explain what user tests were performed. Even though you may not carry out the tests yourself, you still need to monitor them carefully. Document the results, explaining briefly what happened when the tests were performed.

You may wish to add a concluding statement. The concluding statement is not an opinion concerning the results. Rather, it is just a summary of what the test and troubleshooting data have revealed.

A Typical Test Results Document

Figure 10–5 shows a completed Test Results Document for the Variable Power Supply Project. The idea is to record all relevant data, but keep it

FIGURE 10–4
Test Results Document

Test Results Document

Name of Project _____

Person(s) Performing Tests _____

Preliminary Test Results

1. Is all wiring complete and correct?   Init. _____

                                                 Date _____

2. Are all soldering connections good?   Init. _____

                                                 Date _____

3. Are all components in the correct location?

                                                 Init. _____

                                                 Date _____

4. Are all components installed in the correct direction?

                                                 Init. _____

                                                 Date _____

Operational Test Results

1. Any catastrophic failure?   Yes _____ No_____ Init. _____
    Explain: _____
    _____
    _____

2. Any minor problems?   Yes _____ No_____ Init. _____
    Explain: _____
    _____
    _____

6

short and to the point. The Test Results Document basically records what transpired during this important phase of prototype design and construction.

Summary and Recommendations Document

The Summary and Recommendations Document does just what its name implies: It summarizes briefly where the project is at this point. It then offers recommendations as to what direction the project should take.

Here's Where We Are

In one or two paragraphs simply state if the goals and objectives set forth in the Concepts and Requirements Document have been met. Although the Summary and Recommendations Document is usually placed at the end of the Project Report, it is something that could easily be read first by those who want an overall view of the project results.

When the summary is complete, you may offer recommendations as to where the project should go from here. Keep in mind, however, that your

FIGURE 10–4 *continued*

3. Calibration, if necessary.   Yes _____   No_____   Init. _____

_____

_____

4. Is troubleshooting required?   Yes _____   No_____   Init. _____

What is anticipated? _____

_____

_____

5. Is project operation satisfactory at this time?

Yes _____   No_____   Init. _____   Date _____

Troubleshooting Results

1. Define the problem.

_____

_____

_____

2. Steps taken to correct the problem.

_____

_____

_____

3. Is project functioning satisfactorily?

Yes_____   No_____   Init. _____   Date _____

Performance Test Results

1. User test performed. _____

_____

7

recommendations must be based solely on what has taken place during the design and fabrication of the prototype project and should be of a technical nature only. Unless you have a particular expertise to offer, do not get into the sales, marketing, or even production end of things. Leave that for others to ponder. Your job is to offer a recommendation based on how the prototype performed. In that regard, there are four basic possibilities:

1. The prototype worked as expected, no major problems exist, and it is recommended that full-scale production be given serious consideration.

2. The prototype works, but further performance testing is desirable. Perhaps another prototype needs to be built and tested. The point is that more evaluation should be done before a recommendation for full-scale production can be considered.

3. The prototype needs the following modifications before it can be rated a success. Here you list what design and fabrication changes will be required before a reevaluation can be made.

4. The prototype has not performed satisfactorily and is unlikely to do so no matter how many modifications are made. You, therefore, do not recommend that the project proceed to the production stage.

FIGURE 10–4   *continued*

2. User test results. _____
_____

Concluding Comments
_____
_____
_____

8

Whichever decision you make, it is important that you back up that decision with facts. You must, in effect, prove your claim. If the project is ready for production, use the data to show why. You need not repeat all the data at this point; just reference the information. If there is no way to salvage the project, again use the data to show why it shouldn't go any further.

A Typical Summary and Recommendations Document

Figure 10–6 shows the Summary and Recommendations Document for the Variable Power Supply Project. It gives a brief summary of the main developmental points and recommends, based on the test data collected, that consideration be given to full-scale production of the project.

FIGURE 10–5
Completed Test Results
Document

Test Results Document

Name of Project _____ *VARIABLE POWER SUPPLY* _____

Person(s) Performing Tests _____ *AUSTIN BABAYAN* _____

Preliminary Test Results

1. Is all wiring complete and correct?   Init. *A.B.*
                                              Date *6-1*

2. Are all soldering connections good?   Init. *A.B.*
                                              Date *6-1*

3. Are all components in the correct location?
                                              Init. *A.B.*
                                              Date *6-1*

4. Are all components installed in the correct direction?
                                              Init. *A.B.*
                                              Date *6-1*

Operational Test Results

1. Any catastrophic failure?   Yes _____ No *X* Init. *A.B.*

Explain: _____

_____

_____

2. Any minor problems?   Yes _____ No *X* Init. *A.B.*

Explain: _____

_____

_____

6

Something to Think About

Professional, industry-based test results documents can be instructive. Why not gather a few from your industry contacts? Examine them carefully. In what ways are they different from the test results document you are preparing?

10.3 TESTING, TROUBLESHOOTING, AND FINAL DOCUMENTATION FOR THE VARIABLE POWER SUPPLY AND 3-CHANNEL COLOR ORGAN

Let's test and troubleshoot and then write and compile the final documentation for the Sample and Exercise Projects.

FIGURE 10–5 *continued*

3. Calibration, if necessary.   Yes __X__   No _____   Init. __A.B.__

CALIBRATION OF PANEL FACE FOR VOLTAGE READINGS. USED

TRANSISTOR RADIO AS LOAD.

4. Is troubleshooting required? Yes _____   No __X__   Init. __A.B.__

What is anticipated? _____

_____

_____

5. Is project operation satisfactory at this time?

Yes __X__   No _____   Init. __A.B.__   Date __6-1__

Troubleshooting Results

1. Define the problem.

NONE

_____

_____

2. Steps taken to correct the problem.

N/A

_____

_____

3. Is project functioning satisfactorily?

Yes __X__   No _____   Init. __A.B.__   Date __6-1__

Performance Test Results

1. User test performed. FELLOW STUDENT GIVEN PROJECT AND

ASKED TO OPERATE IT UNDER VARIOUS LOAD CONDITIONS.

7

Sample Project: The Variable Power Supply

We'll consider first the testing and troubleshooting procedures, and then the final documentation.

Testing and Troubleshooting the Variable Power Supply

Before the Variable Power Supply prototype can be pronounced a success, it must go through preliminary and operational testing, troubleshooting, cali-

bration (if necessary), and finally, performance testing. Let's examine each in turn.

Preliminary Testing. Preliminary testing begins with a thorough examination of all wired connections, including PC traces. Be sure to check that the positive ends of capacitors C_1 and C_2 are *not* connected together. This is a common mistake. As will be noted on the schematic drawing, the positive ends of these two components are separated, or isolated, from each other by the voltage regulator U_1.

FIGURE 10–5 *continued*

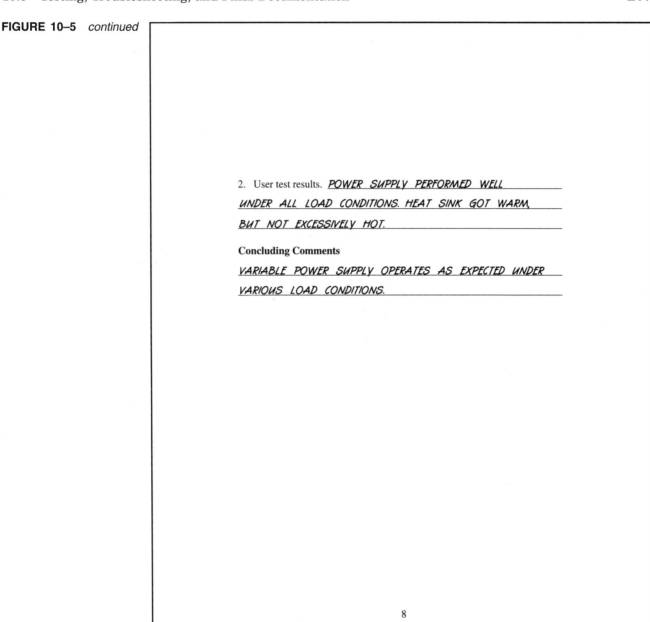

2. User test results. *POWER SUPPLY PERFORMED WELL*
UNDER ALL LOAD CONDITIONS. HEAT SINK GOT WARM,
BUT NOT EXCESSIVELY HOT.

Concluding Comments
VARIABLE POWER SUPPLY OPERATES AS EXPECTED UNDER
VARIOUS LOAD CONDITIONS.

8

Next, check that all components are in the correct location, that polarized components are installed in the right direction, and that component leads are going where they should. Examine carefully diodes D_1–D_4, LED D_5, capacitors C_1 and C_2, and the pins of voltage regulator U_1.

As a final preliminary test, as you did during the breadboarding stage, connect an ohmmeter across the output terminals of the power supply. With the power supply unplugged, turn potentiometer R_3 through its range. As you do so, the meter should indicate a change in resistance of approximately 390 to 5,000 ohms ($\pm10\%$). If there is no change in resistance as the potentiometer is rotated, or if the resistance drops below 200 ohms, recheck all wiring. Once a wiring error is found and corrected, proceed to operational testing.

Operational Testing. Begin by plugging the project in and turning switch S_1 to the "on" position. The LED should light, indicating that a dc voltage is present across one end of R_1 and the negative terminal of the power supply. If the LED does light, place a

FIGURE 10–6
Summary and
Recommendations
Document

Summary and Recommendations Document
Variable Power Supply
by
Austin Babayan

Our goal was to design and fabricate a 1.2- to 15-V Variable Power Supply
prototype project. The goal has been met and the power supply meets all
preliminary project objectives. The initial design has proven valid, and no
unanticipated design or fabrication problems developed during prototype
construction. It is the recommendation of this technician that full-scale
production of the Variable Power Supply be given serious consideration.

Name ___*AUSTIN BABAYAN*___

Signature _____

Date ___*6-15*___

9

voltmeter across the output terminals. While you
are rotating potentiometer R_3, the voltage reading
should go from about 1.2 to 15–18 V dc. If the LED
does not light, or an incorrect or zero voltage reading is obtained, troubleshooting will be necessary. If
the power supply passes these initial operational
tests, skip to the calibration stage.

Troubleshooting. It is necessary to troubleshoot if
the LED does not light and/or an incorrect voltage is
obtained at the power supply's output. Figure 10–7

shows a troubleshooting flowchart for a power supply. The chart indicates the following:

☐ Does the LED light? If the answer is yes, go on
to test for correct output voltages. If the voltages
are right, the operational testing is complete
and you are ready to calibrate.
☐ If the LED does not light, check to see that it is
installed in the proper direction and that it has
not burned out. Again, see if it lights. If the answer is still no, look for 120 V ac at points A–A.

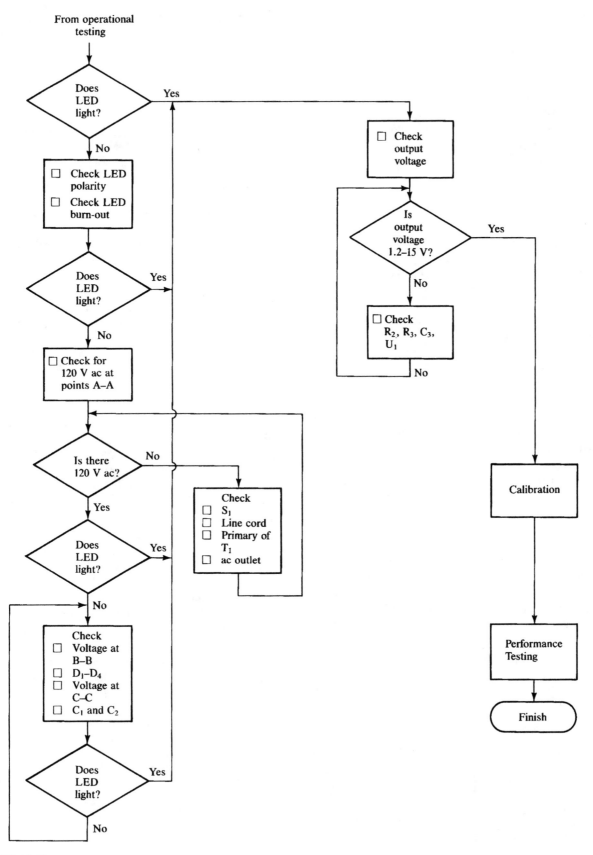

FIGURE 10–7
Variable Power Supply troubleshooting flowchart

☐ If 120 V ac is not received at points A–A, check switch S_1, the line cord, the primary winding of T_1, and even the ac power source—in that order. After each test, go back and check for 120 V ac at points A–A. Eventually, you will get the correct voltage and can check if the LED lights.

☐ If the LED still does not light, examine the voltage at points B–B as well as diodes D_1–D_4, the voltage at points C–C, and whether capacitors C_1 and C_2 are shorted—in that order. Finally, the LED should light.

☐ Once the LED is lit, you can check for correct output voltages. If there are none, or incorrect voltages are obtained, carefully examine resistors R_2 and R_3, as well as capacitor C_2 and voltage regulator U_1. When the correct output voltages are finally produced, proceed to project calibration.

Calibration Procedure. Calibration of the power supply is quite simple. Since you are not using a panel meter, you will mark the cabinet with rub-on lines and letters to indicate voltage as the potentiometer pointer knob is rotated.

With the potentiometer installed and held securely in place, turn it completely counterclockwise. Connect a load to the power supply; a small radio will do fine. Place a voltmeter across the output terminals and turn on the power supply. With a pencil, make a small mark on the cabinet face where the pointer is aimed (Figure 10–8). This will be the 1.2-V mark, as indicated on the voltmeter. Next, turn the potentiometer completely clockwise. Make another pencil mark as before. This will be the maximum voltage, whatever is indicated on the voltmeter. Finally, turn the potentiometer completely counterclockwise again and then slowly advance it. As the meter reads an additional volt, make a new mark. When you have completed the sweep of the pot, go back and rub on lines and numbers, as indicated in Figure 10–9. Spray the face of the cabinet with clear acrylic. You've now completed power supply calibration.

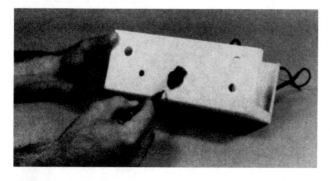

FIGURE 10–8
Calibrating the voltage scale

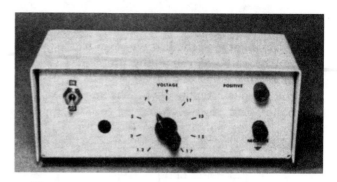

FIGURE 10–9
Completed voltage scale

Performance Testing. To complete performance testing for the Variable Power Supply, begin by leaving the power supply on for 2 to 5 hours without a load across the output terminals. If all is fine at this point, connect a load that will draw $\frac{1}{2}$ to $\frac{3}{4}$ A. A large 9-V radio is a good choice. If the radio plays continuously for a few hours, chances are pretty good you have an excellent prototype Variable Power Supply Project.

Documentation for the Variable Power Supply

We have already examined the Test Results Document for the Variable Power Supply (Figure 10–5). It documents the test and troubleshooting results for one particular prototype power supply. Of course, your results, and thus the Test Results Document, may differ, depending on how your project turns out. The Summary and Recommendations Document for the Variable Power Supply has also been examined (Figure 10–6).

With these two documents complete, the Project Report can be assembled. Add a title page and table of contents and place the report in a folder. If you like, why not take a digital photograph of your finished project and insert it into your document. You now have a working prototype project and Project Report. You should be proud of yourself. Lean back and take a good look. You built it—and doesn't it make you feel great?

The Exercise Project: The 3-Channel Color Organ

For the Exercise Project, we first look at the testing and troubleshooting procedures and then at the final documentation.

Testing and Troubleshooting the 3-Channel Color Organ

Preliminary Testing. Preliminary testing involves standard procedures such as checking circuit wiring, especially off-board wiring, which is extensive; watching for correct component and lead placement; and making sure the polarized components, that is, the SCRs, are installed in the proper direction. It is also a good idea to check each string of light bulbs that is to be used with each channel. Plug them into a 120-V ac wall outlet and make sure all lights turn on.

Operational Testing. When all preliminary tests are complete, you're ready for operational testing: to connect things up and see if the color organ works. Plug a string of lights into each ac socket (SO_1–SO_3) on the color organ case. Plug the color organ itself into the wall outlet. Connect the output from the sound source to phono jack J_1 and open switch S_2. Turn the music on, close switch S_1, and adjust potentiometers R_3, R_5, and R_7 to get the lights to "dance" to the sound of music.

If all lights are responding to the music, turn down the volume of the sound source and close switch S_2. If everything is still operating, you're ready for performance testing. However, if any of the following five conditions exist, immediately turn off the 3-Channel Color Organ and proceed to the troubleshooting section.

1. Neon lamp I_1 does not light.
2. One or more strings of lights do not light.
3. One or more strings of lights come on but do not respond to the sound of music.
4. One or two channels work correctly but not all three.
5. The color organ does not respond well when switch S_2 is closed and the volume of the music source is low.

Troubleshooting. The troubleshooting flowchart for the 3-Channel Color Organ is shown in Figure 10–10. Let's go through it.

☐ Is neon lamp I_1 on? If it isn't, check the components shown in the diagram, in the order suggested. When I_1 does come on, adjust the three potentiometer controls, and ask if all channels seem to be working correctly. If the answer is yes, turn down the volume control, close switch S_2, and ask, again, if everything is all right. If it is, go on to performance testing.

☐ If the answer to the last question is no, examine switch S_2 and recheck the condition.

☐ If when you asked if all channels were responding correctly the answer was no, ask a new question: Do *any* channels work? If the answer is yes, check the components and wiring for the channel or channels that are not working. Each time

you do so, go back and adjust the potentiometers to see if all channels are functioning.

☐ If it turns out that none of the channels work, as mentioned in the previous step, ask the next question: Do the lights for any channel come on, but stay on (do not respond to music)? If the answer is yes, check the wiring between channels and go back to adjust the potentiometers again.

☐ If the answer to the last question was no, there can be only one remaining problem: In one or more channels none of the lights are on. Check the various components for the suspected channel(s), and after each check, go back to see if everything now works correctly. Eventually, it will.

Performance Testing. To carry out performance testing for the 3-Channel Color Organ, simply turn it on and leave it on for at least an hour. Better yet, give it to a friend and let him or her play around with it overnight.

Documentation for the 3-Channel Color Organ

Typical Test Results and Summary and Recommendations Documents for the Exercise Project are shown in Chapter 16, which also contains the Project Report. Use these documents as a guide in completing your own documentation for the 3-Channel Color Organ.

Something to Think About

Industry troubleshooting flowcharts are readily available. Although they may differ in structure from the type studied in this book, examining them will be instructive. Why not obtain as many as possible, pore over them, and hang a few on your laboratory bulletin board?

SUMMARY

In this chapter we looked at the testing, troubleshooting, and final documentation necessary to complete the prototype project. We examined the four basic testing and troubleshooting steps. We saw why preliminary testing, performed before the project is powered up, can, if done thoroughly, save a lot of grief and work. Operational testing allowed us to determine if the project was functioning correctly. If not, we went on to troubleshooting to see what was wrong and how to fix it. Finally, we saw how performance testing is designed to put a project through its paces.

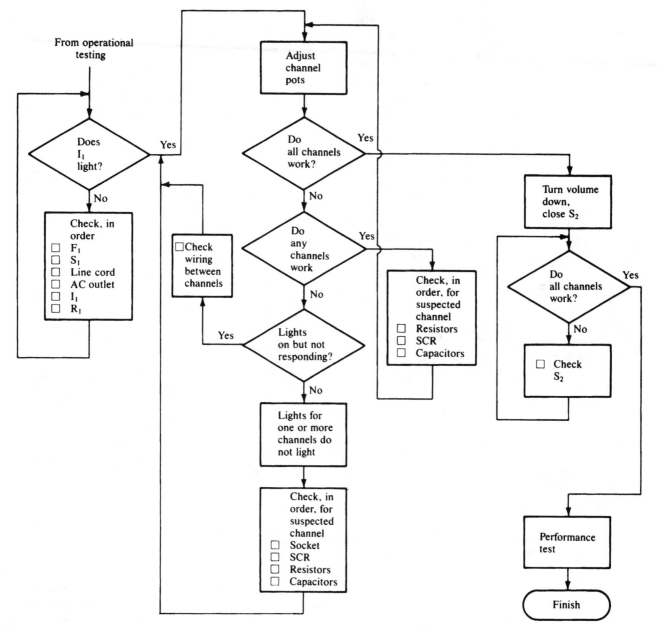

FIGURE 10–10
3-Channel Color Organ troubleshooting flowchart

The two final written documents were introduced and explained: the Test Results Document and the Summary and Recommendations Document. The Test Results Document records the results of the test and troubleshooting procedures. The Summary and Recommendations Document summarizes the important project developments, from the initial design phase through testing and troubleshooting. Finally, all documentation, both written and graphic, was brought together in the Project Report.

Now that your prototype project is complete, it might be nice to take a break and look more closely at a topic touched upon earlier—computer-aided design (CAD). Today, CAD is the way design is being done. It behooves all electronics technicians to know more about computer-based design. Turn the page to Chapter 11 and get the story.

QUESTIONS

1. _____ tests are performed before power is applied to the prototype project.

2. With _____ testing, power is applied to the project for the first time and basic project functioning is determined.

3. _____ is done to determine what is wrong, why it's wrong, and what to do about it.

4. _____ testing is designed to put the prototype project through its paces.

5. Most prototype project errors are a result of a _____ error or a _____ failure.

6. _____ components are much less likely to fail than are _____ components.

7. The four human senses used in troubleshooting are _____, _____, _____, and _____ .

8. Playing the _____ is the best way to approach performance testing.

9. The _____ _____ Document documents, or records, the results of the test and troubleshooting procedures.

10. The _____ and _____ Document summarizes the important project developments.

REVIEW EXERCISES

1. Select various projects or electronic devices and perform preliminary testing on them. Check wiring and component placement against the schematic drawing. This will not only give you good troubleshooting experience but also improve your schematic reading capabilities.

2. Draw the troubleshooting flowchart for any of the projects in Chapter 17.

3. Offer to draw the troubleshooting flowchart for a fellow student's project.

4. List the component failure rate for the prototype project you have been working on.

5. Practice removing old and installing new electronic components from a surplus printed circuit board.

6. Offer to do performance testing on a fellow student's project.

11 Computer-Aided Design (CAD) and the Project Design Process

OBJECTIVES

In this chapter you will learn

- ☐ How CAD schematic drawings are created.
- ☐ About the five elements of a schematic capture program.
- ☐ What netlists are.
- ☐ About the five elements of a PC board design and layout program.
- ☐ How a commercial program is used to create a schematic capture drawing.
- ☐ How a commercial program creates PC board artwork.

When you produced the documentation for your Project Report, we assume you created the written material on a word processor, and you completed the graphic documentation with the aid of CAD software. Today, it is the norm to produce *both* the written and graphic documentation with a computer. CAD programs for creating mechanical drawings, electronic schematics, and PC board design layouts and artwork patterns are, technically and economically, within the reach of personal computer users. Indeed, CAD packages that allow you to produce virtually all the drawings required in your Project Report are currently available for less than $300.

In this chapter, we see how schematic drawings and PC board design layouts and artwork patterns are produced using computer-aided design. Then, we look at a few low-cost, commercial CAD packages that will draw the schematic and lay out the PC board artwork for any project you choose to design.

11.1 MAKING CAD DRAWINGS

Although powerful professional CAD programs such as AutoCAD and P-CAD can produce virtually all the drawings in the 10-drawing set, our primary interest here is discovering how the two most important drawings, schematics and PC board layout designs, are produced; and we want to see how they can be created with low-cost, readily available, and easy-to-use software that does not require the skills of a seasoned drafter. Accordingly, we look first at what it takes to create schematic drawings—in particular, those with schematic capture capability. Then, we will explore the elements of a typical PC board design layout and artwork generation program.

Creating CAD Schematic Drawings

There are two types of CAD programs that let you use the power of the computer to make schematic drawings. The first is simply called a *schematic package;* the second, and more powerful, is known as **schematic capture**. We begin by seeing what differentiates the two, then we will look at the five minimum elements that make up all schematic capture programs, regardless of their cost or sophistication.

Schematic and Schematic Capture

Using a computer to draw schematics substantially reduces the time it takes to produce such drawings compared with doing them manually. Additionally, the drawing elements are of true consistency and high quality (Figure 11–1). Moreover, no drafting skill (though some drafting knowledge) is required.

The real advantage of such programs, however, is in the rework, that is, the ability to revise drawings on screen, rather than on paper. If you've ever done any traditional drafting, you know what an absolute blessing this is. No more dirty, eraser-marked, ink-smudged, and, in some cases, virtually unreadable drawings to pore over, and all the schematic symbols, line work, and component designations are constant and uniform. The days of hand drafting with T-square, triangles, or a mechanical drafting machine are clearly over.

As much a wonder as CAD schematic creation is, however, there is yet one more significant thing that most such programs can do. They can "capture" the data from the created schematic. That data can then be used directly to make printed circuit board layouts. When combined, or merged, with the right PC board layout design program, the two can provide an almost seamless transition from schematic, to PC layout, to PC artwork. That being the case, the only schematic drawing programs worth reviewing are those that contain the schematic capture feature. Accordingly, the following information pertains strictly to schematic capture programs.

The Five Elements of a Schematic Capture Program

A schematic capture program comprises five major elements. Taken in the order necessary to create a complete schematic drawing, they are (1) component placement, (2) line drawings, (3) schematic references, (4) design rule checking, and (5) support documentation. Let's examine each in turn.

All schematic capture programs contain a component library to facilitate *component placement*. Such a library consists of a file containing information describing each component's electrical properties and its representation on the schematic. These libraries have from 350 to over 11,000 devices (Figure 11–2), depending on the package. In addition, you have the power to create your own symbols and add them to the library.

The first step in drawing any schematic is to "pull" from the library the device symbols that form the foundation of your drawing. The selected components are placed on the screen, moved about, rotated, mirrored, and edited as desired (Figure 11–3a). Through program autopanning and zoom features, you move about the drawing at will, "taking in" each component as needed.

Once components are put in place, *lines* are drawn to connect the components together. You usually begin by selecting "New Line" from a menu, then clicking the left button of a mouse at the position where the line is to start. Next, you "rubber-band" (stretch) the line out; then you click again where you want to end or change directions (Figure 11–3b). You continue clicking until you reach the

FIGURE 11–1
High-quality CAD schematic

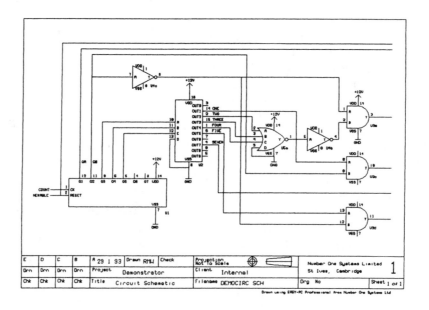

FIGURE 11–2
Schematic symbols library
Courtesy of Number One
Systems, Ltd.

SYMB.SLB

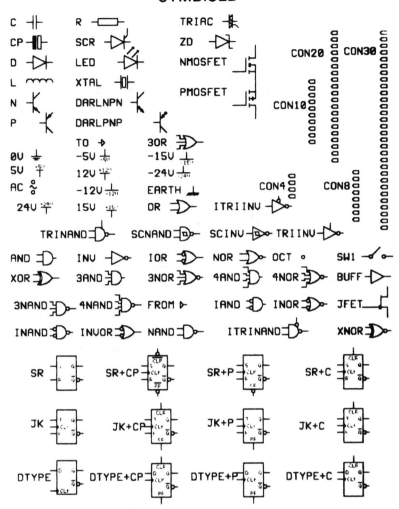

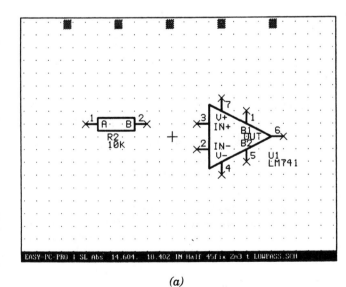

(a)

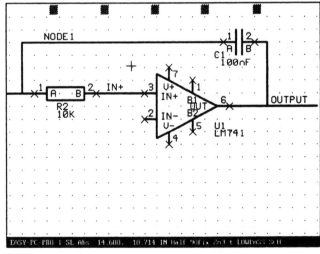

(b)

FIGURE 11–3
Schematic creation
Courtesy of Number One Systems, Ltd.

end of the line. A press of the mouse's right button fixes the line in place.

While components are being placed, or after they are in position and lines are drawn, *schematic reference* nomenclature is brought up. A simple mouse click on pin names and numbers, from a menu, is usually all that is needed to instantly fill the screen with appropriate pin numbers, device IDs, and component values.

In addition to the features just discussed, a schematic capture program will surely have a *design rule checker* routine. It provides, at a minimum, for short circuit, open circuit, and duplicate component description review.

Finally, a flurry of *documentation* is usually available with the better schematic capture programs. Of special importance to designers is the Bill of Materials (BOM) netlist these programs produce. Such a BOM lists all the components used in the circuit and can be employed to create purchase orders.

Creating CAD PC Board Design Layouts and Artwork Patterns

As with CAD-created schematic drawings, there are also five elements to a CAD PC board design package: (1) component placement, (2) trace routing, (3) design rule checking, (4) artwork generation, and (5) documentation. Before we look at each area in detail, however, we need to examine more closely the **netlist** generated by the schematic capture program. It is the netlist, you will recall, that allows us to advance, more or less seamlessly, from schematic to PC board design layout and, eventually, artwork generation.

Netlists

CAD PC board programs let you lay out a printed circuit board on-screen by hand, as you would using decals and transfer tape on a clear acetate base. However, their real power, as we have previously indicated, is their ability to interpret data generated by a schematic capture program and to convert it to components and traces.

The data is contained in files known as netlists. A netlist is simply a grouping of related parameters that are extracted from the schematic. There are BOM netlists, wire netlists, pin netlists, and so on. There is no limit to the number of netlist types, and each schematic capture program has its own repertoire of netlists it can generate. Furthermore, these netlists are in ASCII format, so they can easily be edited by almost any word processing program.

When the appropriate netlists from the schematic capture program are integrated with a compatible PC board design layout program, schematic capture has occurred and we are ready to begin the PC board layout process.

Component Placement

Just as with the schematic capture program, *component placement* is begun by calling up the program's component library (Figure 11–4). As you would expect, the library is nothing more than a collection of device outlines like resistors, capacitors, and ICs. When a component is needed on the printed circuit board, the netlist file calls the specified outline from the component library and makes it available for placement on the board.

Placing components on a printed circuit board can be a challenge. Why? Because components have to be positioned in such a way that electrical connections can be made between any and every component.

Fortunately, there are tools, such as automatic-placement programs, that aid in the layout process. Nonetheless, such programs are rarely able to achieve complete component placement. Much of the layout still needs to be done manually. Manual placement consists of dragging a part from one place to another, using a mouse or keyboard.

To assist you in placing the components, printed circuit board design software provides an interconnectivity pattern on the screen known as a **ratsnest** (Figure 11–5). A ratsnest is a jumble of lines that shows how each component is wired to every other component in the circuit. As a component is moved about on the screen, its attached ratsnest lines follow along with it and, in so doing, expand and contract like rubber bands. After components are permanently in place, their nomenclature can be brought up.

Trace Routing

With the components in place and identified, trace routing, the placement of copper tracks on the PC board, is the next step (Figure 11–6). Most of the work, fortunately, is done with a built-in program known as an **autorouter**. We say "most" because even the best autorouters, known as *Lee routers,* rarely have a completion rate greater than 90%. In the end (actually, toward the end), you will have to resort to some manual routing and editing.

Design Rule Checking

After the PC board is laid out, a design rule check must be made. It is better to find errors here and now than later on the actual PC board.

FIGURE 11–4

PC board component symbols library
Courtesy of Number One Systems, Ltd.

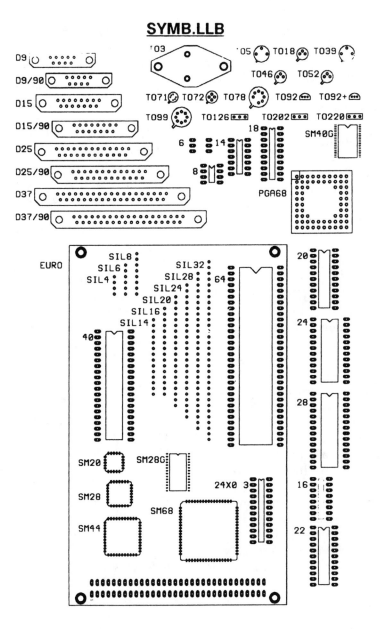

SYMB.LLB

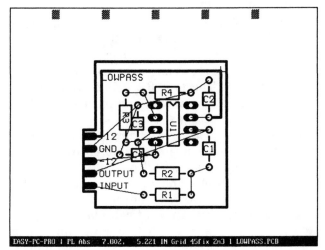

FIGURE 11–5
The ratsnest
Courtesy of Number One Systems, Ltd.

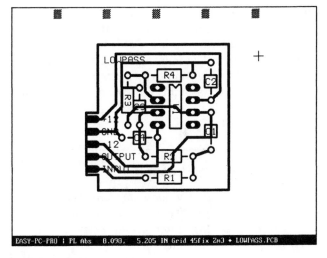

FIGURE 11–6
Trace routing
Courtesy of Number One Systems, Ltd.

The verification process checks to see that pads, vias (plated-through holes), and tracks have been placed according to a set of rules you have established. This is what is referred to as *design rule checking*.

To begin the verification, software first loads the wire netlist, then checks to see if all nodes in each net are connected. If they are not, an error message is sent. Finally, a check is made of the pads, vias, and traces. Pads and traces that are too close together or touching are marked for display. Error messages are also sent.

Artwork Generation

Of course, the ultimate goal of any PC board design layout program is to produce artwork that can be used to make an actual PC board. Toward that end, each layer of a multilayer board is assigned its own separate trace pattern. The various layers are aligned using placement holes drawn on each layer at the beginning of the PC board design process. Other artwork often produced in support of the trace layout includes solder masks and silk-screen masks for board nomenclature.

Any of the final artwork patterns can be printed on an ink-jet or laser printer, pen plotter, or photoplotter. The first two do not always produce a professional-quality output, whereas the plotter always does. Unfortunately, the photoplotter is quite expensive ($6000 and up). Hence, a laser printer (which can be bought for under $400) is probably the best choice for the occasional user of PC board design software.

Documentation

As with CAD schematic software, printed circuit board design programs produce an abundance of support documentation. Most of it is used for manufacturing. Such documentation includes schematics, PC board fabrication drawings, parts lists, and parts placement (assembly) drawings, among other things.

Something to Think About

Using your industry contacts, why not see if you can visit a facility where PC board design is carried out? If granted a tour, prepare a list of questions in advance. When visiting, ask if you can "play" with the software a bit, to get a feel for the real thing.

11.2 LOGIC CREATOR

To better understand how schematic generation and PC board layout are accomplished, we turn to a software package that does both. **Logic Creator** is a low-cost product available from Advanced Microcomputer Systems, Inc. While we do not endorse Logic Creator, per se, we find it accomplishes circuit design and PC board layout quite well. However, there are other excellent products that do likewise.

Here, we examine how a schematic is drawn and its information "captured" using Logic Creator. Then we explore how captured information is used to create PC board artwork.

Keep in mind, however, that what follows is not a complete lesson, page-for-page, screen-for-screen, on how to accomplish the above. The idea is to show how schematic capture and PC layout are done with an actual program.

Schematic Capture Using Logic Creator

Here we create the schematic drawing for a simple **Detector Circuit**. Once produced, we ask Logic Creator to check the schematic for wiring errors. If there are any, Logic Creator will point them out so they can be fixed. We also produce reports: a bill of material, a printed circuit board update file, and a netlist file.

As described earlier, a schematic consists of components, labeled with reference designations and values, wired together. The drawing also includes text, usually in the form of notes. Let's see how Logic Creator performs each task.

Component Placement

The Detector Circuit sketch is shown in Figure 11–7. This is the way schematics normally start out: often on the proverbial napkin or the back of an envelope.

To begin, we select a component from the Logic Creator library. The first IC, a 7420 Dual, Quad-Input NAND gate, is chosen and moved into place. The component actually consists of one part with two identical sections (see Figure 11–8). When the component is placed, its unique reference designator (U_1) and pin numbers are filled.

Next, we select and move the flip-flops (U_2) into place. Then, we add the 330-ohm resistors. Notice that they are automatically labeled R_1 and R_2. Note, also, that the resistors have numbers on their leads like ICs on their pins. These numbers identify the resistors' two leads. Later, Logic Creator uses this information to create a list, or database, for all the component leads and how they

FIGURE 11–7
Detector Circuit sketch

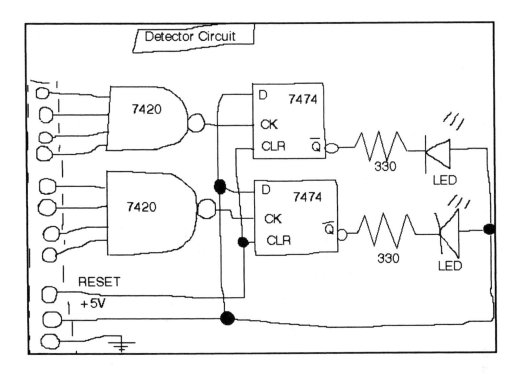

FIGURE 11–8
Component placement
Courtesy Advanced
Microcomputer Systems,
Inc. Circuit Creator

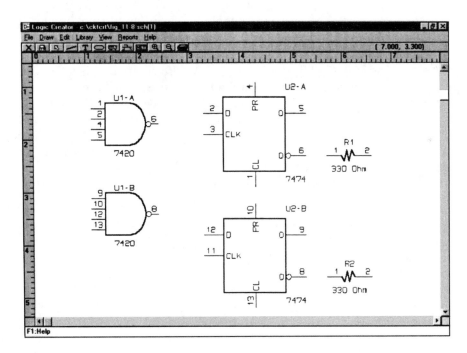

are interconnected. This list is called a netlist. Remember, Logic Creator is not simply a schematic drawing program but a schematic capture package. We must capture component information to be used later.

Finally, we move two LEDs and a connector into place (see Figure 11–9). In doing so, note how easily an LED and its respective resistor are aligned. Leads to both components snap to a "hidden" grid.

This is surely a lot easier than drawing components by hand, one at a time, every time.

Wiring the Components Together

Wiring, as with any schematic drawing program, is straightforward. You select a wire command from the menu, move the cursor to a component lead, and click the left mouse button to begin the wire line, then you drag the line to its next component destination and

FIGURE 11–9
Wiring the components
together
Courtesy Advanced Micro-
computer Systems, Inc.
Circuit Creator

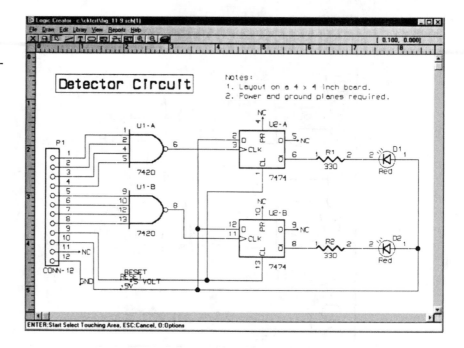

click the left mouse button again. Clicking the right mouse button stops the command. It's that simple (see Figure 11–9).

Component References and Values
Within limits, you can change the components' reference designations and values. However, keep in mind, to capture a schematic and use the acquired information to create a PC board layout, the circuit must "know" a component's characteristics. For example, it makes a big difference when completing the PC board layout if a capacitor, say C_1, is a 0.001

µF or a 1000 µF. Obviously a footprint for the former is different from one for the latter.

Adding Text
To create text is a simple matter. You first select the DRAW/TEXT command from the menu, then you place the cursor at the location where you wish to begin the text string. You now type in text.

The completed drawing is shown in Figure 11–10. It consists of components, with reference designations and values, connecting wires, and appropriate text.

FIGURE 11–10
Complete schematic
Courtesy Advanced Micro-
computer Systems, Inc.
Circuit Creator

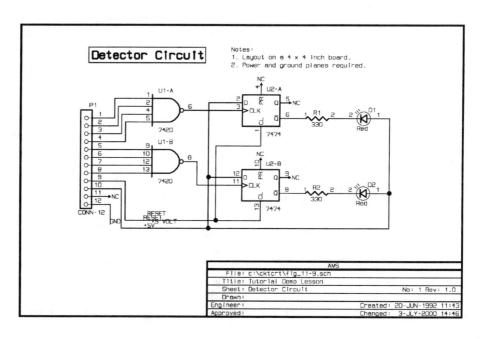

Reports

In addition to the schematic, you can create different reports: a bill of material (BOM), netlist (NET), update list (UPD), and pinlist (PIN). You just choose the FORMAL/EXPORT command from the REPORTS menu. All reports will be created in the same directory from which your schematic was loaded. You then print out a report whenever you like.

Creating PC Board Artwork Using Logic Creator

Now, let's see how Logic Creator takes the schematic just completed and uses it to create a printed circuit board layout. First, we look at various PC board masks and drawings; next, at component placement; third, at actual board routing; and, last, at final checkout.

Masks and Drawings

Several *masks* and *drawings* are used in printed circuit board construction. Let's summarize each type:

Conductor Mask. A conductor mask represents the electrical networks on a single layer of a PC board. There is a different conductor mask for each electrical layer (see Figure 11–11).

Solder Mask. A solder mask paints the solder side of a board. By painting a board in this manner, only the nonpainted areas attract solder during the wave-soldering construction step (see Figure 11–12).

Paste Mask. A paste mask aids in the assembly of PC boards containing surface mount devices. The mask allows only the component pads on a board having no holes to be applied with a solder paste prior to the soldering of the SMD components to the board (see Figure 11–13).

Silk-Screen Mask. A silk-screen mask paints component outlines, part references, part numbers, and other text onto the top or bottom side of a PC board (see Figure 11–14).

Drill Drawing. A drill drawing indicates the location and size of all holes to be drilled in a board (see Figure 11–15).

Assembly Drawing. An assembly drawing shows board outlines, component locations, and other mechanical information (see Figure 11–16).

Power/Ground Plane Mask. A power/ground plane mask makes a conducting layer of a board that is

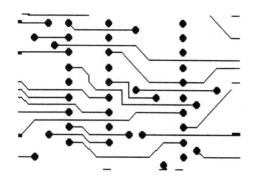

FIGURE 11–11
Conductor mask
Courtesy Advanced Microcomputer Systems, Inc.
Circuit Creator

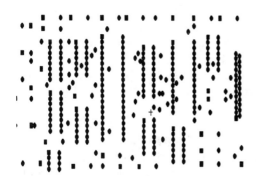

FIGURE 11–12
Solder mask
Courtesy Advanced Microcomputer Systems, Inc.
Circuit Creator

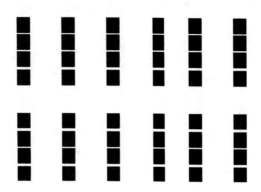

FIGURE 11–13
Paste mask
Courtesy Advanced Microcomputer Systems, Inc.
Circuit Creator

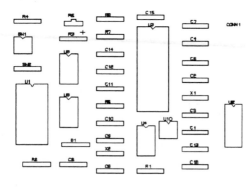

FIGURE 11–14
Silk-screen mask
Courtesy Advanced Microcomputer Systems, Inc.
Circuit Creator

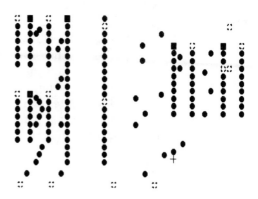

FIGURE 11–17
Power/ground plane mask
Courtesy Advanced Microcomputer Systems, Inc.
Circuit Creator

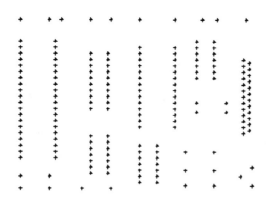

FIGURE 11–15
Drill drawing
Courtesy Advanced Microcomputer Systems, Inc.
Circuit Creator

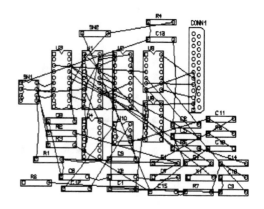

FIGURE 11–18
Schedule drawing
Courtesy Advanced Microcomputer Systems, Inc.
Circuit Creator

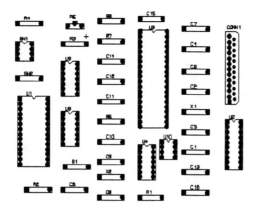

FIGURE 11–16
Assembly drawing
Courtesy Advanced Microcomputer Systems, Inc.
Circuit Creator

dedicated to a single net, usually one of the power or ground signal levels (see Figure 11–17).

Schedule Drawing. A schedule drawing is not required in the building of a PC board; however, it may prove useful during board layout. It shows all connections not yet completed (see Figure 11–18).

Component Placement

The first step in creating a PC board design and layout is to capture, or read, the component and net assignment information from the update-list file produced during the schematic drawing process. The result is a board populated with all the components (see Figure 11–19).

FIGURE 11–19
Component placement
Courtesy Advanced
Microcomputer Systems,
Inc. Circuit Creator

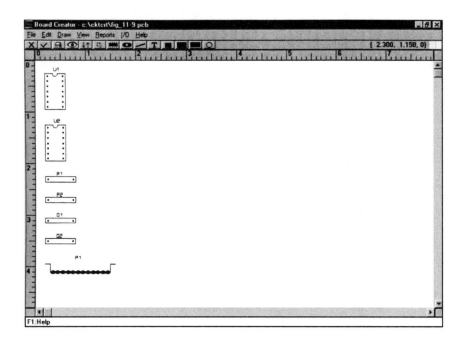

Next, we generate a routing schedule. The resulting *ratsnest* lines show how the components are connected together (see Figure 11–20).

It's time to place components in their most advantageous location for routing. As you would expect, doing this involves clicking and dragging with a mouse. In Figure 11–21 we see the layout after U_2 has been moved. Notice that all the ratsnest lines, with their pin-to-pin connections, have moved with the component.

In Figure 11–22 we have the board layout after all components are placed in a reasonable location for routing the electrical connections.

Setting the board size and number of layers requires completing a pop-up Format Update menu

FIGURE 11–20
Ratsnest
Courtesy Advanced
Microcomputer Systems,
Inc. Circuit Creator

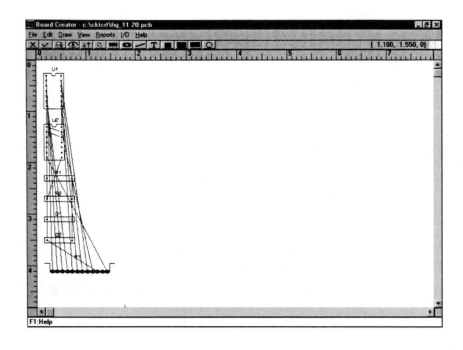

FIGURE 11–21
Moving a component
Courtesy Advanced
Microcomputer Systems,
Inc. Circuit Creator

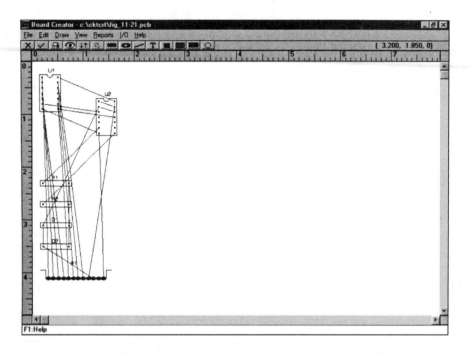

FIGURE 11–22
Components placed in
reasonable location for
routing
Courtesy Advanced
Microcomputer Systems,
Inc. Circuit Creator

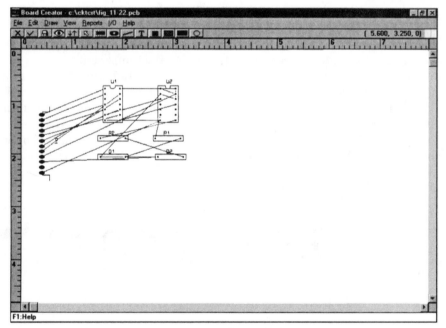

FIGURE 11–23
Format Update menu
Courtesy Advanced
Microcomputer Systems,
Inc. Circuit Creator

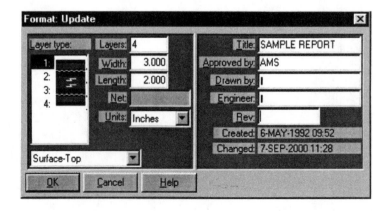

FIGURE 11–24
After two routes
Courtesy Advanced
Microcomputer Systems,
Inc. Circuit Creator

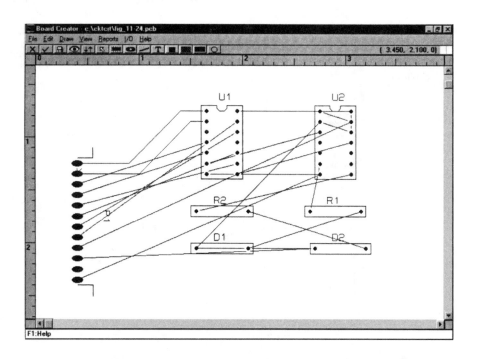

FIGURE 11–25
After nine routes
Courtesy Advanced
Microcomputer Systems,
Inc. Circuit Creator

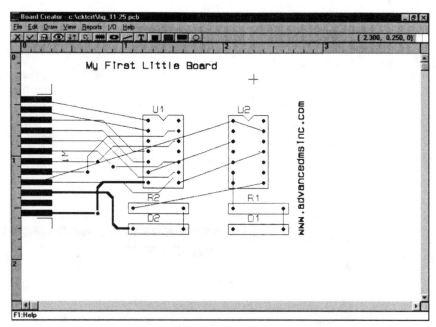

(see Figure 11–23). We create a 3.000″ × 2.000″ board with four layers. The layer types are as follows:

1. Top conductor.
2. Top power layer, "GND."
3. Bottom power layer, "+5V."
4. Bottom conductor.

Routing the Board

To begin routing, we execute the LINE command in the DRAW menu. We will use the component side for mostly horizontal routes and the solder side for mostly vertical routes. We can see from the routing schedule, the ratsnest, which pads need to be connected together.

Routing involves highlighting a ratsnest line and then drawing a route from one pad to another while toggling from component-side to solder-side layers when necessary. Figure 11–24 shows our board after two routes.

In Figure 11–25 we see the board after nine routes.

In Figure 11–26, all routes are shown. This completes the board routing.

FIGURE 11–26
After all routes
Courtesy Advanced
Microcomputer Systems,
Inc. Circuit Creator

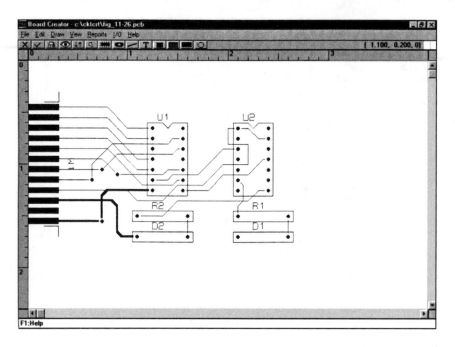

Adding text to the board is straightforward. Logic Creator allows us to place text at any location, horizontally or vertically, in a variety of fonts, and in a "mirroring" format. The latter makes text readable when viewed from the back side of the board (see Figure 11–27).

In Figure 11–28 all the masks and drawings for our PC board are shown. A netlist, parts list, and summary report are in Figure 11–29. The printed circuit board design and layout is complete.

Something to Think About

Now might be a time to gather literature and, perhaps, demonstration packages for a variety of schematic capture and PC board design and layout programs. Where should you begin your search? The Internet, of course.

FIGURE 11–27
Adding text
Courtesy Advanced
Microcomputer Systems,
Inc. Circuit Creator

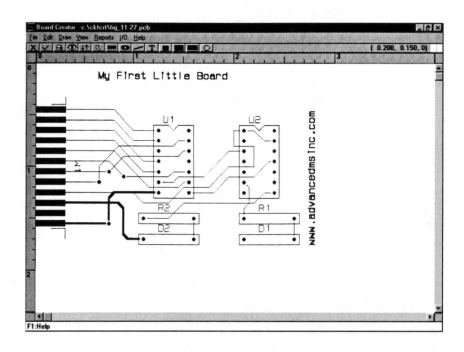

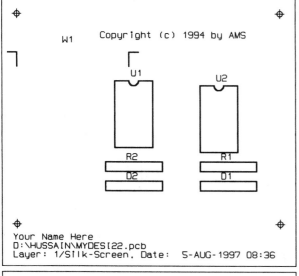

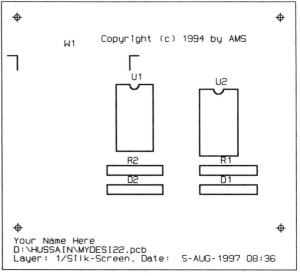

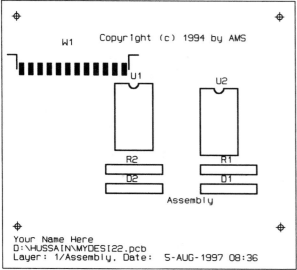

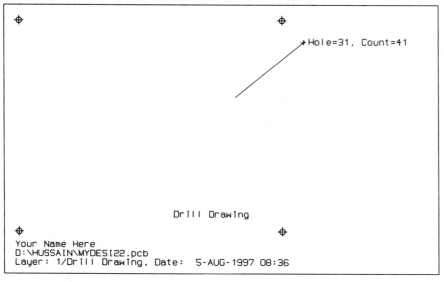

FIGURE 11–28 [a]
Masks
Courtesy Advanced Microcomputer Systems, Inc. Circuit Creator

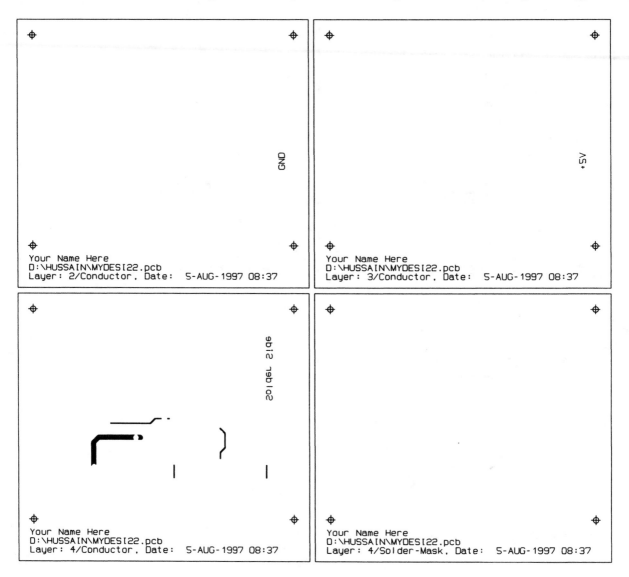

FIGURE 11–28 [b]
Masks

```
            Approved by: AMS
               D awn by: SHAUN DESOUZA
              Engineer: SHAUN DESOUZA
               Program: EZBOARD
                  Date:  1-AUG-1997 16:05
               Created: ...
               Changed:  1-AUG-1997 16:05


            NET_LIST
              NET        3, NC
                PAD (1.200,0.850,0) U1:3
                PAD (1.500,0.950,0) U1:11
                PAD (2.100,1.000,0) U2:4
                 AD (2.100,0.900,0) U2:3
                 AD (2.400,1.100,0) U2:10
                 AD (2.100,0.800,0) U2:2
                 AD (2.100,1.100,0) U2:5
                PAD (2.400,1.200,0) U2:9
                PAD (2.100,1.200,0) U2:6
                PAD (2 100,0.700,0) U2:1
                PAD (1.150,0.450,1) W1:11
                COMMEN1 Net Length= 1.150
              EN _NET
              NE        , EZL#12
                1 \D (1.  )0,0.950,0) U1:4
                P D (0.  50,0.450,1) W1:3
                C  MMENT  Net Length= 10.750
              END_ ET
              NET       1 , EZL#15
                PAL (1.  )0,1.050,0) U1:10
                PAD (0.  50,0.450,1) W1:6
                COMM  NT  Net Length= 0.000
              END_NE
              NET       11, EZL#16
                PAD (1.500,0.850,0) U1:12
                PAD (0.750,0.450,1) W1:7
                COMMENT Net Length= 0.000
              END_NET
              NET       12, EZL#17
                PAD (1.500,0.750,0) U1:13
                PAD (0.850,0.450,1) W1:8
                COMMENT Net Length= 0.000
              END_NET
              NET        5, EZL#13
                PAD (1.200,1.050,0) U1:5
                PAD (0.450,0.450,1) W1:4
                COMMENT Net Length= 0.000
              END_NET
              NET        9, EZL#14
                PAD (1.500,1.150,0) U1:9
                PAD (0.550,0.450,1) W1:5
                COMMENT Net Length= 0.000
```

FIGURE 11-29 [a]
Netlist, parts list, and summary report
Courtesy Advanced Microcomputer Systems, Inc. Circuit Creator

```
                              Your Name Here
-- Net List Report ------------------------------------------ Page: 2 --

 END_NET
 NET      8, EZL#2
   PAD (1.500,1.250,0) U1:8
   COMMENT Net Length= 0.000
 END_NET
 NET      7, GND
   PAD (1.200,1.250,0) U1:7
   PAD (2.100,1.300,0) U2:7
   PAD (1.250,0.450,1) W1:12
   COMMENT Net Length= 1.400
 END_NET
 NET      6, EZL#1
   PAD (1.200,1.150,0) U1:6
   PAD (2.400,1.000,0) U2:11
   COMMENT Net Length= 0.000
 END_NET
 NET      1, EZL#10
   PAD (1.200,0.650,0) U1:1
   PAD (0.150,0.450,1) W1:1
   COMMENT Net Length= 0.000
 END_NET
 NET     13, +5V
   PAD (1.500,0.650,0) U1:14
   PAD (2.400,0.900,0) U2:12
   PAD (2.400,0.700,0) U2:14
   PAD (1.050,0.450,1) W1:10
   COMMENT Net Length= 0.000
 END_NET
 NET      2, EZL#11
   PAD (1.200,0.750,0) U1:2
   PAD (0.250,0.450,1) W1:2
   COMMENT Net Length= 2.600
 END_NET
 NET     15, RESET
   PAD (2.400,0.800,0) U2:13
   PAD (0.950,0.450,1) W1:9
   COMMENT Net Length= 0.000
 END_NET
 NET     14, EZL#3
   PAD (2.400,1.300,0) U2:8
   PAD (2.100,1.500,0) R1:1
   COMMENT Net Length= 0.450
 END_NET
 NET     16, EZL#4
   PAD (2.600,1.500,0) R1:2
   PAD (2.100,1.700,0) D1:1
   COMMENT Net Length= 0.200
 END_NET
 NET     18, EZL#6
   PAD (1.600,1.500,0) R2:2
   PAD (1.600,1.700,0) D2:2
   COMMENT Net Length= 0.200
 END_NET
 NET     17, EZL#5
   PAD (1.100,1.500,0) R2:1
   COMMENT Net Length= 1.550
```

FIGURE 11–29 [b]
Netlist, parts list, and summary report

```
                                        Your Name Here
    -- Net List Report ------------------------------------------ Page: 3 --

      END_NET
      NET      19, EZL#7
        PAD (2.600,1.700,0) D1:2
        COMMENT Net Length= 0.000
      END_NET
      NET      20, EZL#9
        PAD (1.100,1.700,0) D2:1
        COMMENT Net Length= 0.000
      END_NET
      COMMENT Average Length= 0.915
      COMMENT Shortest Net= 0.000
      COMMENT Longest Net= 10.750
      COMMENT Total Length = 18.300
    END_NET_LIST

    End of Report
```

FIGURE 11–29 [c]
Netlist, parts list, and summary report

```
                              Your Name Here
-- Board Summary Report ------------------------------------- Page: 1

              File: D:\HUSSAIN\MYDESI22.pcb
             Title: SAMPLE REPORT
          Revision: AUG-1-19
       Approved by: AMS
          Drawn by: SHAUN DESOUZA
          Engineer: SHAUN DESOUZA
           Program: EZBOARD
              Date:  5-AUG-1997 08:16
           Created: ...
           Changed:  5-AUG-1997 08:16

     Board size: 3000 x 2000

     ____Layer-Type_____
      1: Surface
      2: Power   GND
      3: Power   +5V
      4: Surface

     ____Object_____Count__
      1: Component          7
      2: Text               7
      3: NameNet           20
      4: NamePower          2
      5: Pad               48
      6: Via                5
      7: Line             112

     ____Hole-Size__Count__
      1:       31     41

     Pad definition: 1, Count: 5
      Hole diameter: 31, X offset: 0, Y offset: 0
     Component side: Rounded, 50 x 50
        Solder side: Rounded, 50 x 50
        Inner layer: Rounded, 50 x 50

     Pad definition: 3, Count: 30
      Hole diameter: 31, X offset: 0, Y offset: 0
     Component side: Rounded, 62 x 62
        Solder side: Rounded, 62 x 62
        Inner layer: Rounded, 62 x 62

     Pad definition: 7, Count: 6
      Hole diameter: 31, X offset: 0, Y offset: 0
     Component side: Rectangle, 62 x 62
        Solder side: Rectangle, 62 x 62
        Inner layer: Rectangle, 62 x 62

     Pad definition: 16, Count: 12
      Hole diameter: 0, X offset: 0, Y offset: 0
     Component side: Rectangle, 50 x 125
        Solder side: Rectangle, 50 x 125
        Inner layer: Rectangle, 50 x 125
```

FIGURE 11–29 [d]
Netlist, parts list, and summary report

```
                              Your Name Here
-- Board Summary Report ------------------------------------- Page: 2 --

     ____Line-Width__Count____Length__
      1:        10    103     42750
      2:        50      9      2650

     End of Report
```

FIGURE 11–29 [e]
Netlist, parts list, and summary report

SUMMARY

In this chapter we delved into the CAD drawing process by exploring the five elements of typical schematic capture software. We did the same for a typical PC board design layout and artwork generation program. We then looked at a low-cost, effective, easy-to-use commercial CAD package that does both schematic capture and PC board design and layout.

So far, for project design and fabrication we have assumed the use of insertion mount technology (IMT), in which component leads are inserted in holes in a PC board. However, that is simply no longer the way many circuits are assembled. Today, surface mount technology (SMT) has revolutionized product assembly. To find out more about this critical packaging technology, turn to Part II, Project Design and Fabrication Using Surface Mount Technology.

QUESTIONS

1. Schematic _____ programs "capture" data, then use it to make printed circuit board layouts.

2. List the five elements of a typical schematic capture program.

   1.
   2.
   3.
   4.
   5.

3. Data for schematic capture is held in a _____ file.

4. List the five elements of a typical PC board design layout and artwork generation program.

   1.
   2.
   3.
   4.
   5.

5. To assist in placing the components, printed circuit board design software provides an interconnectivity pattern on the screen known as a _____.

6. As with all schematic capture programs, Logic Creator contains a huge _____ of components.

7. To add text to a drawing created using Logic Creator, you simply place the cursor at the location where you wish to begin the text _____ and begin typing.

8. Four reports that can be created using Logic Creator are (1) _____
_____ _____ (BOM), (2) _____ (NET), _____
_____ (UPD), and _____ (PIN).

9. List five masks Logic Creator can create.

   1.

   2.

   3.

   4.

   5.

10. List the four PC board layers used by Logic Creator in creating the Detector Circuit.

   1.

   2.

   3.

   4.

PART II

Project Design and Fabrication Using Surface Mount Technology

12 Introduction to Surface Mount Technology

OBJECTIVES

In this chapter you will learn

- ☐ What the SMT packaging revolution is all about.
- ☐ About the history of packaging technologies.
- ☐ About the many advantages of surface mount technology.
- ☐ About discrete passive and active surface mount components.
- ☐ About surface mount integrated circuits.
- ☐ About automatic SMT assembly using reflow soldering.
- ☐ About automatic SMT assembly using flow soldering.
- ☐ About an emerging career path, that of SMT technologist.

"They" have been around for sometime now, having exploded onto the scene in the late 1980s. Break into any electronic device, particularly a consumer electronics product, and you see them everywhere—that is, if your eyesight is keen enough. They are, to be sure, extremely tiny, which is why, in part, they have permeated, penetrated, pervaded, infused, and saturated today's electronic circuits. They are found not only on one side, but in many cases on both sides, of a printed circuit board. "They," of course, are **surface mount devices**, **SMDs**, and they represent a component packaging revolution. As a result, both your job—prototype electronics technician—*and* the electronics industry have changed forever.

Surface mount devices are so small and weigh so little, a teaspoon of assorted components would contain hundreds. If you had a healthy sneeze a few inches from the spoon, you would probably lose half your supply forever—blown about to mix and get lost among the tidbits and scraps on your work bench. The components have no leads in the traditional sense, that is, wires intended to be inserted into holes drilled in a printed circuit board. Instead, what leads, or terminals, there are do not go *through* the PC board but rather are soldered to pads on *top* of the board—on its surface. Thus the name, **surface mount technology (SMT)**. It's a technology, as we've indicated, that is sweeping electronics and resulting in lower-cost, ever-more-reliable products. It's an exciting technology that every electronics technician, particularly one working with prototype project design and fabrication, must know more about. Part II of this book will start you on this exciting learning journey.

In this chapter we present an introduction to surface mount technology. If you have time to read but one chapter on the subject, this is the one. In it we explore the technology in general, examine surface mount components in detail, and look closely at industrial SMT assembly practices. In sum, it's an overview of the subject.

Chapter 13, however, takes you one significant step further, providing not only SMT knowledge but an SMT experience. It's all about project design and fabrication using surface mount components.

Chapter 14, on SMT rework (removal and replacement of surface mount devices), covers the technology as most electronics technicians will encounter it—in commercial and industrial electronic products. Here we'll see how to work with the new technology in the "real world."

12.1   SURFACE MOUNT TECHNOLOGY—A PACKAGING REVOLUTION

Surface mount technology is a packaging revolution making a powerful impact on the electronics industry in the new millennium. By the mid-1990s, 70% to 80% of all electronic assemblies were already surface mount. Today, the percentage is even higher. To examine this phenomenon further, we begin by defining the technology, looking at how it evolved, and exploring its many advantages while touching on a few limitations.

Surface Mount Technology Defined

Let's begin by seeing just what SMT is and what remaining role there will be, if any, for insertion mount technology (IMT) as the century unfolds.

SMT: What Is It?

Surface mount technology uses very small, "leadless" components, often one-quarter to one-tenth the size of standard components, that are soldered to the *surface* of a specially designed PC board. Both discrete components (resistors, capacitors, diodes, transistors, etc.) and integrated circuits come in surface mount packages (Figure 12–1).

FIGURE 12–1
Surface mount components

The new technology came about primarily in response to the density limits imposed on current IC packaging. The typical DIP (dual-in-line package) IC has lead spacing of 0.100″. That has worked well enough for most small- to medium-scale integration, but today's VLSI (very-large-scale integration) chips need more pins to connect to the outside world. When you're talking 96 to 244 pins, the DIP just can't handle it. The chip would be more than half a foot long!

With a surface mounted IC, the lead spacing is only 0.050″ (and, in some cases, 0.025″ and even 0.020″), half that of the DIP IC. That factor alone results in a 30 to 60% size reduction. Furthermore, many surface mount ICs have pins on all four sides, giving even more chip connections in a given space. One of the first commercial products to make extensive use of surface mount components, particularly ICs, was the IBM 286 AT personal computer, introduced in the mid-1980s. Its motherboard was almost completely populated with surface mount devices.

IMT: What's Left for It?

Does the rise of SMT mean the fall, and possible demise, of IMT, insertion mount technology? A reduction maybe, but certainly not oblivion—IMT will be around for some time to come. For one thing, there are component packages that, because of their need to dissipate large amounts of heat, will probably never be seen in surface mount configurations. Large power transistors, heavy-duty SCRs and triacs, and many relays are examples that come quickly to mind. Nonetheless, IMT components are getting harder to find. SMT is simply more profitable for electronics manufacturers. For many, there just isn't enough business in IMT components anymore to justify continual manufacturing.

What you are likely to see, with regard to IMT, is what is already taking place—a mixture of both technologies on a single printed circuit board. For instance, Micree Corporation of Torrance, California, is an electronic contract manufacturing firm that assembles printed circuit boards for outside vendors. Micree is equipped to handle both IMT and SMT assemblies; however, it has yet to assemble a purely SMT board of moderate-to-large size. All such boards have at least a few IMT components. It looks like SMT and IMT are destined to coexist for a few more years.

Development of Packaging Technologies

How did this surface mount packaging revolution come about? What were its antecedents? Let's find out by briefly examining the development of packaging technologies, starting with the infamous vacuum tube and its related hand-wired chassis.

Vacuum Tubes and Hand-Wired Chassis

The vacuum tube, the first control component, ushered in the electronics age at the turn of the 20th century. For over 50 years, this relatively large, expensive, power-hungry, and breakable component remained the heart of all radios, televisions, industrial controls, and even the first computers. Its biggest drawback was its limited life. Like a light bulb, it burned out and had to be replaced on a regular basis. All vacuum tubes were installed in sockets that were in turn wired to other components (resistors, capacitors, etc.) in the circuit. All this wiring, of course, was done by hand.

We had a glimpse of what that hand wiring meant when we examined the cut-slash-and-hook (CSH) method of breadboarding in Chapter 6. If you want to see the results of this approach on a production basis, however, try to get hold of an old 1930s, '40s, or '50s radio and peek under the metal chassis. It's amazing that so many got built. Hand wiring on a metal chassis, using terminal strips, is a laborious, time-consuming task, yet even today, some form of hand wiring is required to connect off-board components to a printed circuit board (see Chapter 9).

Transistors and Printed Circuit Boards

The commercial advent of the transistor in the late 1950s, with its small size, low power consumption, and, most importantly, high reliability (it didn't burn out and thus did not have to be replaced) fundamentally changed the way we built electronic circuits. This solid-state component, tiny and with leads like any other component, needed no socket—it could be installed just like a resistor or a capacitor. What a perfect match for a parallel emerging technology, the printed circuit board.

The marriage of transistor and PC board in the early 1960s meant rapid circuit production at much lower cost. Component leads were first inserted into holes drilled in the printed circuit board. The board was then turned over and the individual leads soldered by hand using wire solder and a soldering iron. The ubiquitous five-transistor radio of the 1960s best reflects this midcentury packaging technology.

DIP Integrated Circuits and Automatic Insertion

The dual-in-line package (DIP) IC, developed in the late 1960s and early 1970s, was also fully compatible with PC board technology. In addition to the packaging of whole circuits on a silicon chip, there also occurred at this time two important advancements in assembly technology. First, **automatic insertion equipment** found its way into the larger circuit assembly houses. The machinery, which was and still is quite complex and expensive, can automatically pick up an IC from a magazine or tube and insert the pins precisely into predrilled holes in a PC board. Discrete components can be "handled" in much the same way.

Second, the *flow solder machine* appeared. With it, a stuffed PC board is passed over a bath of molten solder. Presto! All the leads and pins are soldered to the printed circuit board in one quick, automatic operation. Except for an occasional touch-up and the attachment of special components (certain connectors, for example), hand soldering is all but eliminated.

The packaging and assembly technology just described—automatic insertion of IMT components, soldered, for the most part, by machine, not by hand—is essentially how circuit assembly has taken place in the United States over the past three decades. Surface mount technology, however, has changed all that.

Surface Mount Technology: The Next Step Has Arrived

As we have indicated, surface mount technology is the new packaging paradigm. We'll see how the new component packages are attached to printed circuit boards a little later in the chapter. It is worth noting here, however, that the technology itself is not new, having emerged from *hybrid circuits* as far back as the 1950s. Actual surface mount ICs, known as *flatpacks*, saw their debut in military equipment in the 1960s. In the following decade, other kinds of surface mount devices, LEDs for example, were also developed. Meanwhile, advances in Europe (particularly Switzerland) and Japan were forcing the technology along at a rapid pace. By 1985, half of all electronic components used in Japan were SMCs. This Japanese success proved a stimulus for U.S. industries, and now surface mount technology has spread swiftly, not just in the United States, but throughout the world.

Advantages and Limitations of Surface Mount Technology

Obviously, if SMT is permeating every facet of the electronics industry, it must possess significant advantages over IMT. Here we'll explore a half-dozen of the more important ones. We must also, in all fairness, look at limitations on implementing this latest technology. The latter, we will discover, are due more to human than to technological restrictions.

Why SMT?

Following are six reasons why surface mount technology has changed the face of electronic circuitry.

1. Size Reduction. SMT components are, on average, one-quarter to one-eighth the size of their bigger IMT cousins. When you consider the much lower profile of SMT components, however, the size advantage is even more noticeable (Figure 12–2). Furthermore, when surface mount components are placed on printed circuit boards, we see major overall circuit size reduction. Why? First, the components can be placed on both sides of the PC board. Second, circuit traces are thinner for SMT circuits. Third, since lead and pin holes are eliminated, the space formerly taken up by holes and their pads is now available for more components and traces. In total, PC board real estate savings with SMT average 50%. As a result, you can put more circuitry in the same space or the same circuitry in less space. Either way you win.

2. Weight Reduction. The weight of SMT components is a mere one-fifth that of traditional electronic components. As a consequence, circuit boards are much lighter and, thus, with lower component mass, less subject to the negative effects of vibration. Often, the resulting "credit card–size" boards are so light they can be glued to surfaces rather than attached with traditional fasteners. The weight and size advantages evident with SMT can combine to create still more benefits, as we shall see in a moment.

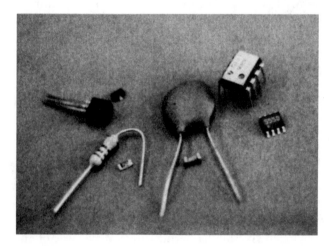

FIGURE 12–2
SMT and IMT components compared

3. Improved Performance and Reliability. Because SMCs are so small and essentially leadless, we find they exhibit better propagation speeds, greater noise immunity, less cross talk, and fewer solder joint failures.

4. Improved Manufacturing of Circuit Boards. Surface mount components were designed for automatic insertion from the start. It was never intended that SMT circuits would be mass-produced by hand. Fully automated assembly processes mean less manufacturing floor space, less inventory, and circuits that can be assembled on customer demand in a few hours, not weeks or months.

5. Cost Reductions. The SMT advantages discussed so far can often lead to major cost reductions. There is less PC board to buy, there are no drilling costs, and fewer expensive layers in multi-layer boards are necessary. Furthermore, a mini–circuit board can mean a smaller housing or cabinet, which reduces costs further. If everything is smaller and lighter as a result of SMT, manufacturers pay less for handling and shipping. The entire assembly plant can be smaller if for no other reason than that less inventory space is demanded. Hence, leasing costs are lower. Actually, these cost savings can become a happy never-ending story. More-reliable SMCs require fewer inspectors, quicker production means faster accounts payable, and so on. I am sure you can continue the list more or less indefinitely.

6. New Products. When transistors replaced vacuum tubes, a whole range of electronic products emerged that were not possible or practical to produce with vacuum tube technology. The same thing happened as integrated circuits superseded transistors. Although SMT does not represent a *functional* advance over transistors or DIP IC technology, it does signify a major *packaging* improvement. As a consequence, many new products, such as the MicroWand II in Figure 12–3, owe their very existence to SMT. Believe me, if it's small enough to put in your wallet, along with your credit cards, it is probably SMT-dependent. Watch for such items in the months and years to come.

SMT: Some Limitations

Generally, the limitations on implementation of surface mount technology fall into two categories: (1) those having to do with the technology and (2) those involving the attitudes of individuals responsible for putting SMT into effect.

Concerning the first group, there are some legitimate apprehensions. To begin with, there is the

FIGURE 12–3
An SMT product
Courtesy Hand Held
Products, Inc.

cost of surface mount components. During the late 1980s, SMCs were 20 to 30% higher in price than equivalent IMT components. Today, however, pricing, at least in industrial quantities, is at parity with traditional components. In fact, many SMCs are today less costly than their insertion mount equivalents.

Component *availability* is, or was, another concern. Although virtually every major component manufacturer offers an SMC line, availability and delivery times are not what most users in the industry would like, but the situation is moving in the right direction. Every month more distributors, with improved service, come on board with a line of surface mount components.

Standards, or the supposed lack of them, is another issue that is still mentioned in the industry and one that still causes some to hold back from entering the world of SMT. In the beginning, a lack of agreement on component case styles and sizes did generate considerable havoc throughout the electronics industry. To be sure, work is still necessary to merge European, Japanese, and American standards, but, as with cost and availability, the industry is making strong headway. This issue should become a nonissue in the very near future.

On a more technical note, it is true that the *design* of SMT printed circuit boards requires special knowledge and equipment. Although you can design simple SMT PC boards with traditional design methods (we'll do so in Chapter 13), more complex layouts need to be established using computer-aided design techniques. The SMT PC board designer clearly must have knowledge of and access to CAD PC layout tools.

The *repair* of surface mount circuits also sets off alarm bells for some in the industry. True, it is difficult, if not impossible, to repair complex SMT boards in the field, yet bench repair work involving SMT, with the right tools used in the correct way, is *actually* easier to accomplish than similar work with IMT. We'll see why this is so in Chapters 13 and 14.

When it comes to people's willingness to adopt surface mount technologies, the concern usually boils down to the "learning curve problem." Learning anything new is a challenge, but it is also an exciting opportunity. What was mentioned in Chapter 3 with regard to TQM is true here. Fortunately, the electronics industry and the people in it are used to change—indeed, many thrive on it. If there is any group of individuals prepared to meet the challenge of a new technology, it is those working in electronics. SMT is being eagerly embraced throughout the industry.

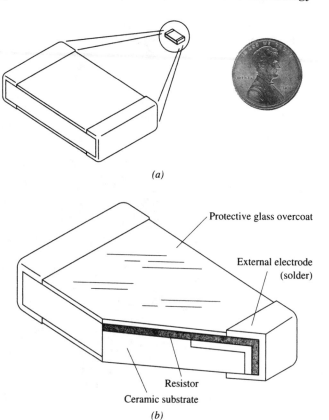

FIGURE 12–4
Leadless chip resistor

12.2   SURFACE MOUNT COMPONENTS

In the previous section we introduced you to the world of surface mount technology and explored its many advantages. It is now time to turn attention to the building blocks of SMT, **surface mount components (SMCs)**. In doing so, we might take any of three approaches to component classification. We could categorize SMCs on a functional basis: passive and active, for instance. Or we might discuss these components by first dividing them into discrete devices and integrated circuits. Finally, we could even break things down strictly according to packaging styles: two-terminal, three-terminal, and so forth. Our best approach might be to combine the three, since all factors are relevant. Accordingly, we will look first at discrete passive components, next at discrete active components, third at integrated circuits, and last, as a catch-all, at a few SMCs that do not readily fall into any of the other categories.

Keep in mind, however, that space limitations prevent us from discussing every possible surface mount component configuration. Even if space were not a factor, the field is changing so rapidly—new packaging styles are appearing monthly—there is little we can do here but cover the basics. We will, therefore, examine only the most common component packages.

Discrete Passive Components

Discrete passive components are individual components that do not change their basic character when an electrical signal is applied. Resistors and capacitors are prime examples. Let's examine each in turn.

Surface Mount Resistors

Not surprisingly, the *leadless chip resistor* is the most popular surface mount component (Figure 12–4a). It is available in standard values from 10 ohms to 10 megohms. Wattage ratings are as low as $\frac{1}{16}$ W and as high as $\frac{1}{4}$ W. Figure 12–4b shows how an SMC chip resistor is constructed.

The resistors are extremely small. The 0805 type is but 0.080″ long, 0.050″ wide, and 0.050″ tall (Figure 12–5). Dimensions for the other standard chip resistors, 1206 and 1210, are also shown in the figure.

Note, in Figure 12–4a, the resistor's external electrodes or terminals. A terminal can be one-, two-, three-, four-, or five-sided, depending on the degree of bonding demanded. The five-sided terminal chip is becoming the norm, since it provides more soldering surface.

SMC resistors do not, of course, have colored bands to indicate their value in ohms as do standard leaded resistors. So how can we tell their value? A simple three-digit system is in general use, with the numbers located on either the substrate side or the resistance element side. The first two digits indicate the first two significant numbers of the resistance value; the third digit, the number of zeros.

For example, a code number 220 indicates a 22-ohm resistor. A code number 221 would designate a 220-ohm resistor. If the resistance value is less than 10 ohms, an "R" is used to indicate a decimal point. Therefore, 2R2 specifies a 2.2-ohm resistor.

FIGURE 12–5
Chip resistor sizes

EIA Standard Chip Resistor Sizes				
EIA Size Code	Wattage	Length	Width	Thickness
0805	$\frac{1}{10}-\frac{1}{16}$	0.080"	0.050"	0.050"
1206	$\frac{1}{8}$	0.126"	0.063"	0.059"
1210	$\frac{1}{4}$	0.125"	0.095"	0.065"

Surface Mount Capacitors

The *ceramic chip capacitor* is the second most widely used surface mount component (Figure 12–6a). It comes in both nonpolarized and polarized versions. The former range in value from 1 pF to 1 μF; the latter (most often tantalums) are available from 0.1 μF to 100 μF. Voltages go as high as 1000 V.

In Figure 12–6b we see the inside of a ceramic chip capacitor. Essentially, the device is a sandwich of interleaved metal film and dielectric layers.

SMC capacitors look just like SMC resistors. They are the same size (some tantalums are larger),

and the terminals are identical. There is one difference in appearance, however. In most cases, no value designation is printed on the capacitor—there isn't room. Instead, the thin cardboard backing that runs the length of a strip of packaged capacitors has a value printed on it every inch or so.

Discrete Active Components

Now let's turn to two active discrete SMT components, those that switch or amplify. We'll examine diodes and transistors.

Surface Mount Diodes

Two-terminal diodes come in a **MELF** (metal-electrode face) leadless cylinder (Figure 12–7a). The package is also referred to as an SOD, for

(a)

FIGURE 12–6
Ceramic chip capacitor

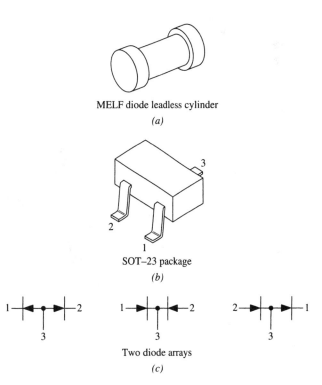

MELF diode leadless cylinder
(a)

SOT–23 package
(b)

Two diode arrays
(c)

FIGURE 12–7
Surface mount diodes

small-outline diode (or SOD-80). The cylinder is but $\frac{1}{10}''$ in diameter and $\frac{1}{5}''$ long.

Diodes have begun to appear in three-terminal SOT-23 (small-outline transistor) packages as well (Figure 12–7b). Though originally developed for transistors, SOTs are available with two diodes in a common-anode, common-cathode, or series connection (Figure 12–7c).

Surface Mount Transistors

Low-power transistors come in SOT-23 cases, the package designed for them (Figure 12–8a). The device is but 0.118″ long, 0.05″ wide, and 0.04″ tall. Note, in Figure 12–8a, the **gull-wing** lead configuration. With this shape it is much easier to solder and desolder the leads, as you will discover in Chapter 13. The gull-wing configuration will appear again when we examine surface mount ICs.

Transistors required to dissipate more power than the general-purpose type appear in the SOT-89

package (Figure 12–8b). Observe that the center (collector) tab extends under the device body and emerges as a small tab on the other side. The greater surface area of the tab enables it to dissipate more heat. SOT-89 cases are 0.18″ long, 0.102″ wide, and 0.063″ tall.

The four-terminal SOT-143 is also a popular SMC case style (Figure 12–8c). Measuring 0.118″ long, 0.05″ wide, and 0.04″ tall, it is used for bridge rectifiers, field-effect transistors, and a pair of diodes. Note that one lead is wider than the others to provide a convenient index mark.

Surface Mount Integrated Circuits

Surface mount ICs are available in a number of packages with a variety of lead configurations. Here we will concentrate on the three most popular. We will examine the **small-outline integrated circuit (SOIC)**, the **plastic-leaded chip**

FIGURE 12–8
Surface mount transistors

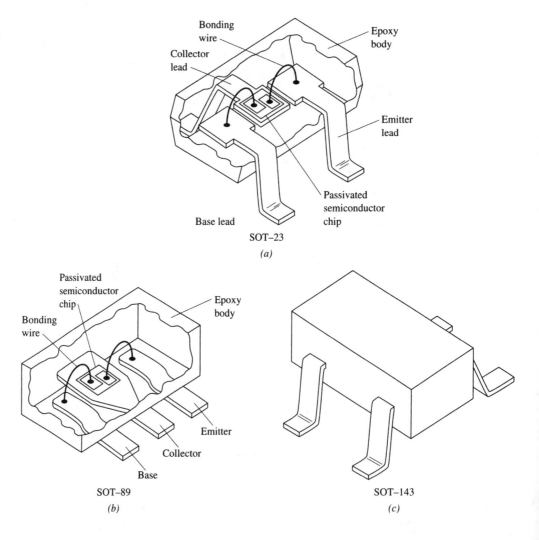

carrier (PLCC), and the **leadless ceramic chip carrier (LCCC)**.

Small-Outline Integrated Circuit (SOIC)

An SOIC looks like a miniature DIP integrated circuit (Figure 12–9a). However, these surface mount ICs, with from 8 to 28 pins, are but one-fourth the size of their DIP cousins. Actually, when you consider the SOIC's extremely low profile, the total mass of the IC package is around one-eighth that of an equivalent DIP.

SOICs are almost exclusively gull-winged (Figure 12–9b), though J-lead versions are available. J-leads, unfortunately, are more difficult to inspect once in a circuit. They are also harder to put in place and remove (solder and desolder).

SOICs come in regular and wide-body forms. The former, 0.150″ wide, are confined to 8, 14, and

16 pins (SO-8, SO-14, and SO-16). The latter, 0.300″ wide (often designated SOLIC), are for 16-, 20-, 24-, and 28-pin ICs. Lead spacing for SOICs is only 0.050″ (50 mil), half that of the traditional DIP.

Plastic-Leaded Chip Carrier (PLCC)

If an IC needs more than 22 pins, the plastic-leaded chip carrier (PLCC) is the choice (Figure 12–10a). With the PLCC, terminals are placed along all four sides. ICs with 24, 28, 44, 68, 84, and 124 terminals are common in the PLCC package. Terminal spacing, for the standard version, is 0.050″, the same as with the SOIC case style.

Unlike the SOIC, however, the PLCC has J-leads, not gull-wing leads (Figure 12–10b). The J-lead has important advantages and at least two significant drawbacks. On the positive side, J-leads are quite flexible, they take up less space than gull wings, and they are easier to handle—the leads don't get bent or tangled. On a negative note, as was mentioned earlier, J-leads are harder to inspect when in a circuit, and the flux residue that accumulates around the terminals is more difficult to wash away. The J-leaded PLCC finds its greatest use in microprocessors.

A fine-pitch version of the PLCC is beginning to receive wide acceptance (Figure 12–10c). With this chip, lead spacing of 0.025″ is the norm, though 0.020″ spacing is fairly common. The Europeans are even experimenting with a lead pitch of 0.017″.

The leads of the fine-pitch PLCC can be so easily damaged in handling (bumping up against one another in a tubular storage package, for example) that tabs are extended on all four corners to act as pin blockers (Figure 12–10c).

Leadless Ceramic Chip Carrier (LCCC)

The leadless ceramic chip carrier (LCCC), though rarely used in consumer electronics because of its high cost, is nonetheless found extensively in military electronics (Figure 12–11). It consists of a square tray with a lid on top. Into the tray is placed an assortment of semiconductor chips (and, in many cases, passive components) that are in turn installed on a conductive interconnecting pattern inside. After the components are installed, the lid is bonded to the top of the carrier.

As you can see from the figure, an LCCC has no external leads at all. Instead, the terminals of an LCCC are bonded directly to recessed slots found on all four sides of the carrier. The terminals are internally connected to the circuitry inside.

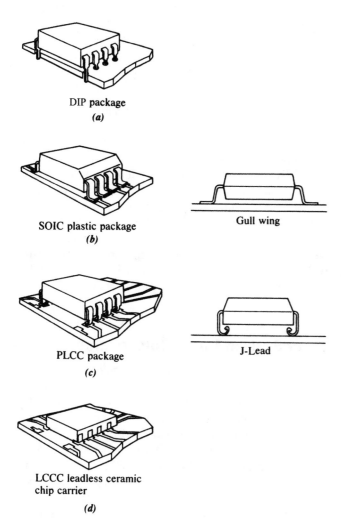

DIP package
(a)

SOIC plastic package   Gull wing
(b)

PLCC package   J-Lead
(c)

LCCC leadless ceramic
chip carrier
(d)

FIGURE 12–9
Small-outline integrated circuit (SOIC)

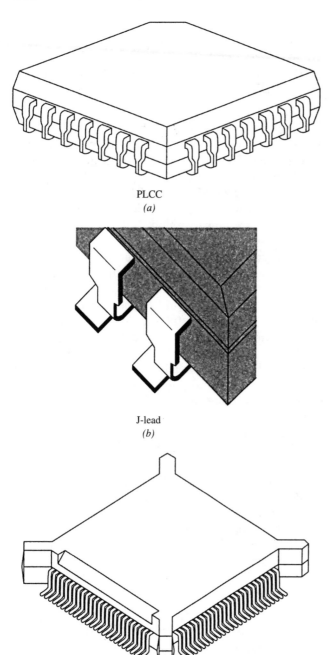

PLCC
(a)

J-lead
(b)

Fine pitch PLCC
(c)

FIGURE 12–10
Plastic-leaded chip carrier (PLCC)

They are externally connected to the PC board by solder columns.

One major advantage of the LCCC is that it has no leads to extend beyond the walls of the carrier. The resulting savings in PC board space can be as high as 50% over an equivalent PLCC. LCCCs are

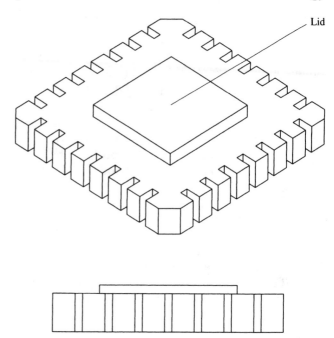

FIGURE 12–11
Leadless ceramic chip carrier (LCCC)

also capable of an extremely high terminal count, up to 244 pins as of this writing.

A major drawback with LCCCs concerns the *coefficients of thermal expansion*, or CTE, factor. The LCCC terminals are bonded directly to PC board solder columns; there are no flexible leads as with gull-wing and J-lead components. Since the terminals and columns are often made of different materials, they expand at different rates when heated. As a result, solder joints may fracture. The CTE problem is another reason why LCCCs are seldom found in commercial electronics equipment.

Other Surface Mount Components

As mentioned at the beginning of the chapter, there is certainly no dearth of surface mount components. Except for very large and high-power devices, virtually all components are now offered in SMT styles. Space does not permit us to detail even the majority of them here. Nonetheless, we should take a moment to look at three SMT components not yet discussed. *Trimmer resistors, trimmer capacitors,* and *light-emitting diodes* (LEDs) are likely to show up frequently in SMT projects that you design and build. Let's turn our attention to these three as we conclude this section on surface mount components.

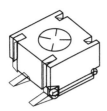

FIGURE 12–12
Surface mount trimmer resistor

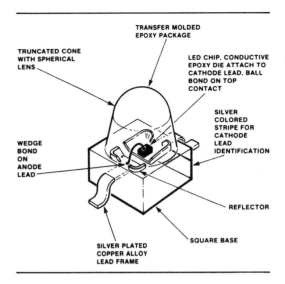

FIGURE 12–14
Surface mount LED
Courtesy Hewlett-Packard Co.
Reproduced with permission.

Surface Mount Trimmer Resistors

A large number of SMT-style trimmer resistors are available, both in sealed and nonsealed versions (Figure 12–12). The former are used with flow soldering, the latter with reflow (wave) soldering. (These automated soldering processes are discussed later in the chapter.) Resistance values range from 100 ohms to 2 megohms. The trimmers are either linear or audio taper, and either single or multiturn. It should be pointed out, however, that most trimmer resistors are not designed for repeated adjustments: up to 10 turns is about all they can handle. Adjustments are accomplished with a miniature screwdriver or special tool. Both the sealed and nonsealed types come in leadless or stubby pin configurations.

Surface Mount Trimmer Capacitors

Surface mount trimmer capacitors, like their trimmer resistor cousins, come in sealed (enclosed) and nonsealed (caseless) versions (Figure 12–13). Typical trimmer capacitor values are 1.4 pF to 3.0 pF, and 7.0 pF to 50 pF. As with SMT trimmer resistors, these tiny capacitors are not designed to be repeatedly adjusted. Once in a circuit and set for a particular value, they are rarely "tweaked" again.

Surface Mount Light-Emitting Diodes (LEDs)

Subminiature surface mounted LEDs are now finding wide application in electronic circuits. Construction of a typical SMC LED is shown in Figure 12–14. Two lead configurations are popular: gull-wing and yoke. The gull-wing, Figure 12–14, we have seen before with SOT and SOIC packages. This lead configuration allows an LED to sit flat on the surface of a PC board. The special yoke lead configuration, Figure 12–15, permits the LED to be placed inverted through a hole in the PC board. This type is also referred to as a panel-mount SMT LED. Both kinds of LEDs come in the standard red, yellow, orange, and green colors. LED arrays are also available in surface mount packages.

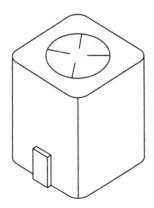

FIGURE 12–13
Surface mount trimmer capacitor

Something to Think About

Surplus SMCs are often available for the asking—you just have to know whom to ask. Why not locate a user of SMCs and explain your need? You want samples to have around and to show around. Many circuit board assemblers have reels of overruns. A donation is not difficult to obtain. After all, what are a few hundred SMC resistors, capacitors, and transistors—a spoonful—to them?

FIGURE 12–15
Surface mount LED
lead configurations
Courtesy Hewlett-Packard
Co. Reproduced with
permission.

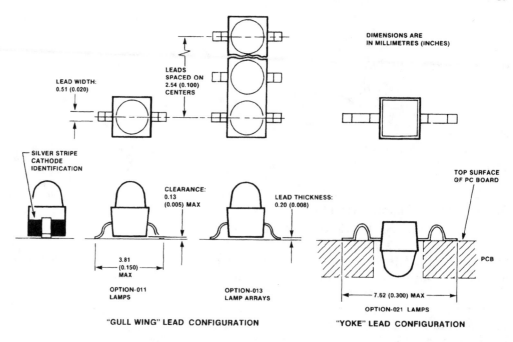

12.3 INDUSTRIAL SMT ASSEMBLY PROCESSES

From its origin, surface mount assembly was designed to be fully automated. True, SMT proto-types, as we shall see in Chapter 13, are assembled by hand, as are a few low-population, very-low-volume production boards. Anything more than that, however, requires the use of automatic pick-and-place machines and flow or reflow soldering equipment. Essentially, there are two types of automated assembly, both identified by the method they use to solder SMCs to a printed circuit board: the reflow process and the flow (wave) process. The two can be broken into six fundamental stages, or steps. As we will discover shortly, Stages 1, 3, 5, and 6 are almost identical for both types. Stages 2 and 4 are what set the two processes apart. We will investigate first the reflow method and then the flow procedure. Finally, we'll conclude this introductory chapter, appropriately, with the examination of an emerging career path, that of SMT technologist.

Automatic SMT Assembly Using Reflow Soldering

The six stages in the **reflow soldering** process are diagrammed in Figure 12–16. *First*, incoming SMT PC boards are visually inspected for defects. *Second*, solder paste is applied to SMC pads (foot-prints) by a silk-screening (printing) operation. *Third*, individual surface mount components are automatically picked up and placed on the PC board. *Fourth*, the entire printed circuit board is heated until the solder melts, or reflows. *Fifth*, the assembled board is visually inspected to determine if good component bonding has taken place. If problems are identified, rework may be required. (SMT rework is discussed in Chapter 14.) Finally, in the *sixth* stage, the completed circuit is cleaned to rid it of flux residue.

The reflow soldering approach to SMT assembly just outlined offers numerous advantages, many of which we'll discover as we turn to a detailed discussion of each stage. We will then conclude with a peek at the entire process as it is actually done in an electronics contract manufacturing facility—Micree Corporation.

Stage 1—PC Board Inspection

Whether a company fabricates its own printed circuit boards or contracts their manufacture out to a PC house, the boards must be carefully inspected prior to the application of solder paste in stage 2. It is generally accepted that SMT PC boards demand closer scrutiny than boards designed for insertion mount technology. The smaller pads, the thinner traces, the emphasis on multilayers, and related factors combine to make close visual inspection under the magnifying glass a must. The inspector can take a "random check" approach, making spot checks at arbitrary locations, or concentrate on a particularly troubling area: extremely thin traces, for instance. Either way, the inspector looks for

FIGURE 12–16
Reflow soldering

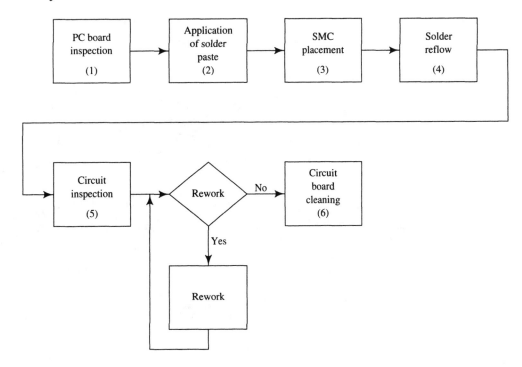

physical defects in the board laminations, solder bridges, trace breaks, and for an even, smooth solder coat. If problems are identified, the board must be channeled to a rework station or rejected outright. If a bad board is allowed to proceed to stage 2, and beyond, the result can be prohibitively expensive and time-consuming, as whole circuit boards are resigned to the recycling heap.

Stage 2—Application of Solder Paste

In stage 2, solder paste is applied to printed circuit board pads. Solder paste is a mixture of tiny spherical solder particles, flux, and a carrier vehicle. When applied to the board's pads, it retains its sticky, "tacky" characteristic for many hours. This feature allows a surface mount component placed on the board to remain in position; no adhesive or glue is required to prevent the component from shifting around or falling off the PC board.

The paste is applied using a semiautomatic screen printer (Figure 12–17). An image of the PC board pads is first photographically "burned" into a stainless steel, fine-meshed "silk" screen held in a frame. As a result, open spaces on the screen are formed at each pad location. The rest of the screen remains opaque. Next, solder paste is mixed and lightly spread over the screen. The PC board is then placed under the screen and carefully aligned. The paste is now "squeegeed" over the screen. As you would expect, solder paste is forced through openings in the screen onto the pads

immediately below. The screen is next lifted and the PC board removed. Finally, a new printed circuit board is put in place and the whole process repeated. When a different circuit is to be assembled, its screen is inserted into the printer and a fresh board run is begun.

The process just described is not quite as straightforward and automatic as it might appear; considerable operator skill is involved. To produce a succession of well-"pasted" boards in a reasonably short period of time demands plenty of hands-on experience.

Stage 3—Surface Mount Component Placement

Today, automatic pick-and-place equipment can install anywhere from 1000 to 500,000 SMCs per hour. Whether it's toward the former or the latter depends on a number of factors, machine intelligence and the range of components that can be handled being the two most relevant. Prices for the units vary considerably, from a low of $20,000 to over $1 million.

Three categories of placement equipment are in wide use. The huge, expensive *mass placement machines* place a whole batch of SMCs on a PC board at one time (Figure 12–18). Such machines require extensive programming and setup time. They are ideal, however, for complex boards and extremely high production runs.

With *in-line SMC placement machines*, circuit boards pass under a series of placement heads,

FIGURE 12–17
Semiautomatic screen
printer

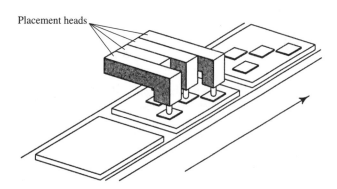

FIGURE 12–18
Mass placement SMC machine

each of which places a single SMC on the board (Figure 12–19). When comparatively few SMCs are to be installed, this approach works well.

X-Y pick-and-place machines are the most popular choice for automatic SMC assembly. Such devices are essentially specialized computer-driven pick-and-place robots (Figure 12–20). In operation, a moving pick-and-place vacuum head fetches components one at a time and places them on the PC board. Components come from a dispensing reel, magazine, or bin dispenser.

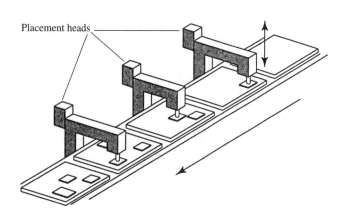

FIGURE 12–19
In-line SMC placement machine

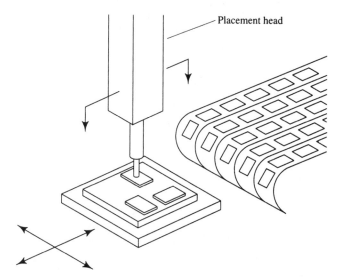

FIGURE 12–20
X-Y pick-and-place machine

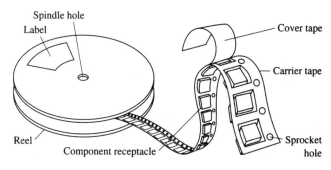

FIGURE 12–21
SMC dispensing reel

The reel illustrated in Figure 12–21 is the most widely used component delivery method for X-Y pick-and-place machines. As you can see from the drawing, hundreds of SMCs are stored in a plastic tape that is wound around the reel. As the plastic tape is fed to the pick-and-place machine the cover tape is automatically peeled away. The exposed SMCs are then picked up and put in place by the vacuum head. In operation, several reels, each with a particular component, are set in position to feed the X-Y pick-and-place machine.

Stage 4—Solder Reflow

In stage 4, the stuffed PC board, with its surface mount component terminals settled in their cushion of solder paste, is placed on a conveyor belt and transported into an oven. The solder paste is melted (reflowed), the board is removed and allowed to cool, and no one has lifted a component or a soldering iron. It sounds simple, and in theory it is. In actuality, the solder reflow process is a bit more involved, depending on which of three heating methods is used.

The *convection oven* is the simplest, easiest, and least expensive way to reflow solder in an industrial setting. It's basically a forced-air heating system similar in concept to the heater that warms your house. The PC board sits in an oven, warm air is circulated, and the solder paste melts. No chemicals or complex processes are involved. About the only disadvantage to this approach is the need for critical temperature control.

In the *infrared reflow soldering system*, the solder paste is melted by infrared heat lamps. Because it is quick, the process works well for large circuit runs. Temperature control, however, is even more essential than with the convection oven.

Vapor phase reflow soldering represents a third way to reflow solder paste. A dense blanket of vapor is created in the oven by boiling liquid fluorocarbons.

The hot vapor condenses on the PC board with its SMCs and solder paste, causing the latter to melt. The main advantage here is that it is impossible to overheat the surface mount components. On the down side, the fluorocarbons are very expensive, often costing hundreds of dollars per gallon. The oven itself is also quite costly, making this method the most expensive of the three just discussed.

Stage 5—Circuit Inspection

After solder has been reflowed and allowed to solidify and cool in stage 4, the printed circuit board is ready for circuit inspection. Given the small size of SMCs, they are always inspected under a magnifying glass. The inspector looks first for the standard problems that can appear with any PC board assembly: solder bridges, solder balls, trace breaks, and so forth. There are, however, two difficulties unique to SMT that require the inspector's singular attention. When SMCs are reflow or flow soldered, they tend to exhibit what is known as the **drawbridge effect** or the **tombstone effect** (Figure 12–22). Both effects are evident when only one terminal of a two-terminal chip is actually soldered to the board. The other terminal lifts completely off its pad.

In the drawbridge, or "popping wheelies," effect, the rise is less dramatic than in the tombstone effect, where the SMC assumes a vertical position. In both cases, however, the result is completely unacceptable — the chips are effectively out of the circuit.

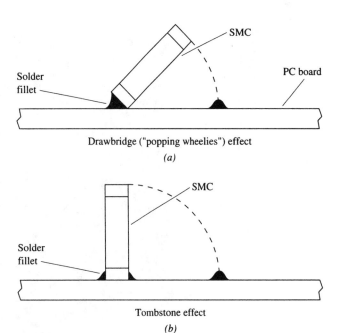

FIGURE 12–22
Drawbridging and tombstoning effects

Two phenomena cause drawbridging and tombstoning. Either an uneven quantity of solder paste is applied to a chip's pads, or the solder under one terminal melts before it does under the other. Attention to detail and operator skill during stages 2, 3, and 4 will minimize these effects.

Stage 6—Circuit Board Cleaning

If drawbridging, tombstoning, or related component misalignment occurs in stage 5, circuit board rework will be required (Figure 12–16). Such rework can be fairly complex and involved. We will hold off discussing the subject here, since we devote all of Chapter 14 to the topic.

The final stage in the reflow soldering process involves circuit board cleaning. Virtually all but the least expensive consumer electronics PC boards are washed of flux residue and other contaminants. Doing so results in better board appearance and less likelihood of circuit failure later on. Two cleaning procedures are in wide use. One is *water-based* (aqueous), the other uses *organic solvents*.

The water-based cleaning method is accomplished much in the same manner that your dishes are washed at home. A hot spray of iodized water is directed, by hand or by machine, over the circuit board. A scrubbing action, again by hand or by machine, helps remove the flux residue. The PC board is then placed in a low-temperature oven to evaporate all the water. The aqueous process is inexpen-sive and environmentally safe. Nonetheless, it is generally not as effective as solvent-based cleaning.

Solvent cleaners, when used in industry, are usually CFC (chlorofluorocarbon)-based. Such cleaners do a good job of removing contaminants, particularly rosin-based fluxes. CFCs, however, are believed to be reducing ozone concentrations in the stratosphere. They are an environmental no-no.

Fortunately, a natural solution to the dependence on CFCs may be at hand. It has been claimed that ordinary lemon juice does as good a job of removing rosin fluxes as do dangerous CFCs. If that is true, it will be a win-win-win situation. The circuit manufacturers will be happy (lemon juice is relatively cheap), the environmentally conscious will rejoice, and, of course, the lemon growers won't want to look a gift horse in the mouth. Oh, but if all our environmental/technological problems could be solved so easily!

The Reflow Process Illustrated

As mentioned earlier in the chapter, Micree Corporation is an electronics contract manufacturing facility located in Southern California. In addition to its traditional insertion mount technology assembly line, Micree has a complete SMT reflow soldering setup, pictures of which are shown in Figures 12–23 to 12–29.

In Figure 12–23 we see blank circuit boards that have just arrived from an outside vendor. After

FIGURE 12–23
Inspecting incoming printed circuit boards

they are carefully unwrapped, they will be visually inspected under a magnifying glass.

Figure 12–24 pictures Micree's semiautomatic screen printer in an open, or raised, position. Note the metal screen and its frame locked in place.

Figures 12–25 and 12–26 highlight the SMC placement equipment. The Dynapert machine shown in Figure 12–25 is an X-Y pick-and-place unit. In Figure 12–26, José Hernandez, Micree's technical manager, adjusts one of the many component dispensing reels used for the next product run.

The TurbAir Infraflo 500c machine shown in Figure 12–27 is, of course, used to reflow the solder paste. It employs a convection-type heater.

FIGURE 12–24

Micree's semiautomatic screen printer

FIGURE 12–25

An X-Y pick-and-place machine

FIGURE 12–26
Adjusting component
dispensing reels

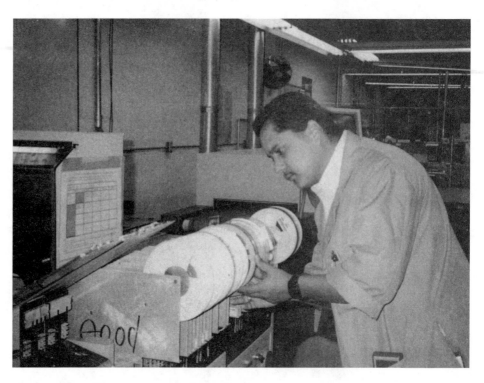

FIGURE 12–27
A convection-type heater

In Figure 12–28 an inspector examines a completed circuit board through a magnifier. Note the microscope to his right, used to examine finer details.

Finally, in Figure 12–29, we see circuit boards being cleaned with iodized water. Notice the scrub brush hanging from the basin in the center of the picture.

Automatic SMT Assembly Using Flow (Wave) Soldering

Automatic SMT assembly using flow soldering, like the process previously outlined, requires six stages (Figure 12–30). In examining flow soldering, however, we won't have to discuss stages 1, 3, 5, and 6

FIGURE 12–28
Circuit board inspection

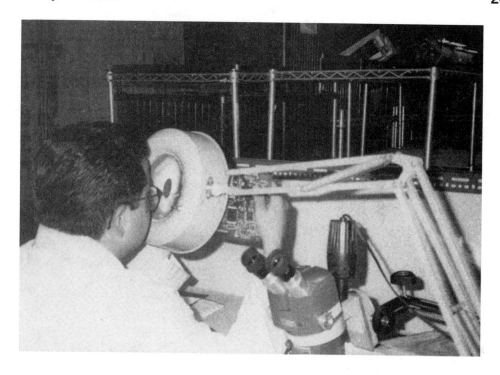

FIGURE 12–29
Circuit board cleaning

again. PC board inspection, surface mount component placement, circuit board inspection, and circuit board cleaning are the same for both processes. Stages 2 and 4 are fundamentally different, however. It is these two stages, as they apply to flow soldering, therefore, that we need to explore in some detail.

The Basic Flow Soldering Procedure

Flow soldering is not new; it has been used with insertion mount technology for years. Recall that in IMT, components sit on top of the component side of a PC board and leads and pins protrude through holes to the solder side of the board (Figure 12–31a).

FIGURE 12–30
Six stages in the flow
soldering process

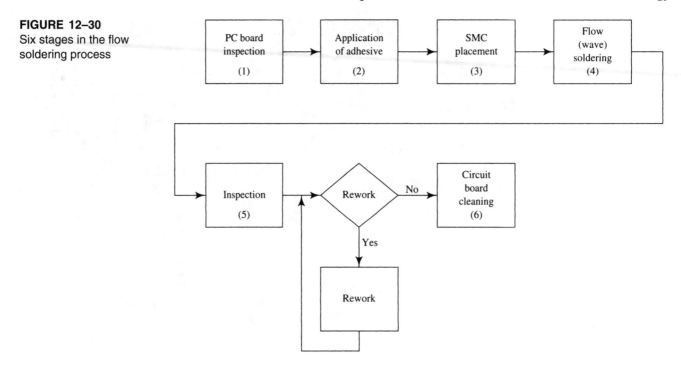

With the reflow method, the solder side of the board is passed over a wave of molten solder. All the leads are quickly soldered to the PC board pads. Note that in this process the component body never touches the liquid solder.

With SMT flow soldering, the procedure is similar but with one significant difference. As shown in Figure 12–31b, the inverted SMT PC board, components and all, is passed over the molten solder. Solder, of course, does not adhere to the components, only the

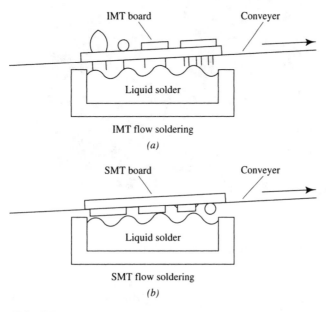

FIGURE 12–31
Flow soldering

terminals and pads. And yet, looking at this method, two concerns come quickly to mind. One, will the SMCs be damaged by the hot solder? And two, how are the components held in place? How do we prevent them from dropping off into the pool of solder?

Although the first point is cause for some anxiety, if SMCs are designed with the heat problem in mind, if they are preheated before entering the solder bath, and if they are not overexposed to the molten solder, they will survive the thermal shock of the solder wave. As to the second issue, obviously the SMCs must be bonded to the PC board before the entire assembly is flow soldered. The bonding takes place in stage 2, application of adhesive. At stage 4, the flow solder process actually occurs. Let's turn our attention to both stages.

Stage 2—Application of Adhesive

With the flow solder method, an adhesive must be used to hold SMCs in place until wave soldering is completed. A drop of cement is placed at the component site, by automatic syringe or screen printing (stage 2), and the surface mount component put in place with automatic pick-and-place equipment (stage 3). The dot of adhesive is not, of course, placed on the circuit board pads but between them on the blank board (Figure 12–32).

The adhesive must be cured, usually with UV light, immediately after application to remove its tackiness. Furthermore, the adhesive must be strong enough to hold the SMC in place while it is heated by molten solder but not so strong that the component cannot be removed during rework.

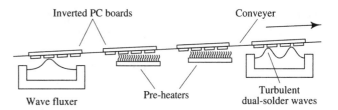

FIGURE 12–33
SMT flow soldering

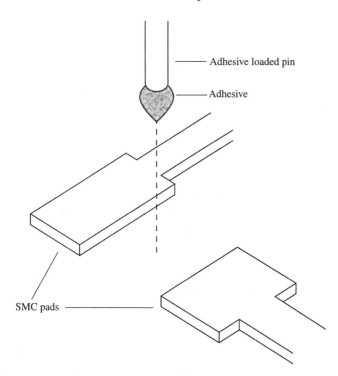

FIGURE 12–32
Placement of adhesive

Various glues, cements, and epoxies are used to adhere components to their PC boards. Most are *thixotropic* (they have the ability to liquefy). They can be applied by all the methods used with solder paste. Nonthixotropic adhesives, on the other hand, are syringe- or pump-dispensed.

Stage 4—Flow (Wave) Soldering

Flow soldering of IMT components uses a single-wave system. When SMT components are involved, a double-wave system is required.

An overview of the process is shown in Figure 12–33. The inverted, stuffed PC board is first passed over a wave fluxer that applies a thin coating of flux to the components and PC board. Next, the circuit is preheated to reduce the negative effects of thermal shock. The board then enters the dual-wave molten solder bath. The first wave is quite *turbulent*; it forces solder into hard-to-reach places. The second wave is *smooth*; its purpose is to smooth out the connections just made by the turbulent wave.

Flow soldering is a proven technology that works well with boards that are mixed (use both IMT and SMT components), as well as with those that are pure SMT. It is not without its disadvantages, of course—missed connections caused by the "shadow effect" being just one, yet flow soldering will continue to see wide application in the industry as surface mount technology explodes on the electronics packaging scene.

The SMT Technologist: An Emerging Career Path

If what you have read so far about the surface mount technology revolution inspires you, read on for a few more paragraphs. You may want to become an SMTT—surface mount technology technologist.

What Does an SMT Technologist (SMTT) Do?

The SMT technologist is an *emerging* career. If you try to look it up in the U.S. Department of Labor's *Occupational Outlook Handbook*, you won't find it. In the department's eyes, it doesn't exist—yet. In actuality it does, though there are some electronics technicians and engineers working full time with SMT who don't know that they are, in effect, SMT technologists. True, an SMTT isn't just someone who works in the field: those who simply design SMCs, lay out circuit boards for SMT components, and operate a specific piece of SMT equipment are not SMT technologists. The SMT technologist has a much broader knowledge—not to mention experience—than any surface mount specialist.

A surface mount technology technologist must know SMT from A to Z. That is, he or she needs to understand the complete SMT process: what equipment to purchase and why; how to install and maintain the equipment (and, in some cases, repair it); all about SMCs; how to design SMT circuit boards; and so on. The SMTT works for a particular company or hires out as an SMT consultant. Duties include holding seminars, training personnel, and, in general, getting companies going with SMT. He or she must not only be familiar with the technology itself but understand the economic and personnel implications of introducing SMT. The SMTT, with considerable know-how and years of industrial experience, will be the point person for the new packaging revolution of the 21st century.

How Do You Prepare for a Career as an SMTT?

At this time there are no two- or four-year college degree programs specifically designed to train surface mount technology technologists, though certificate programs are beginning to appear. Maybe that's just as well. If there was ever a career that emphasized practical experience, it would be the SMTT. In other words, you don't go to school to become an SMTT; you work with and around the technology, sometimes for years, before you can declare yourself an SMT technologist. Nonetheless, there is much to be learned in school about surface mount technology, and you've begun the process by reading this introductory chapter. Whether you see SMT as a subfield of electronics that every would-be electronics technician needs to understand, or as an eventual career path all its own, there is a lot more to know. You can take the next step by proceeding on to Chapter 13: Project Design and Fabrication Using Surface Mount Technology.

Something to Think About

Why not find a surface mount technology technologist and conduct an information interview? Go prepared with questions specific to the field. Ask if you can observe him or her on the job. Try to determine if this is indeed an emerging career path.

SUMMARY

In this chapter we began by exploring a new electronics packaging revolution—surface mount technology (SMT). We defined SMT and discussed the future of IMT. We looked at the history of packaging technologies and examined the many advantages (and a few limitations) of surface mount technology.

Next, we investigated surface mount components. We scrutinized discrete components, both passive and active. We studied surface mount integrated circuits: SOICs, PLCCs, and LCCCs. We also delved into surface mount trimmer resistors and capacitors and SMC LEDs.

It was then on to an examination of industrial SMT assembly processes. We analyzed the six-stage reflow soldering procedure in detail. We saw how the flow soldering approach differs from the reflow method. We concluded the chapter with a peek at an emerging career path, that of surface mount technology technologist.

Are you eager to get your hands on SMCs and build an actual circuit project? The design and fabrication of such a project is possible, even with traditional materials and tools. To find out how, turn to Chapter 13.

QUESTIONS

1. Surface mount components, unlike IMT components, are soldered to the _____ of a specially designed PC board.

2. List six advantages of surface mount technology over insertion mount technology.

    1.

    2.

    3.

    4.

    5.

    6.

3. List two limitations of surface mount technology.

    1.

    2.

4. Surface mount components are but _____ to _____ the size of IMT components.

5. Most surface mount components have no leads; they have _____ instead.

6. The small-outline integrated circuit (SOIC) looks like a miniature _____ IC.

7. List the six stages in the SMT reflow soldering process.

    1.

    2.

    3.

    4.

    5.

    6.

8. Reflow soldering is characterized by the reflow of solder _____.

9. List the six stages in the SMT flow soldering process.

    1.

    2.

    3.

    4.

    5.

    6.

10. The flow soldering method requires SMCs to be _____ to the PC board prior to their immersion in molten solder.

13

Project Design and Fabrication Using Surface Mount Technology

OBJECTIVES

In this chapter you will learn

☐ Where to get surface mount components.

☐ How to breadboard surface mount circuits.

☐ How to design PC boards for surface mount components.

☐ How to produce the PC board artwork for surface mount components.

☐ How to fabricate an SMT PC board.

☐ How to hand solder SMCs to a circuit board.

☐ How to do surface mount assembly using solder wire.

☐ How to do surface mount assembly using solder paste.

☐ How to design and fabricate a surface mount sample project.

There are electronics people, many of whom have been in the industry for years, who will laugh in your face when you tell them what you want to do; and when you say that you want to do it with traditional tools, particularly with an ordinary soldering iron, they'll just shake their heads, turn away, and refuse to have anything more to do with you. After all, you must have been staring at the little buggers for too long—it's affected your sanity.

This sort of thing has happened to me, more than once, and it may happen to you too. Believe me, for many the idea of hand-assembling surface mount components into a working prototype project seems ludicrous. After all, as even we have noted

(Chapter 12), SMT was developed with automatic assembly in mind. Finding, grasping, arranging, and somehow soldering those tiny, leadless SMT components by hand is ridiculous. Right? Wrong—delightfully wrong! Not only is it possible, it can be accomplished easily, quickly, and, most importantly, for the most part with the tools you now have on your workbench. Furthermore, as you are about to discover, prototyping with SMT is fascinating and fun. Don't get the wrong idea. There's much new here to be learned—you can't breadboard and prototype SMT projects in the same way that you do IMTs, but by using familiar tools, albeit with different techniques, you can become a skilled SMT prototype technician. You will be lifted to a higher level of assembly sophistication, where you'll feel, and in some cases, act, like a surgeon in an operating room. Then you'll be ready to whip that SMT prototype project out of your shirt pocket, smile, stretch out your arm, and show it to any moribund electronics cynic.

Chapter 13 is one of the two "do it" chapters of Part II; it gives you not only more SMT knowledge but an SMT experience. We will begin with *SMT project design*. We'll look at SMT design factors, a whole new method of breadboarding, and examine closely ways to create printed circuit board design layouts and artwork patterns exclusively for surface mount component projects.

Next, we will turn to *SMT project fabrication*. We will see how SMT PC board fabrication differs from the more traditional approach. We will exhaustively examine SMC assembly using solder

273

wire and solder paste. Moreover, we'll do all this while designing and building an SMT sample project: the 555 Dual LED Flasher.

13.1 SMT PROJECT DESIGN

To get the most from this chapter, you must first read Chapters 1–10, or at least be familiar with the basics of project design and fabrication using insertion mount technology (IMT). In Chapter 13 we do not begin these topics anew just because we are dealing with a new technology—surface mount. Rather, we use what we have already learned with more traditional IMT as a point of departure for examining SMT. Both technologies have much in common with regard to project design and fabrication. It's what sets SMT apart from IMT in dealing with these subjects that we concentrate on now.

In this section, we examine SMT design factors, a fascinating breadboarding approach known, appropriately, as "Surfboarding," SMT PC design layout, and SMT PC artwork generation. By the time we are through, we'll have created a photographic negative for our 555 Dual LED Flasher sample SMT Project.

SMT Design Factors

Before we jump right into SMT project building, we need to pause a moment and answer two important preliminary questions: (1) When is it appropriate to design and fabricate with SMT as opposed to IMT? and (2) Where do we get SMT components? When we have dealt with these two questions, then it will be time to get started on the actual project design.

When Do We Go SMT?

In making the decision to go with either an SMT or IMT design, three factors are critical. *First*, does the project demand an SMT solution? Could it even be designed, mass-produced, and sold with traditional components? If it needs to be very small, very thin, or very light, and it doesn't require too many off-board components (switches, panel lights, etc.), then SMT may be the *only* way to go. *Second*, are SMCs available, at competitive prices, to enable us to build the product? If the answer is yes, we've moved another step further toward a decision to use SMT. *Third*, is the equipment and expertise on hand to do the job? Assuming the answer is also yes, we are ready to jump on the SMT bandwagon—our next project will be SMT-based.

Where Do We Get SMT Components?

Suppose you want to build an SMT project. Where do you get the components? Most of your traditional suppliers have an SMC line. Such suppliers are definitely the first ones to check with. Begin by looking at a catalog's table of contents for a special listing of surface mount items. If a list is there, great! If not, they still may be selling SMCs, just not placing them in a separate category. Look for the component you want, say a 74LS02 Quad 2-Input NOR Gate IC, and see if it is listed under TTL ICs as a 74LS02D. If the "D" is there, you've found a surface mount SOIC. One popular vendor (JAMECO) lists no SMCs in its catalog, yet it carries a rather impressive line. You just have to specify your needs; then the vendor will let you know if it's available in a surface mount version.

As you search around, contacting vendors, distributors, and retail outlets, you will notice that some SMCs are definitely more available than others. Chip resistors and capacitors are the easiest components to find, followed by transistors, LEDs, signal diodes, and trimmer pots. Integrated circuits (linear and digital), unfortunately, are the most challenging to locate. A vendor will have a dozen or so, often not many more. Until this situation changes, and it will, you are going to be somewhat circumscribed in what SMT projects you can build.

To aid you in finding SMCs, a list of suppliers willing to provide such components in single-lot quantities, at moderate prices, is presented in Appendix F. The author has contacted all the companies listed and has bought SMCs from most. They are all eager to sell to schools and individuals in low-volume quantities. Of course, the list is in no way complete—it represents just one person's effort to locate surface mount components for project building. Expand on it by searching around locally, by mail order, and, of course, on the Internet.

An SMT Sample Project: The 555 Dual LED Flasher

In choosing a sample SMT project for this chapter, we insisted that it be simple, useful, easy to build, contain typical (but readily available) SMCs, and give us a feel for what SMT design and fabrication are all about. The 555 Dual LED Flasher, built on a $1'' \times 1\frac{3}{4}''$ PC board taped to a 9-V battery (its power source), meets these criteria (Figure 13–1). The schematic for the project is shown in Figure 13–2. In addition to the sample project, the design and fabrication of which will be carried out in this chapter, two additional SMT projects, all built on the same size $1'' \times 1\frac{3}{4}''$ printed circuit board and operated from a 9-V battery, are presented in Appendix E. Each project is displayed on a single sheet that contains a project schematic, description, picture,

FIGURE 13–1
555 Dual LED Flasher Project

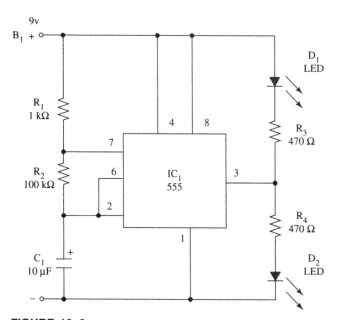

FIGURE 13–2
Schematic drawing for 555 Dual LED Flasher Project

parts list, PC artwork pattern, and assembly drawing. Both are simple, useful, and interesting projects made with readily available surface mount components. After working your way through this chapter, we trust you'll be inspired enough to build them.

SMT Breadboarding

You have your project schematic, you've obtained the necessary SMCs to build the device, yet you don't want to jump right into a PC board layout. You want to breadboard the circuit first. But how? Of the five breadboarding approaches discussed in Chapter 6, four are clearly out of the question here. There is no way you are going to breadboard an SMT circuit using the cut-slash-and-hook, point-to-point, wire wrap, or universal PC boards. The only approach that seems even remotely possible is *solderless circuit board;* however, you're working with components that for the most part have no leads, or if they do, the leads are not spaced or shaped "right." Clearly the solderless circuit board technique, by itself, isn't going to work. Someone has to come up with a better idea.

Fortunately, someone has! He's Rob Laschinski, president of Capital Advanced Technologies, Inc., and his company has developed a unique breadboarding method specifically designed for surface mount components. It's called *Surfboards,* and you don't have to live in the home state of the Beach Boys (like the author does) to appreciate how easy and effective this approach is. We'll explore the basic Surfboard concept, examine the different types of Surfboards, analyze how circuits are assembled using this novel procedure, and look at a Surfboarded 555 Dual LED Flasher.

The Surfboard Concept

The Surfboard concept involves the use of specially designed universal PC boards with single in-line pins (SIPs) spaced 0.100″ apart (Figure 13–3). The pins fit into a solderless circuit board. Thus, this method elegantly combines the universal PC board and solderless circuit board concepts.

The boards themselves incorporate foil patterns that accommodate a wide variety of component mounting footprints, both discrete and IC. Several device sizes can be mounted on the universal foil pattern. Figure 13–4a shows the mounting of 0805, 1206, and MELF packages. Larger components, such as inductors, tantalum capacitors, and crystals can often be straddle-mounted across foil patterns (Figure 13–4b). The boards are quite "forgiving," too. As shown in Figure 13–4c, the large foil land patterns allow for considerable component orientation latitude, an especially useful feature for beginners. Three-terminal components, such as transistors, some LEDs, potentiometers, and trimmers, can be mounted on the three-terminal universal mounting zones found on many board models (Figure 13–4d). All in all, it's a great way (if not the

FIGURE 13–3
Surfboards

only way) to breadboard surface mount technology projects.

Surfboards for All Seasons

Surfboards come in two series: the 6000 series for discrete devices and the 9000 series for ICs. Although we don't have space to delineate every one of the many designs in each series, we'll take a few moments to highlight the more widely used boards.

Of the 6000 series, the *Basic Design* model shown in Figure 13–5a is very popular. It consists of vertical foils with a single horizontal "common" foil across the top. The board accommodates two- and three-terminal devices.

The versatile *General Purpose Design,* Figure 13–5b, is specifically intended for three-terminal devices, though it can accommodate components with only two terminals as well. In this version, the same basic pattern is repeated in multiples of three foils, with the first foil in each group being an inverted L-shape that forms a mounting footprint for three-terminal devices such as SOT-23 transistors and trimmer components. Other discrete boards are available, too, such as the 6306 model, Figure 13–5c, designed specifically for SOT-89 components.

The 9000 series can handle both SOICs and PLCCs. An example of the former (the 9163), used to breadboard the 555 Dual LED Flasher sample

SMT project, is shown in Figure 13–6a. Note that a direct pin-out for each IC pin is provided. The 9163 accepts 8-, 14-, or 16-pin SOICs. Discrete components are cross-mounted between foils.

A 20-terminal PLCC Surfboard, the 9301, is shown in Figure 13–6b. Every terminal on the IC is brought out to its own SIP pin. Discrete devices can be mounted across adjoining traces.

Building SMT Circuits with Surfboards

Regardless of which Surfboard models you use, the tiny SMT components have to be soldered to the board's foil patterns. Capital Advanced Technologies, Inc. recommends gluing components in place with a spot of adhesive, then hand- or dip-soldering them. Since we will be reviewing this and similar approaches to attaching SMCs to circuit boards in Section 13.2, we will not discuss the topic here. If you simply must know now how it is done, skip to the appropriate pages in the text.

Surfboarding the 555 Dual LED Flasher

The Surfboarded 555 Dual LED Flasher is shown in Figure 13–7. Note two things. First, we are using only one Surfboard (the 9163), not two. We could have used the 9082 board that accommodates an 8-pin SOIC, but then we would have had to use a 6000 series discrete board, like the one shown in Figure 13–5a, to hold our discrete components (a capacitor, two LEDs, and four resistors).

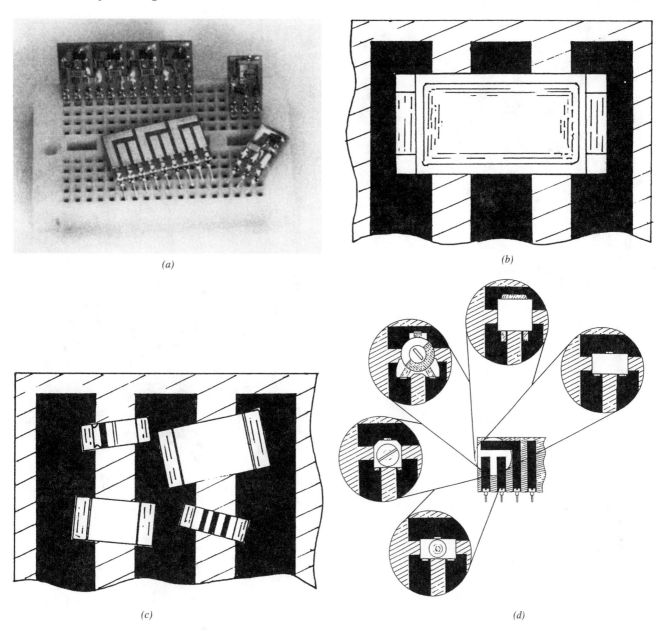

(a)

(b)

(c)

(d)

FIGURE 13–4
Placing SMCs on Surfboards

With our single 9163, however, we have no such problem. As you can see in the figure, by using the lower eight pins (four on each side) of the DIP outline to accommodate the 555 IC, we are left with enough vertical foils to mount our discrete components.

Second, observe that some "external" wiring on the solderless circuit board is necessary to complete the circuit. This is to be expected. After all that's why we plug the Surfboard into a traditional solderless circuit board in the first place.

SMT PC Board Design Layout

OK, you've breadboarded (or Surfboarded) your project and are now ready to design the PC board. To help you do so, we'll examine some preliminary considerations, the issue of scale, and how the design layout is accomplished. We will conclude with a look at the PC Design layout for our 555 Dual LED Flasher.

Although it is certainly possible, even desirable, to do SMT project design and artwork generation

FIGURE 13–5
6000 series discrete device Surfboards

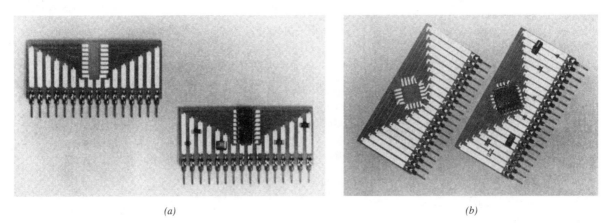

FIGURE 13–6
9000 series integrated circuit Surfboards

FIGURE 13–7
Surfboarded 555 Dual
LED Flasher Project

FIGURE 13–8
SMT PC board design
layout materials

☐ Graph paper ($8\frac{1}{2}''$ × 11″), 20 squares to the inch (0.050″
    spacing)
☐ Magnifier (5× or 10×)
☐ Pencil (2H)
☐ Ruler (0.10″)
☐ Surface mount device patterns (dry transfers) 1×, assorted

with CAD, here we will introduce you to the more traditional tape-up method; however, should CAD be available to you, you may simply skip this section and refer to Chapter 7 for suggestions and advice on creating CAD-generated PC board designs for IMT projects.

Preliminary Considerations

To begin with, you're going to need some materials. In Figure 13–8 we list the basics. Note that you must purchase graph paper with lines spaced only 0.050″ apart, the same pin spacing for SOICs. It's available at any good stationery store. You will also need a magnifier. The type shown in Figure 1–20 works great, not only for PC design layout and artwork creation, but for actual project assembly as well. Finally, in addition to the standard materials, you'll need dry transfers (rub-ons) of surface mount device patterns. Although they will, of course, be required for the actual tape-up, you will want them at the design stage, too—especially if you are working full scale, or 1×. Some typical patterns are shown in Figure 13–9.

Next, you need to gather the SMCs you will be using. Notice how small the components are and how tiny are the individual terminals and pins. Study the devices under a magnifying glass.

As with traditional PC board design, you will need to determine which components are to be onboard and which off-board. When designing with SMT, there are usually very few off-board components. Our 555 Dual LED Flasher, for instance, has only one, the 9-V battery snap.

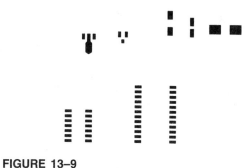

FIGURE 13–9
Typical SMT device patterns

Should your PC board be single- or double-sided? That's easy. Unless you plan to use both sides of the board for circuit components, you'll use a single-sided board for all your SMT designs.

In determining the size and shape of your board, lay the components out on a sheet of paper to see how much space they occupy, then add about one-third additional space for traces. Don't forget to add room for nomenclature: name of project, part numbers, polarity designators, and so on. The PC board for our sample project is $1'' \times 1\frac{3}{4}''$. Those dimensions were chosen so that the board could be taped against a 9-V battery.

Finally, is the board to be viewed from the component or trace side during the layout process? Actually, for SMT you will be working exclusively on the trace side, since that is where your components will be placed. You will simply lay out the pads and traces, then plop the components down on top. It's the easiest possible way to lay out a printed circuit board.

An Issue of Scale

As with traditional PC board design (see Chapter 7), you must decide whether to work full scale (1×), or double or even quadruple (2× or 4×) the final size. There are certainly advantages to working at 2× or 4×. For one thing, it's easier to see what you are doing, without working under a magnifying glass. Also, SMT PC board design templates are readily obtainable in 2× and 4× versions. There are not, however, any such templates available at 1×. However, working at greater than full scale presents one very serious drawback. The artwork at the larger scales must be photographically reduced. If you have the facility to do so—great. You will probably want to design your SMT PC board at 2× or 4×. However, if you cannot reduce your artwork to make a full-scale negative, you must work at 1×.

Can it be done? Can one even design and tape up an SMT board at full scale? Of course! The fact that 1× surface mount artwork patterns are made at all is an indication that it's doable. You will need to work under a magnifier, you will have to be extremely careful in laying down tape, some of which will be as thin as 0.015″, and you will have no PC board design template to aid you in the initial layouts. If you can accept the challenge, however, the results are often impressive. Not only was our 555 Dual LED Flasher sample project laid out by hand at 1×, so were the projects presented in Appendix E. It can be done!

Doing the Design Layout

To begin a PC board design layout at 1×, tape down a sheet of graph paper having 20 squares to the

FIGURE 13–10
Freehand trial layout for 555 Dual LED Flasher Project

inch (0.050″ line spacing). On the sheet, draw a perimeter defining the borders of your PC board. For our sample project, a $1'' \times 1\frac{3}{4}''$ rectangle is laid out (Figure 13–10). Since you don't have a PC design template to work with, you will want to make use of the dry-transfer SMD patterns that you'll use later at the artwork stage. Actually, it's just as well; the patterns are accurate and easy to work with. Start by laying down a key component pattern, one for an IC perhaps. The IC pads should be centered on the graph paper grid lines. Transfer the pattern with a ball burnisher, then burnish it hard with a board burnisher or the back of a pocket comb. As you burnish, protect the pattern with the blue backing paper that comes with each rub-on sheet.

Now, following the schematic drawing, lay down a few discrete component terminal pads. Begin drawing lines (straight or curved), representing traces, from one pin or terminal to another. For the straight lines, you need not use a straightedge at this point: remember, you're only establishing a freehand trial layout sketch during this first iteration (Figure 13–10). Don't forget to include a pad for every component terminal or pin. If there are to be any PC board mounting holes, now is the time to identify them. The completed sketch for the 555 Dual LED Flasher is shown in Figure 13–10.

Rarely will you be able to complete the PC board design layout with one pass. A second, final design layout is almost always necessary. Redraw the perimeter of your board near the first iteration, then begin the layout anew, taking your cues from the first try. Lay down new patterns, improving the original arrangement as needed. This time you might draw connecting straight lines with a straightedge. Avoid any acute (less than 90°) angles. One thing nice about working 1× in surface mount is that in many cases your line widths are going to be quite close to the tape width used on the PC board artwork. The final design layout for the sample project is shown in Figure 13–11.

If you are satisfied with the layout, check and recheck it against the schematic drawing. Remember, errors caught at this stage will be much less costly than those discovered later on.

FIGURE 13–11
Final design layout for 555 Dual LED Flasher Project

SMT PC Artwork

We will start by examining the artwork materials you'll need. We'll then see how SMT dry transfers are applied and how tape is laid down. We will conclude with a look at artwork touch-up.

Artwork Materials

To begin with, you'll need a clear acetate sheet, $8\frac{1}{2}'' \times 11''$, either 0.005″ or 0.007″ thick. You will want an X-Acto knife with a #60 blade and a burnishing tool of some sort. A light table and magnifying glass are also required.

Currently, your best source for surface mount device patterns is Datak Corp. It sells "Direct Etch" dry transfer packages for a wide range of device patterns. Each package, selling for about $2, contains enough patterns to make about a dozen simple PC board layouts. In Figure 13–12a are listed the packages you'll need to get started. The part number, SMD component designation, and device pattern are shown.

Black pressure-sensitive tape is readily available from companies such as Datak. You will need a roll or two of tape 0.015″, 0.031″, and 0.040″ wide to get started (Figure 13–12b).

In addition to these materials, a set of right-angle board delineation marks would be useful. Of course, you could use 0.080″-wide tape instead. Sets of numbers, symbols, and letters will also be needed. They are available in stationery stores.

Applying SMD Dry Transfers

When applying dry transfers along with pressure-sensitive tape, you must work slowly and carefully, somewhat like a surgeon. Patience and practice are the key. Take it nice and slow.

Begin by placing your PC design layout on a light table. Lay a sheet of acetate over the graph paper and tape both securely in place. Swing your magnifying glass into position.

Your first step will be to lay your board delineation marks in place. Remember, the inside edges of the tape identify the outer border of the PC board. If you have any mounting holes to locate, do so now.

You have already used SMD dry-transfer patterns when laying out your original PC design. You

DATAK part no.	SMD component designation	SMD component pattern
DE 20	DIP	
DE 22	1805	
DE 23	1206	
DE 24	1210	
DE 26	3216	
DE 29	7243	
DE 32	SOT–23	
DE 33	SOT–89	

(a)

Tape size	
0.015″	
0.031″	
0.040″	

(b)

FIGURE 13–12
Artwork materials

now apply them onto the acetate sheet in the same way, but here you must be extra careful. Avoid allowing the dry-transfer sheet to touch any portion of the artwork. If it does, the sticky surface may adhere to the previously applied patterns and accidentally pull them up.

There is another problem to be mindful of. If you are not careful, individual terminal pads will move relative to each other. Although the spacing of the pads is precisely determined, they are not held in common. Each tiny pad is an individual dry-transfer rub-on. Keep this in mind.

In Figure 13–13 we see the board delineation marks and SMD dry-transfer patterns set in place for our 555 Dual LED Flasher. We are now ready to lay down tape.

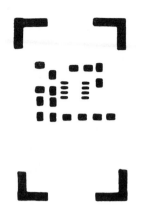

FIGURE 13–13
Board delineation marks and SMD patterns for 555 Dual
LED Flasher Project

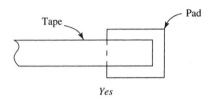

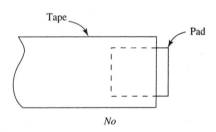

FIGURE 13–14
Tapes and pads

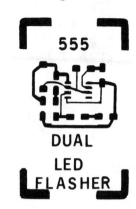

FIGURE 13–15
Artwork for 555 Dual LED Flasher Project

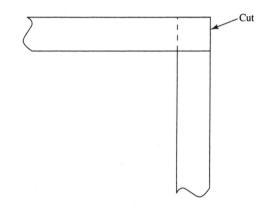

FIGURE 13–16
Overlap and cut

Laying Down Tape

Keeping in mind that your tape width must be smaller than the pad it intersects (Figure 13–14), it is, nonetheless, good practice to work with the widest tape possible. We do this not so much to save copper or to carry more current but because thicker tape is easier to work with, and the resulting trace will have less chance of acid undercut. Note that in our completed 555 Dual LED Flasher tape-up, Figure 13–15, two tape widths are used. For the most part, 0.031″-wide tape is laid down between discrete component pads, while the much thinner 0.015″ tape is reserved for traces run to IC pads that are only 0.025″ thick. The concept is simple: start wide and work down to the thinner tapes only as needed.

Tape routing can be either curved or straight, or a combination of the two. Curved traces are per-

fectly acceptable, and since the tape is so thin, rather tight curves are possible.

To lay down tape on angled curves you'll want to use the "overlap and cut" method. As shown in Figure 13–16, you simply lay one strip of tape over the other and cut as close to the edge as you can.

Note that as you apply tape over dry-transfer pads, then pull it back against the X-Acto knife blade to make the cut, the tape tends to pull the pad up off the acetate (Figure 13–17a). There are at least two ways to combat this negative effect: (1) Place the tape completely over the pad, press down firmly, and cut the tape directly at the end of the pad (Figure 13–17b). It's a variation of the "overlap and cut" approach. You do not pull the tape up to cut, thus avoiding the pulling effect. (2) You can do it the traditional way, that is, pull the tape back against the X-Acto knife blade. As you do, however, do not press the tape into place; instead, lay it down (float it) gently over the pad (Figure 13–17c). Before taping up the actual board, experiment with both of these methods. With a little practice, you'll find the technique that works better for you.

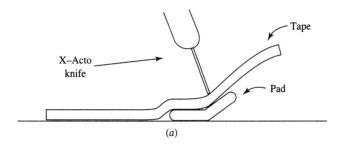

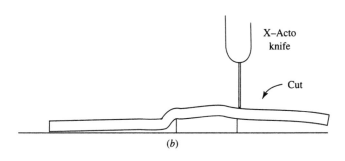

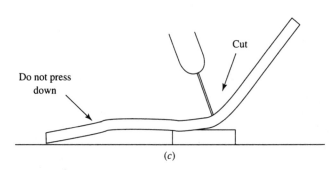

FIGURE 13–17
Working with tape and dry transfers

Finalizing the Artwork

Once dry-transfer pads and pressure-sensitive tape are laid in place, it's time to apply any nomenclature. Using the underlying graph paper as a guide, rub numbers, symbols, and letters onto the acetate as desired. If you need to remove a dry transfer for any reason, don't attempt to lift it with your X-Acto knife or other sharp tool. Simply press a strip of clear tape over the offending letter, number, or symbol and pull up. The dry transfer will lift right off the acetate sheet.

Now, examine the completed artwork carefully under a magnifying glass. Identify any X-Acto knife cuts in the tape or pads. These transparent slits, if they exist, must be made opaque before a negative is produced. If they are not, they will appear as black lines on the negative. If left on the negative, breaks in circuit traces may result when the PC board is fabricated. To cover the tape slits, use a special pen designed for the purpose (found at stationery stores) or a thick black felt-tip pen. Apply in a dabbing motion to cover the slit. Work slowly, applying only a little ink at a time.

With the artwork complete, it's time to make the negative (see Chapter 8). If there will be a delay between artwork generation and negative production, protect the original artwork by placing it, face down, on a sheet of cardboard (a business card works well). Secure the artwork in place with clear tape.

Something to Think About

All of today's PC board design- and artwork-generation software packages include SMT layout tools—they have to. Why not see if you can work with one, at school or in industry? Are there differences in approach from working with IMT design?

13.2 SMT PROJECT FABRICATION

With your project artwork negative in hand, you are now ready to begin the SMT project construction process. Here we will explore PC board fabrication and surface mount assembly using solder wire and solder paste. We will conclude with a look at the finished 555 Dual LED Flasher Sample Project. After completing this and the previous section, you'll be able to build your very own SMT project—from initial design to final assembly.

Making the Printed Circuit Board

In examining the PC board fabrication process, we'll first briefly review the traditional method, then we will note how the fabrication of an SMT PC board is different.

The Traditional PC Board Fabrication Process Reviewed

The traditional, through-hole, PC board fabrication process was discussed extensively in Chapter 8. We will touch only on the highlights here, by way of review.

To begin with, a PC board fabrication drawing must be prepared. Such a drawing, for the sample project, is shown in Figure 13–18. Note, in contrast to the more traditional drawing in Figure 8–10, that our SMT illustration is much simpler. That is because it has no hole drilling data.

FIGURE 13–18
PC board fabrication
drawing for the 555 Dual
LED Flasher Project

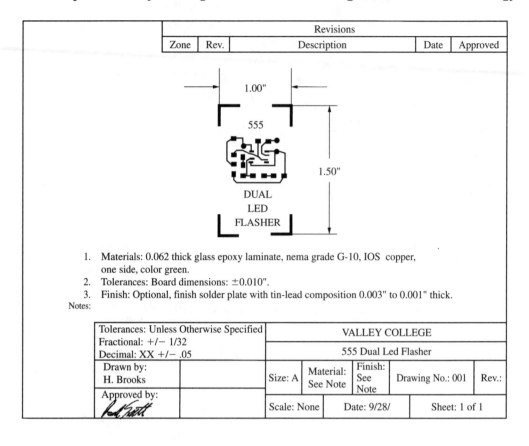

		Revisions		
Zone	Rev.	Description	Date	Approved

1.00"

555

1.50"

DUAL
LED
FLASHER

Notes:
1. Materials: 0.062 thick glass epoxy laminate, nema grade G-10, IOS copper, one side, color green.
2. Tolerances: Board dimensions: ±0.010".
3. Finish: Optional, finish solder plate with tin-lead composition 0.003" to 0.001" thick.

Tolerances: Unless Otherwise Specified Fractional: +/− 1/32 Decimal: XX +/− .05		VALLEY COLLEGE				
		555 Dual Led Flasher				
Drawn by: H. Brooks		Size: A	Material: See Note	Finish: See Note	Drawing No.: 001	Rev.:
Approved by:		Scale: None	Date: 9/28/		Sheet: 1 of 1	

You will recall from Chapter 8 that there are 10 steps to the fabrication of a PC board. They are listed in Figure 13–19.

(1) The blank board is cleaned of contaminants and oily residue with fine steel wool. (2) A liquid photoresist (KPR) is applied with a paint brush. (3) The photoresist is dried with a hair dryer. (4) The sensitized board with the negative on top is exposed to ultraviolet light. (5) The image is developed.

Ten Steps to Fabricating a PC Board

Step 1   Clean the blank PC board.
Step 2   Apply liquid photoresist.
Step 3   Dry the photoresist.
Step 4   Expose the board to ultraviolet light.
Step 5   Develop the image.
Step 6   Rinse and dry the board.
Step 7   Etch the board.
Step 8   Wash and dry the board.
Step 9   Remove the remaining photoresist layer.
Step 10 Drill the board.

FIGURE 13–19
Ten steps to fabricating a PC board

(6) The board is rinsed and drip-dried. (7) The board is immersed in acid, usually ferric chloride. (8) When etching is complete, the board is thoroughly washed and then dried with a paper towel. (9) The remaining photoresist layer is removed by lightly rubbing the board with steel wool. Finally, the board is drilled.

The SMT PC Board Fabrication Process

Fabrication of an SMT PC board is, step-for-step, identical with that for a traditional board, with one major exception—there is no step 10. Assuming that there are to be no PC board mounting holes, then obviously step 10, drilling component lead or pin holes, is unnecessary with an SMT board.

There are, nonetheless, two minor steps, or cautions, to be observed when making your SMT PC board. First, since some of your traces may be extremely thin, 0.015", for instance, it behooves you to check the etching progress frequently. Pull the board out of the acid bath often and examine it closely for acid undercuts. Those narrow traces are delicate; leaving your board in the acid bath any longer than necessary to etch away the unwanted copper is asking for trouble.

Second, once the board is complete, you will want to give it a good inspection under a magnifying glass, but that may not be enough to locate any possible trace breaks. Therefore, you should check each trace (and pad) with a continuity checker (ohmmeter). Only then can you be sure if the board is ready to be stuffed with surface mount components.

Attaching SMCs to the PC Board Prior to Hand Soldering

In prototype work, surface mount components are soldered to the PC board by hand, but with SMT, unlike with IMT, the solder joint must supply not only the necessary electrical connection between component and pad but the mechanical bond as well. The question is, How do we even hold the SMCs in place in order to do the soldering? It seems one would need three hands: one to hold the solder, one to grasp the soldering iron, and one to grip the component. With IMT, we can get away with two hands because the component leads do the "holding." But with SMCs, there are no leads, or at least none that are inserted into the PC board and bent to secure the component. A "third hand" is needed. How to supply that third hand is what we will discuss shortly. Before we get started, however, let's look at the tools and materials required for the various approaches we will be examining.

Tools and Materials for Attaching SMCs

In a few moments we will examine three methods used to hold SMCs in place while hand-soldering with solder wire. Since you will undoubtedly be experimenting with all three approaches, you should obtain all of the tools and materials listed in Figure 13–20.

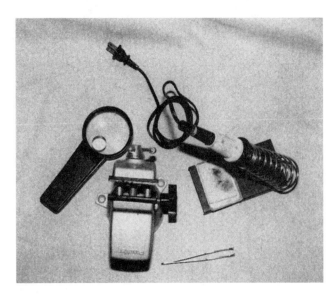

FIGURE 13–21
Tools for attaching SMCs

The soldering iron needs to be from 25 to 40 W and have a tinned tip with a conical shape, $\frac{1}{16}''$ or less in diameter. A tweezers, like the one shown in Figure 13–21, is mandatory for picking up the tiny SMCs. It should have forceps-style tips. A vise is a must for securing the PC board while components are being attached. The type displayed in Figure 13–21 works well. It is assumed that you already have a magnifying glass. Nonetheless, the small, handheld unit shown in the figure is also useful and worth obtaining.

As to the materials listed in Figure 13–20, they are fairly self-explanatory. Solder should be 63/37 or, better yet, silver-bearing solder at 62/36/2. A diameter of 0.020″ is OK; 0.015″ is better. A noncorrosive liquid flux and drop dispenser are necessary, as is a light-duty spray defluxer. A general-purpose plastic cement, the kind for plastic, wood, or metal, is fine. When choosing solder wick, select a diameter of 0.030″. Ordinary clear tape, $\frac{1}{2}''$ in diameter, is fine. The rest of the materials shown in Figure 13–22 are probably lying around your house. Now is a good time to gather the items up and place them on your workbench.

Taping SMCs Down

One way to provide a "third hand" to temporarily hold an SMC in place is with ordinary clear tape. As seen in Figure 13–23a, you begin by simply grasping a short strip of the tape and touching it to the SMC, picking up the latter. You then tape the SMC onto its PC board pads. Using clear tape instead of masking tape allows you to see through to align SMC pins and

---

Tools
☐ Soldering iron
☐ Tweezers
☐ Vise
☐ Magnifying glass (handheld)

Materials
☐ Solder
☐ Liquid flux
☐ Drop dispenser
☐ Defluxer
☐ Adhesive (plastic cement)
☐ Solder wick
☐ Clear tape

FIGURE 13–20
Tools and materials for attaching SMCs

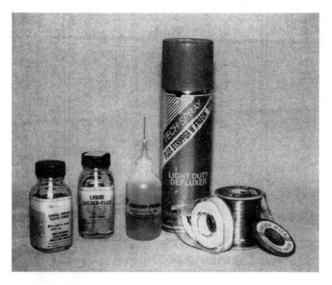

FIGURE 13–22
Materials for attaching SMCs

The tape method is quick, easy, and inexpensive, plus there is no messy adhesive to deal with. Its only drawback has to do with board crowding. The tape method doesn't work well when the PC board is already crammed with SMCs. It is hard to work the tape in between the components to provide a secure adhesion.

Using Adhesive

Another way to hold an SMC to the circuit board in preparation for final soldering is to glue it in place. A general-purpose glue, available in a jar, or special epoxies, dispensed from a syringe, are used. Of the former, the General Purpose Plastic Cement, 10-324, from GC Electronics works well. If you wish to use an epoxy, the SME-12 Epoxy, packaged in a preloaded syringe, from Capital Technologies, Inc. is probably your best choice, since it has been formulated specifically to adhere SMCs.

When working with a glue, the best way to dispense the material is with a toothpick. Dip the pick into the liquid, stopping just as it touches the surface, then place a dot of glue on the PC board, being careful not to get any of it on the copper pads (Figure 13–24). Gently press the component into

terminals onto the board pads. With the component held securely in place by the tape, you now solder the exposed terminal (Figure 13–23b). Remove the tape and solder the remaining terminal(s) to their pads (Figure 13–23c).

FIGURE 13–23
Taping SMCs down

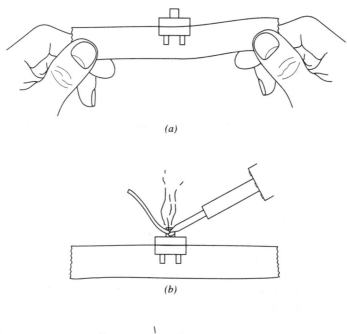

(a)

(b)

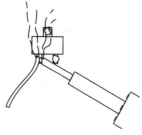

(c)

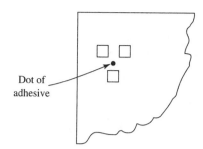

FIGURE 13–24
Dispensing adhesive

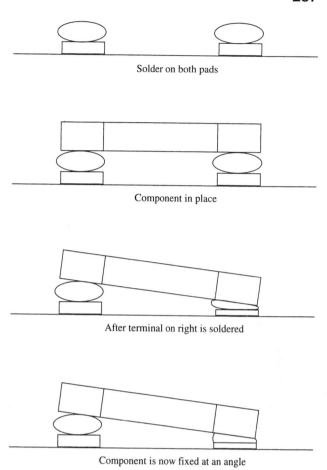

Solder on both pads

Component in place

After terminal on right is soldered

Component is now fixed at an angle

FIGURE 13–25
The drawbridge effect revisited

place and let the glue dry for about 30 minutes. Be careful not to apply too much glue—remember, its purpose is only to secure the SMC for soldering. The rule is, "A little dab'll do you."

Using syringe-dispensed epoxy is a little more difficult until you get the hang of it. You must first set up the syringe for dispensing, dispense the material, place the component, then cure the epoxy. The last step can be done by placing the PC board in a toaster oven, set at 300°, for 10 to 20 minutes. If you are using the SME-12 Epoxy from Capital Technologies, Inc., simply follow the instructions that come in the package.

Tag Soldering

One of the most effective ways to attach SMCs is by tag soldering. It's easy, quick, and involves no additional materials such as tape or adhesives. All you need is the solder wire and soldering iron that you will be using anyway.

First, apply a dab of liquid flux to *one* PC board pad, then create a pool of solder on that pad. Allow the solder to solidify.

Next, with the SMC held in a tweezers, rest the component on its PC board pads and hold it in place. Using a soldering iron held in your other hand, reflow the solder, causing the component terminal to "sink" close to the board surface. Remove the iron, allow the solder to again cool, and release the tweezers. Now with the component "held" in place, solder the other terminals or pins in a traditional manner.

When tag soldering, you must avoid the temptation to first apply solder to *all* component pads, then put the component in place, and finally reheat each terminal. If you do this, a variation of the drawbridge effect, discussed in Chapter 12, will occur. As the first terminal sinks to the PC board when heated, the component will tip slightly (Figure 13–25). Heating subsequent terminals will not allow the component to "flatten out," since it is already held rigid at an angle. The result is unacceptable.

Which Way Is Best?

Three ways to secure an SMC to a circuit board prior to hand soldering have been addressed. Which one is best? That depends! All have their advantages and disadvantages. Some may work better in one situation than another. Furthermore, on a given board, more than a single method may be required. In the end, you may want to come up with your own customized means of providing a "third hand." In doing so, be flexible, inventive, and experiment a bit. You'll discover that there is definitely more than one way to hold a surface mount component to a printed circuit board.

Surface Mount Assembly Using Solder Wire

Once an SMC is held in place by the methods just described (or by a procedure you have devised), it's time to solder the remaining terminals or leads in place. You will need a soldering iron and tip, solder, solder wick, liquid flux, flux remover, nailbrush, vise, magnifying glass, and lots of patience—"A slow hand with an easy touch."

First, we will examine how to solder discrete SMCs, then how to attach surface mount ICs. We'll conclude with a brief discussion of flux and component removal.

Soldering Two-Terminal Discrete SMCs

To begin with, we assume that your PC board is in a vise, your magnifying glass is swung into place, and your tools and materials are in easy reach. To solder a chip resistor or capacitor (or a similar two-terminal SMC) to the PC board, follow these four steps:

1. Clean the PC board pads. Using steel wool or some other fine abrasive, be sure the copper pads are clean, shiny (deoxidized), and ready to accept solder.

2. Tag-solder one component terminal to its pad. (We pick the tag-solder method to give you a chance to practice soldering.) Apply liquid flux and flow a puddle of solder onto the pad. When the solder has cooled, position a component with your tweezers and hold it in place. Now place the tip of your soldering iron so that it touches both the pad and edge of the component terminal (Figure 13–26a). The solder should reflow in less than 2 seconds. As soon as you see the solder melting, remove your iron. Allow the joint to cool,

then extract the tweezers. Your component is now in place.

3. Solder the other terminal. Do so by first applying a dab of liquid flux. Next, with solder wire in one hand and a soldering iron in the other, solder the terminal to the pad in a manner similar to that discussed in step 2. It's basically the four-step soldering process you learned about in Chapter 7: apply heat, apply solder, remove heat, remove solder. The only difference is the time element. With SMCs, heat can be a real killer. The whole soldering process should take no more than 2 seconds—3 at the most. You'll have to practice a bit, but you will be able to do it.

4. Inspect your work. Using the magnifying lens attached to your workbench or a handheld unit, examine the connection carefully. Make sure there is enough solder, yet not too much. Figure 13–26b shows good and bad SMT two-terminal solder joints.

Soldering Three- and Four-Lead Discrete SMCs

Most three- and four-lead SMCs have gull-wing leads. The steps necessary to solder these components in place are the same as those for two-terminal devices: (1) clean the board; (2) tag-solder

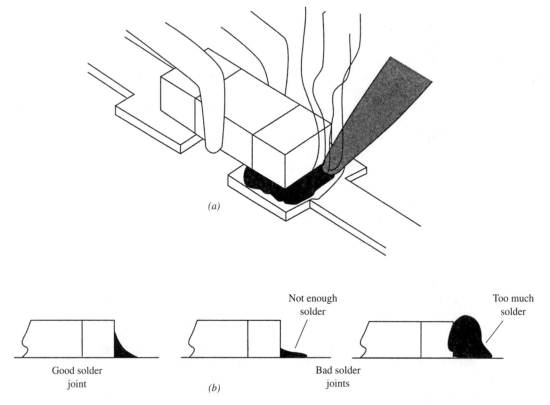

(a)

Not enough solder

Too much solder

Good solder joint

(b)

Bad solder joints

FIGURE 13–26
Soldering two-terminal SMCs to a PC board

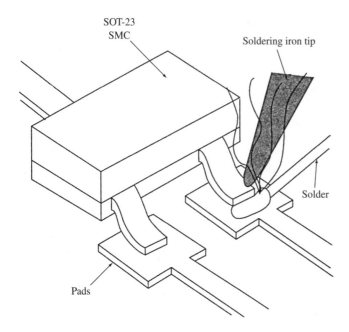

FIGURE 13–27
Soldering gull-wing leads

4. Place the IC onto the copper pads. Be sure to center the IC right-to-left as well as top-to-bottom. Hold the IC in place.

5. Apply the soldering iron tip to the pretinned pad. Hold the iron in place just long enough to flow the solder.

6. Pivot the IC about the soldered pin as necessary to again align all pins with their respective pads.

7. Solder a second pin kitty-corner to the first. In doing so, don't forget to apply a small amount of liquid flux.

8. Bathe one line of pins with liquid flux.

9. Solder the line of pins, one at a time. Move quickly, yet adhere to the four traditional soldering steps: apply heat; apply solder; remove solder; remove heat. Try to spend no more than 2 seconds on each pin. Apply liquid flux to the opposite row of pins and solder as just described.

10. Inspect your work under a magnifying glass.

Flux and Component Removal

With soldering completed and the PC board inspected, it's time to remove flux residue. Using a light-duty defluxer in a well-ventilated area (preferably outdoors), spray the board in a back-and-forth motion from top to bottom. Wait a few seconds and brush the board vigorously with a nailbrush or similar abrasive. Don't worry about knocking off any components. If the SMCs can't withstand the brushing, then they weren't well soldered to begin with. Now is the time to find that out. Inspect your work one more time under a magnifying glass.

Unfortunately, there will be times when you'll need to remove a component (wrong component, component in backward, component severely misaligned, etc.). The subject of SMC removal is discussed in detail in the next chapter. As a general procedure, you simply dip a solder wick into liquid flux, place it over the lead or pin, and apply heat with a soldering iron (see Chapter 6). When all possible solder has been removed, the component should lift up with a slight pull. (See Chapter 14 for an extensive discussion of SMC removal and replacement.)

Surface Mount Assembly Using Solder Paste

It is possible to completely assemble your SMT prototype project without ever having to pick up a piece of solder or a soldering iron. You can do it with solder paste. It's a method that lets you reflow the entire circuit in one operation. The paste is first applied to all PC board pads, then individual SMCs are placed

one terminal; (3) solder the other terminals; and (4) inspect your work. It is particularly important with gull-wing leads, however, to do the actual soldering as quickly as possible. Don't linger with your soldering iron in place. Get off the lead and pad as quickly as possible.

With a three-lead package, such as an SOT-23 case, tag-solder the lone lead first, then use your finger or a tweezers to shift the other pads into alignment, if necessary. Now solder the remaining leads as per step 3. When soldering gull-wing leads, place the solder wire where the edge of the lead touches the copper pad (Figure 13–27).

Four-lead SMCs are handled in much the same way. Start by tag soldering a small corner lead. Then proceed as with the three-lead SMC.

When soldering any multilead component, be sure to use plenty of liquid flux: it makes for a much more effective solder connection. Finally, don't forget to inspect your work closely under a magnifying glass.

Soldering SMT Integrated Circuits

Hand soldering DIP gull-wing SOICs to a circuit board is not difficult. Just follow these 10 steps:

1. Make sure all pads are clean and ready for soldering.

2. Create a pool of solder on a corner pad in preparation for tag soldering.

3. Hold the IC with a tweezers or grip it between your fingers.

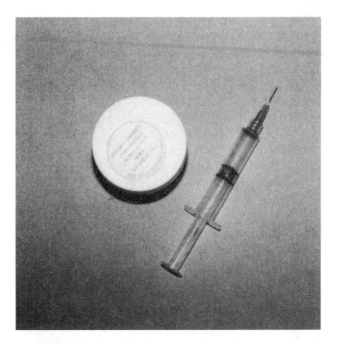

FIGURE 13–28
Soldering paste

on the board. Third, the entire assembly is heated until the solder melts. Finally, the circuit is allowed to cool, and any remaining flux is removed. Let's see what the paste consists of, how it is dispensed, and how to reflow it using an ordinary toaster oven.

Solder Paste Characteristics

Solder pastes, or creams, consist of minute prealloyed spherical solder particles, much smaller than a grain of salt, suspended in solder flux and activators. A 63/37 mixture with a melting point of 360°F (183°C) is common. Even better is the 62/36/2 mixture with a melting point of only 355°F (179°C). The lower the reflow temperature, the more desirable the paste.

The paste is available in a jar (bulk) or a prepackaged syringe. Examples of both are shown in Figure 13–28. Costs vary and are subject to change, so consult your distributor for the latest prices.

It is vital to be extremely safety conscious when working with solder pastes. Remember, approximately 40% of the mixture is lead, a deadly substance if ingested. A typical warning label is worth repeating here:

> **Harmful or fatal if swallowed**. Contains lead and/or tin. If swallowed induce vomiting immediately and call physician. Vapor harmful, use with adequate ventilation. Do not use in open or unsealed container. Do not reuse container. **Keep out of reach of children.**

It is a good idea to wash your hands (and particularly under the fingernails) frequently when work-ing with solder paste. Actually, it's even smarter to wear latex gloves.

Solder Paste Application and Component Placement

Although you can apply solder paste by scooping it out of a jar with a toothpick and dabbing it onto the PC board pads, it's best to dispense the material with a prepackaged syringe. If you are able to purchase paste only in bulk, then you can buy your own syringes and load them with paste. Sets of inexpensive, disposable syringes with various tip sizes are available at most electronics or hobby stores.

If you have never used a syringe to dispense solder paste, you'll want to practice a bit first. To begin, be sure the paste is thoroughly mixed. It's a good idea to roll the syringe barrel against a hard surface from time to time to mix up the ingredients.

You may want to rehearse on a sheet of paper or a spare PC board. As you do, notice that the paste tends to run out of the tip after you release the plunger. Also, at times a watery, oily substance may exude from the syringe, leaving the solder paste behind. When that happens, it's a good indication that additional mixing is required.

After you get the hang of paste dispensing, turn to your project PC board and try to cover each pad with a thin layer of the cream. When you are ready to place the component, grasp it with a tweezers and lay the device on top of its pads. Do not press the component to the board surface. Push down just enough to sink the SMC slightly into the paste, yet not so much that the paste oozes out. When all components are in place, you're ready to reflow the solder by placing the circuit in an oven.

Solder Reflow with a Toaster Oven

Using an ordinary toaster oven (Figure 13–29), commandeered exclusively for the purpose (you don't

FIGURE 13–29
Toaster oven used to reflow solder paste

want to mix solder bits with your morning toast), it's a simple matter to reflow your prototype circuit. Follow these five steps:

1. Place the circuit in the toaster oven. Set the oven temperature to 200°F and leave it there for 5 minutes. You are now curing the solder paste and preheating the board.
2. Increase oven temperature to 400°F, where it should remain for 8 to 10 minutes. The solder will reflow during this step.
3. Remove the circuit from the oven using a tweezers or pliers. Do not touch the circuit; it will be quite hot.
4. Let the circuit cool at room temperature for 5 minutes.
5. Inspect the circuit carefully under a magnifying glass for cracks. Pull on the components to check if they are secured in place.

If all seems OK, clean the circuit with a light defluxer, as described earlier. If a solder joint has not reflowed satisfactorily, rather than reheating the entire circuit, touch up the offending joint with solder wire and a soldering iron.

The Completed Sample Project

As we conclude this chapter on SMT project design and fabrication let's look at our completed 555 Dual LED Flasher Sample Project and the SMT practice board.

The 555 Dual LED Flasher

Our completed sample project is shown in Figure 13–1. Notice that the board is attached to a 9-V battery with double-sided tape. The components were tag soldered into position and the project completed with solder wire and a soldering iron. The battery snap is the only non–surface mount component. Leads were attached by first tag soldering their pads, then reflowing the solder with a lead held in place.

SMT Practice Board

The artwork for an SMT practice PC board is shown in Figure 13–30. It is also available in Appendix E. As you can see, it contains the pads, or footprints, for a variety of SMCs. Traces 0.031″ and 0.015″ wide have been added at random.

If you fabricate the board, you'll gain experience manufacturing an SMT PC board with very thin traces. You can then use the board to practice surface mount assembly using solder wire or solder paste.

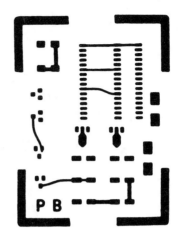

FIGURE 13–30
SMT practice board

Something to Think About

Rob Laschinski, president of Capital Advanced Technologies, the maker of Surfboards, has a most interesting Web site full of great information on SMT fabrication. Why not visit it at www.capitaladvanced.com? (For further Web sites to visit, see Appendix K.)

SUMMARY

In Chapter 13 we learned how to design and fabricate surface mount circuits. We began with a look at SMT design factors: when to build with SMT and where to purchase SMCs. We then explored a fascinating way to breadboard surface mount components using Surfboards. Next, it was on to SMT PC board design layout considerations. We concluded Section 13.1 by investigating SMT PC artwork, with emphasis on applying dry transfers and laying down extremely thin tape.

In Section 13.2, SMT Project Fabrication, we first saw how a surface mount PC board differs from the more traditional through-hole type. We then investigated ways, such as taping, using adhesive, and tag soldering, to initially attach SMCs to a circuit board. We examined two methods of surface mount assembly: one using solder wire, the other using solder paste. We concluded with a final peek at our sample project, the 555 Dual LED Flasher.

Working with prototype SMT is one thing, reworking stuffed production boards is another. Removing and replacing SMCs that were automatically inserted is an art in itself. To find out more, turn to Chapter 14.

QUESTIONS

1. The Surfboard breadboarding concept combines the _____ PC board and the _____ circuit board approaches.

2. Surfboards come in two series, one for _____ components and the other for _____ circuits.

3. What is the main disadvantage of working 2× or 4× when creating a PC board layout?

4. In designing a PC board, a freehand _____ layout sketch and a _____ design sketch are usually required.

5. In producing artwork for your SMT PC board design layout, _____ transfers and pressure-sensitive _____ are used.

6. With SMT PC board artwork, tape as thin as _____ is often used.

7. Fabrication of an SMT PC board is identical with that of a traditional board with one important exception—no _____ need to be drilled.

8. List the three common methods used to supply a "third hand" when attaching SMCs.

    1.

    2.

    3.

9. When soldering SMCs to a circuit board using solder wire, plenty of liquid _____ should be used.

10. Solder paste used for SMT assembly is best dispensed with a _____.

REVIEW EXERCISES

1. Referring to Review Exercise 3 in Chapter 7, produce a 1× freehand trial layout sketch for the circuit shown. Produce either curved or straight traces, or a combination of both. Be sure to account for any off-board components.

2. From the freehand trial layout sketch in Exercise 1, draw the final design layout. Use dry-transfer surface mount device patterns, a pencil, and a straightedge.

3. Referring to Review Exercise 7 in Chapter 7, produce a 1× freehand trial layout sketch for the circuit shown. Produce either curved or straight traces, or a combination of both. Be sure to account for any off-board components.

4. From the freehand trial layout sketch in Exercise 3, draw the final design layout. Use dry-transfer surface mount device patterns, a pencil, and a straightedge.

5. Using the following artwork, lay traces between various component pads with 0.031″-wide tape. Route both curved and straight traces. Continue to lay down tape until you are reasonably proficient.

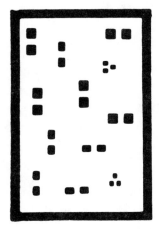

6. Using the following artwork, lay traces between various component pads with 0.015″-wide tape. Route both curved and straight traces. Continue to lay down tape until you are reasonably proficient.

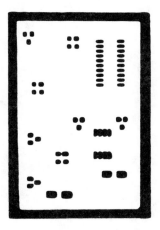

7. Fabricate the SMT Practice Board. Check it carefully for underetching of the thin traces.

8. Using the SMT Practice Board (or a similar board of your own design), attach various SMCs using three methods: (1) taping; (2) adhesive; and (3) tag soldering. Compare the results. Determine which method you prefer.

9. Continuing with Exercise 8, complete the component attachment process by hand soldering with liquid flux, solder wire, and a soldering iron.

10. Using the SMT Practice Board (or a similar board of your own design), attach various SMCs using solder paste.

14 SMT Rework—Removal and Replacement of Surface Mount Devices

OBJECTIVES

In this chapter you will learn

- [ ] How traditional soldering tools are used for SMC removal.
- [ ] How direct heat attachments are used to remove SMCs.
- [ ] How handheld air devices are used to remove SMCs.
- [ ] How SMCs can be replaced with direct heat attachments.
- [ ] How SMCs are replaced with handheld air devices.
- [ ] What inspection and rejection techniques are used to determine solder joint integrity.
- [ ] How to determine acceptable SMT solder joint workmanship.

Today's electronic circuits, as any consumer knows, are exceedingly reliable. The vast majority of them never fail; they seem to last forever, or at least until an improved version comes along. Yet some circuits, even those employing SMDs, do at times cease to function; they experience physical damage, environmental degradation, or even component mortality. When that happens, two choices present themselves: you can throw the circuit board away, or you can try to repair it. Although the first option predominates in the low-priced consumer electronics field, "board chucking," contrary to popular impression, is by no means universal within the industry. Today's multilayer SMD PC boards can be quite expensive. When they do fail, they are more likely to be repaired than recycled. That's where the SMT electronics technician comes in.

Traditionally, circuit repair followed troubleshooting in one more-or-less continuous operation. Once the problem was discovered, the actual repair—the removal and replacement of an electronics component—was quickly effected. You clipped the failed component out and soldered a new one in. Not so with today's SMT circuits. Troubleshooting is one thing, circuit rework, quite another. With SMT, removal and replacement (R & R) is an art and skill unto its own, requiring special tools and training. As an electronics technician you'll probably spend more time reworking SMT circuits than building their prototypes. Since that is to be the case, in this last chapter on surface mount technology, let's take a look at how such circuit repair is accomplished.

14.1 SMD REMOVAL: TAKING IT OFF

When a surface mount component fails, when it is installed incorrectly, or when it is simply the right component but in the wrong place, it has to be removed. Doing so requires special tools and techniques in order to avoid what has traditionally been referred to as "destructive repair," where a "little" PC board damage was acceptable. With SMT rework, however, protecting the printed circuit board is paramount. Thus a "nondestructive repair" approach is mandatory. Although component removal may not save the component, the procedure *must* save the PC board pads and traces. The reworked board must then be as ready as a new board to accept a surface mount device. With that in mind, let's now examine the tools and techniques used to make SMD removal as "painless," or nondestructive, as possible.

Tools for Disengagement: Methods of SMD Removal

There are a variety of tools and materials available for removing SMDs. Which ones you choose will depend on a number of factors, such as the cost of equipment, quality of rework demanded, volume of component removal desired, speed of removal necessary, types of components and boards used, and personal preference. If you do only an occasional component removal, traditional soldering tools, such as the ones described in Chapter 13, will often suffice. On the other hand, if high-volume, truly professional rework (perhaps involving military specifications) is necessary, a complete professional rework station is usually demanded. Here we will examine four options with regard to tools for SMD removal: (1) traditional solder tools; (2) direct heat attachment soldering tools; (3) handheld hot-air devices; and (4) professional rework stations. As you explore these options remember that most of the tools and equipment used are for component replacement as well as component removal.

Traditional Soldering Tools for SMD Removal

Traditional soldering tools, such as the ones discussed in Chapter 13, can be used for SMD removal. Their cost is low, actually nonexistent if you have

acquired the tools for other purposes. The speed of SMD removal using this method is slow, of course, but the quality of rework can be quite acceptable. Furthermore, since you are already familiar with the tools, you should find them comfortable, convenient, and easy to work with.

As shown in Figure 14–1, you will need a 25- to 40-W soldering iron with a conically shaped tinned tip, $\frac{1}{16}''$ or less in diameter. You will want lots of liquid flux with a nail polish brush for dispensing (a drop dispenser is also handy). Have plenty of solder wick, 0.030″ and 0.060″ wide, on hand. Of course, you will want a tweezers. An inexpensive dental pick, like the one shown in the figure, is quite useful. You will need the pick to lift off "uncooperative" components, especially those that have been glued in place. Naturally, some sort of vise to hold the circuit board is mandatory. You will be pulling up on components as you heat them for removal, so the board must be secured when you apply the upward tug. Finally, you may also want a magnifying glass, depending on the keenness of your eyesight. Using the glass now, however, even if you don't think you need it, might save your eyesight for future work.

Direct Heat Attachment Soldering Tools: Removal Through Heat Conduction

With the advent of surface mount components, tools and attachments specifically designed to remove and replace them have been developed. Of particular interest, at the low-cost end, are direct heat attachments that screw onto the end of a soldering

FIGURE 14–1
Traditional tools for SMD removal

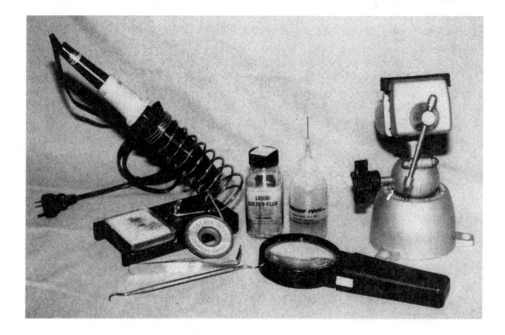

FIGURE 14–2
Direct heat attachments

iron (Figure 14–2). In effect, they are specially formed soldering tips designed to conform to the geometry of various SMCs. Regardless of the shape and style of the tips, all have one thing in common: they heat all component terminations simultaneously. The attachments are readily available, inexpensive, and easy (even fun) to use. They represent a great way to do excellent SMT rework with a minimal financial investment.

Three direct heat attachment tip styles are in wide use (Figure 14–3). The *slotted-spade tips* are designed primarily for use from the side or top of a component, depending on the space available adjacent to the specified component. *Tunnel tips* are designed for use on dense, highly populated boards where approach would be made from above the component. *Quad tips* are used to remove four-sided IC packages. All three tip styles can usually be screwed into a standard soldering iron.

Handheld Hot-Air Devices: Removal Through Heat Convection

Soldering irons, whether equipped with conventional tips or direct heat attachments, melt solder through heat conduction—the transfer of heat through a conductor (the soldering tip). It is also possible to transfer heat by the circulation or movement of air. When the latter approach, known as

convection heating, is used, solder is quite easily reflowed. A handheld hot-air torch, like the one shown in Figure 14–4, is ideal for this purpose. The implement sells for under $100, and of all the ways to remove SMDs, it's supposedly the easiest, though not necessarily the best.

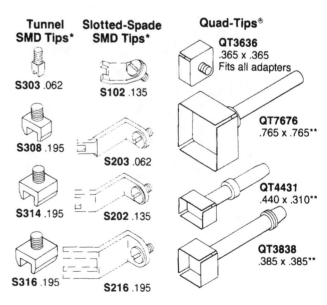

FIGURE 14–3
Direct heat attachment tip styles
Courtesy Hexacon Electric Co.

FIGURE 14–4
Handheld hot-air torch

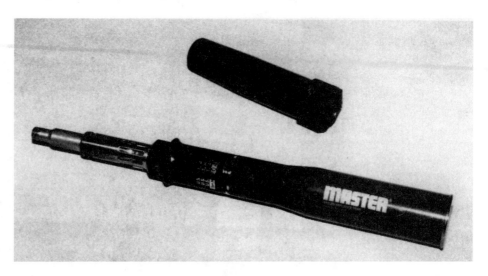

The hot-air butane torch shown in Figure 14–4 provides a variable *flameless* heat source of from approximately 500° F to almost 2400° F, more than enough to melt solder. The tool is safe and requires no batteries or line cord. It's totally portable, thus ideal for in-field service. We will examine its use as a desoldering tool shortly.

Professional Rework Stations

To replace the complete range of SMDs, particularly those with high pin counts, the professional rework station is the equipment of choice. Although expensive ($1000 to $5000 or more) and requiring special operator skills, such tabletop stations are ideal in large-volume production environments. A typical "upper-end" unit from Pace, Inc., is shown in Figure 14–5.

The station shown in the figure is a self-contained, completely equipped system for rework of both surface mount and insertion mount circuits. With it one can align, install, remove, and replace the full spectrum of fine-pitch and standard surface

FIGURE 14–5
Professional rework station
Photo courtesy of PACE, Inc.

mount devices. Because of space limitations and the need for specialized training, however, we will not be discussing the use of professional rework stations in this text.

Effecting the Pullout: Dislodging SMDs

It's now time to see how the tools and materials just discussed are applied to the removal of SMDs. Regardless of which method is being reviewed, it is important throughout to "keep your eye on the prize"—that is, the circuit board, not the surface mount component. Saving the SMD is not only difficult (almost impossible with ICs), it is rarely necessary. Previously used chip resistors and capacitors, as well as three-lead transistors, are better discarded altogether and replaced with fresh components. In the long run you'll save both time and money—and end up with a far more reliable product.

The PC board is another matter, of course; it simply must not be damaged. It has to be saved. If you start pulling up its tiny pads and traces while attempting to remove a surface mount component, the resulting damage may be irreversible. It would be most unfortunate to have to discard an entire board just because the pads or traces for one or two components have been lifted. To avoid such an eventuality, follow the procedures outlined for SMD removal closely—and take it slow and easy. Try thinking like an Egyptologist does as she carefully and methodically unwraps the gauze from an ancient mummy.

In examining the dislodging of SMDs, we'll look first at how it is done with traditional soldering tools. Next, we will see how direct heat attachment soldering instruments are used. Finally, we'll examine SMD removal with the handheld hot-air torch.

Using Traditional Tools to Remove SMDs

Using traditional tools and materials, it is easy to remove SMDs, if you are careful and patient. Let's see how it is done for two- and three-terminal devices and for SOICs.

In removing a two-terminal SMD, such as a chip capacitor or resistor, proceed as follows:

1. Using a drop dispenser, apply a dab of liquid flux to each terminal.
2. Grip the component across its body, not its terminals, with a tweezers (Figure 14–6).
3. Using a soldering iron, alternately and rapidly heat one terminal, then the other, continuing back and forth until the solder on both terminals is liquefied. While doing this, maintain an upward pull on the component.
4. When the component breaks free, set it aside to be discarded later.

5. Clean the PC board pads, restoring them to pristine condition so that they are ready to accept a component replacement. Begin by applying a new dab of flux to each pad.
6. Using solder wick, remove any remaining solder from the pads. Inspect the board carefully.

To remove a three-terminal SMT component, do the following:

1. Apply liquid flux to all terminals.
2. Use solder wick to remove as much solder as possible from each terminal.
3. With a dental pick, pull up the single lead while heating it with a soldering iron (Figure 14–7).
4. Now grasp the component with a tweezers and while simultaneously heating the remaining two terminals with a soldering iron, pull upward. The component should break free.
5. Clean the pads of remaining solder with flux and solder wick. Inspect your work carefully.

When removing an SOIC, progress slowly and cautiously. Remember, the PC board pads are tiny and are not plated through to the other side. Such pads will lift easily with excess heat and pressure. Don't plan on saving the IC; consider it unsalvageable from the start. Concentrate on preserving the board pads and traces. Proceed as follows:

1. Dip a strip of solder wick into liquid flux, then "lay" it across one row of IC pins as you press the soldering iron down on top. Do not "drag" the wick across the pins; move the iron to adjacent spots to soak up solder.
2. Repeat as necessary to remove as much solder as possible.
3. Do the same thing for the other row of IC pins.
4. Using a dental pick, select one row and gently lift each pin from its pad, one at a time (Figure 14–8). The IC will eventually break away.
5. Clean the pads of remaining solder with flux and solder wick. Inspect the board pads carefully.

Remember, removing the IC is easy. Extracting it without damaging the PC board is the challenge. The trick is to remove all the solder you possibly can before attempting to lift the IC.

Using Direct Heat Attachments to Remove SMDs

Direct heat attachments are designed specifically to remove and replace SMDs. The tips are made to match the precise dimensions of currently available surface mount components. They heat all terminals, leads, or pins of a given component at the same time. Generally, slotted-spade tips are used to remove discrete devices, and tunnel tips work best to

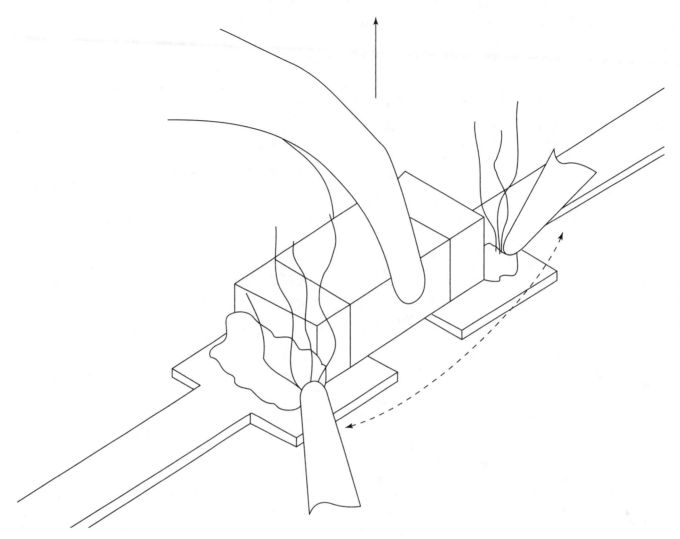

FIGURE 14–6
Removing a two-terminal device

dislodge SOICs. Quad tips, of course, are made to lift PLCCs and LCCCs.

When removing discrete 1206 and SOT-23 type packages, apply the slotted-spade tip so that its surface touches the solder fillet between the component leads and the board pads (Figure 14–9). As soon as the solder completely melts, remove the component with a slight twisting action of the tip. Due to surface tension, the component should adhere to the tip. If it does not, use a tweezers to lift the component from the board while the solder is molten. Do not press on the PC board while heating; let the heat of the tip do the work. Clean pads with flux and solder wick, then inspect your work carefully.

SOICs are removed in much the same manner as discretes, only a tunnel tip is used. Make sure the tip is in contact with all leads, then press gently and firmly until solder melts. Twist slightly while lifting the tip slowly. The IC should adhere to the tip due to surface tension of the solder. If the package does not adhere to the tip, slide a tweezers under one end of the IC and gently lift. Remember, it is the delicate pads and traces, not the IC, that you want to protect. Clean the pads of excess solder with flux and solder wick. Finally, inspect your work carefully.

Using a Handheld Hot-Air Device to Remove SMDs

SMDs can be removed with a minimum amount of damage to the PC board by the use of a handheld hot-air torch. The tool is quite easy to use. You simply fire it up, adjust the hot-air-temperature output, and aim the torch at the component to be removed. The only caution is to keep the hot air off adjacent components. With a little practice, you can easily avoid dislodging unwanted components.

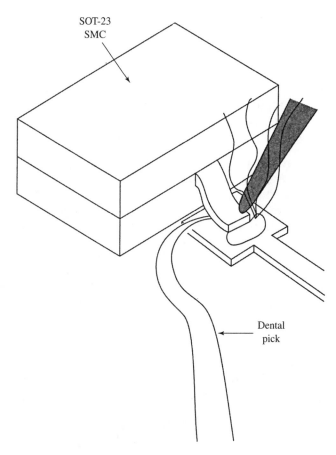

FIGURE 14–7
Removing a three-terminal device

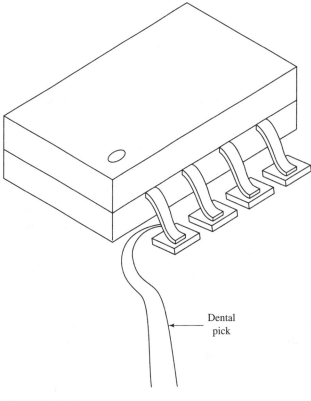

FIGURE 14–8
Removing SOICs

When removing discretes, keep the torch temperature low as you aim hot air at a tiny SMD. As the solder joints melt, lift the component with a tweezers.

With SOICs, follow the same procedure, only raise the air temperature a bit and rotate the torch slightly so that hot air blows over the entire component, body and pins. When solder begins to reflow, use a dental pick to gently lift the IC off the board. As with the other methods of component removal discussed earlier, clean the board pads with flux and solder wick, then inspect your work carefully.

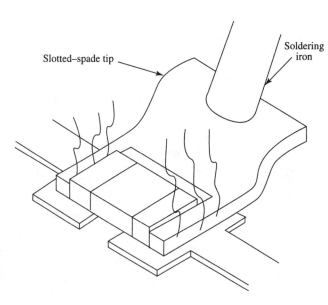

FIGURE 14–9
Removing SMDs with a slotted-spade tip

> ### *Something to Think About*
>
> *Why not grab a few surplus SMT circuit boards and start tearing them apart? With traditional tools or, if you are lucky enough, a professional rework station, practice removing SMCs. Remember, though, when removing components, avoid damaging the PC board itself.*

14.2 SMD REPLACEMENT: PUTTING THEM IN THEIR PLACE

Virtually all surface mount components removed from a PC board will be replaced with an identical or similar component, and just as there are techniques to be mastered in carefully dislodging an SMD from its pads, there are skills to be acquired in setting new ones in their place. In this section we will concentrate on SMD replacement with direct heat attachments and with handheld hot-air devices. (For a discussion of the placement of SMD using traditional hand tools, refer to Chapter 13.) We will also examine SMT solder joint workmanship standards, identifying the factors that make for a good, professional-quality solder connection.

SMD Replacement with Direct Heat Attachment

There are two ways to use direct heat attachments to install surface mount components. You can employ the *tinned-pad* method or the *solder-paste* procedure. Let's examine each in turn.

The Tinned-Pad Method

The tinned-pad method is simple and direct, at least for discrete SMDs. First, you apply a thick coat of solder to the component pads, tinning them. You do this with a traditional soldering iron tip. Next, you hold the component in place while heating its terminals and tinned PC board pads with a slotted-spade tip. Finally, you remove the direct heat attachment and allow the solder joint to solidify. Here are the steps in more detail:

1. Install the appropriate slotted-spade tip.
2. Using a standard soldering iron and tip, melt a thick layer of solder onto each terminal pad.
3. Using tweezers, place the SMC on its tinned pads.
4. Place the slotted-spade tip so that both forks simultaneously touch the tinned pads and SMD terminals.
5. As the solder melts, use the tweezers to gently press the SMC flat against its pads.
6. Remove the slotted-spade tip and wait a few seconds for the solder to solidify, then remove the tweezers.
7. Inspect your work carefully.

To replace SOICs, proceed as outlined above, but use the appropriate-size tunnel tip. Since the SOIC will have gull-wing leads, not body terminals, heat from the tunnel tip must be carried through the component leads to the solder underneath.

Thus, additional time will be needed to reflow the previously tinned solder pads.

The Solder-Paste Method

Direct heat attachments can be used with solder paste to put SMDs in their place. (For a discussion of solder paste, see Chapter 13.) Here is the procedure for replacing two-terminal SMCs:

1. Install the appropriate slotted-spade tip.
2. Using a syringe, apply a small amount of solder paste to the pad surfaces.
3. Using tweezers, gently place the component onto the solder paste. Do not apply pressure; use the tweezers for alignment only.
4. Place the slotted-spade tip on the component pads and hold until solder begins to melt.
5. While continuing to hold the tweezers in place, lift the tip off the connection.
6. When solder has solidified, remove the tweezers.
7. Inspect your work carefully.

To replace leaded-terminal components, such as SOT-23s and SOICs, proceed as follows:

1. Using a syringe, apply a small amount of solder paste to each pad. Use a minimum amount of solder, particularly with SOICs, to prevent solder bridges.
2. Using tweezers, place the component onto the solder paste. Do not apply pressure; use the tweezers for alignment purposes only. If installing SOICs, hold the package on the very end to allow access of the tunnel tip.
3. Place the tip, slotted-spade or tunnel, over the component or package. Make sure the tip is in contact with all leads. Hold briefly until the solder begins to melt.
4. While continuing to hold the tweezers in place, lift the tip off the connection.
5. When solder has solidified, remove tweezers.
6. Inspect the connection carefully.

SMD Replacement with Handheld Hot-Air Devices

Just as you can remove surface mount components with a handheld air device, you can replace them with the same tool. Let's examine first how the PC board surface is prepared, then how a hot-air torch is used to stick SMCs in their place.

PC Board Surface Preparation

As with any method used to attach SMCs to a PC board, the component pads must first be cleared of all solder and contaminants. Use liquid flux and solder wick to restore pads to pristine condition. If necessary, spray the board with a light-duty flux remover to complete the preparation task.

Apply solder paste to individual pads with a syringe, as discussed in Chapter 13. It is extremely important to avoid solder bridging. The best approach is to squeeze the syringe slightly, release the pressure, then quickly move from pad to pad, dispensing a small amount of paste as you go. Practice on a surplus PC board until you have perfected the technique.

After applying the paste, grip the component to be attached with a tweezers and deposit it gently onto its pads. Align the component carefully and press down lightly. Do not force the component to the pads: press just enough to sink it slightly into the paste. Continue to hold the component in place with the tweezers.

Making the SMDs Stick

To melt solder paste, you must apply heat with a hot-air torch. In doing so, keep one caution in mind. The solder paste on all pads of a given component must melt at the same time. If uneven melting occurs, poor solder joint formation will result.

The best way to make this procedure work is to keep the torch in constant motion as you bring it down toward the PC board from a height of 6″ to 8″. When you are within 2″ to 3″ from the board, go no farther, but keep the torch moving back and forth or in a circular motion. It will take from 10 to 15 seconds before the solder begins to melt. As it does, the paste will change colors, going from gray to silver. When that happens, remove the heat source and then the tweezers. The component should now be "stuck" in place.

Solder Joint Integrity

Regardless of which method is used to initially install or eventually replace SMCs, there are solder joint workmanship standards that must be adhered to, particularly in a professional production environment. Although we cannot hope to delineate the full range of industry workmanship standards here, we can at least illuminate what is acceptable and what is not. Let's look first at inspection and rejection techniques, then at solder joint workmanship.

Inspection and Rejection Techniques

Inspection of SMT assemblies in a production environment is done visually under a magnifying glass. The magnifier should be 5× to 10× and have a light source for shadowless illumination of the area being viewed. The magnifier should also be suitable to permit inspection of each solder connection in its entirety.

If on inspection any of the following characteristics are present, product rejection must be considered:

1. Spattering of solder on adjacent connections or components.
2. Voids or holes that are not in accordance with SMT standards criteria.
3. Solder that obscures the connection configuration, that is, excess solder.
4. Fractured solder connections.
5. Unclean connections with lint, solder, residue, splash, dirt, and the like.
6. Dewetting.
7. Insufficient solder.
8. Lands connected by interfacial connections that show evidence of failure to wet the metallic surfaces.

In the final analysis, all metal that comes in contact with solder should exhibit proper wetting. Wetting, you will recall, is the formation of a uniform, smooth, unbroken, and adherent film of solder to the base metal.

Solder Joint Workmanship

With regard to solder joint workmanship, we'll examine the criteria for five popular SMT components.

Five-Face Chip Carrier Components. The solder fillet is concave with a positive wetting angle to the land pattern and the metallized areas of the component (Figure 14–10a). The preferred fillet is one-third to three-fourths component height with fillet extending the full width of the metallized area in contact with the land pattern.

Cylindrical Components. The solder fillet is concave with a positive wetting angle to the land pattern and the metalized areas of the component (Figure 14–10b). The preferred fillet is one-third to three-fourths component height with fillet extending the full width of the metallized area in contact with the land pattern.

Gull-Wing Components. The solder fillet is concave with a positive wetting angle to the lead and land pattern (Figure 14–10c). Solder fillet should form under heel bend.

J-Lead Component. The solder fillet shows a positive wetting angle to lead and land pattern (Figure 14–10d). Solder fillet should form under heel bend with fillet concave or straight.

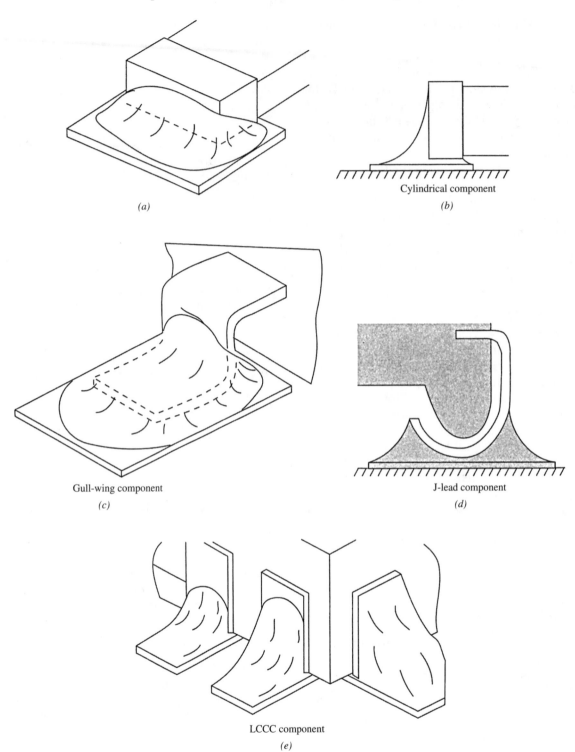

(a)

Cylindrical component

(b)

Gull-wing component

(c)

J-lead component

(d)

LCCC component

(e)

FIGURE 14–10
Solder joint workmanship

LCCC Components. The solder fillet shows a positive wetting angle to metalized areas of component and land pattern (Figure 14–10e). Minimum fillet extends one-third of the metallized area height. Maximum fillet height extends to the top of the metallized area of the component.

Try to achieve the these solder joint workmanship standards with all surface mount assemblies you work on, be they production units or your own prototypes.

Something to Think About

OK, now how about putting the SMCs back on the PC board? Actually, you don't want to reuse the ones you just removed. Never install used SMCs; always place new ones on the board.

SUMMARY

In Section 14.1 we learned about the removal of SMCs in a production environment. We began by looking at the tools for disengagement: traditional soldering tools, direct heat attachment soldering tools, handheld hot-air devices, and professional rework stations. We then saw how the first three are actually used to remove surface mount components.

In Section 14.2 it was on to replacing SMCs, first with direct heat attachments and then with handheld hot-air devices. We concluded with a look at solder joint integrity.

QUESTIONS

1. Direct heat attachment soldering tools are specially formed soldering _____ designed to conform to the geometry of various SMCs.

2. Handheld hot-air devices remove SMCs through heat _____.

3. To replace the complete range of SMCs, the professional _____ station is used.

4. When removing SMCs, the first priority is to save the _____, not the component.

5. When using traditional soldering tools to remove an SOIC, the trick is to remove all the _____ you can before attempting to lift the IC.

6. There are two ways to use direct heat attachments to install SMCs: the _____ _____ method and the _____ _____ procedure.

7. Direct heat attachment tips come in three configurations: _____-_____ and _____.

8. When applying solder paste, you should use a _____ with an appropriate-size tip.

9. SMT assemblies in a production environment are visually inspected under a _____ glass.

10. With regard to solder joint workmanship, in most cases a solder _____ with a positive wetting action is formed.

PART III
Student Construction Projects

15 Sample Project Report/Variable Power Supply

Project Report

Variable Power Supply

by

Austin Babayan

June 15

1

Contents

Written Documentation

* Concepts and Requirements Document

* Experiment Results Document

* Test Results Document

* Summary and Recommendations Document

Graphics Documentation

* Design Drawings

* Working Schematic Drawing

* Breadboard Drawing

* Printed Circuit Board Design Layout Drawing

* Printed Circuit Board Artwork Drawing

* Printed Circuit Board Fabrication Drawing

* Printed Circuit Board Assembly Drawing

* Sheet Metal Drawing

* Wiring Diagram

* Final Packaging Drawing

2

Concepts and Requirements Document
Variable Power Supply
by
Austin Babayan

Project Goals

This Concepts and Requirements Document proposes that students in Electronics 150 design and build a working prototype for a Variable Power Supply that may later be mass-produced. Electronics students need an inexpensive 0- to 15-V dc power supply to power such projects as transistor radios, burglar alarms, digital counters, and a host of solid-state-based experiments. A search has failed to uncover a low-cost commercial power supply of the type required. Furthermore, undertaking such a project will provide electronics students with a valuable electronics design and fabrication experience.

Project Objectives

The project is to have the following objectives: inexpensive to produce (less than $25), reliable, safe, attractive, and easy to use. These objectives will be met in the following ways:

* **Cost.** By using noncritical, readily available hardware and electronic components, bought in bulk quantities, the cost will be kept low.

* **Reliability.** By using solid-state active components and printed circuit board construction techniques, a high degree of reliability will be obtained.

* **Safety.** Use of an internal fuse, thermal shutdown voltage regulator, insulated output terminals, and a completely sealed metal enclosure, will assure safe operation.

3

* **Appearance.** A metal enclosure, either painted or covered with self-adhesive vinyl paper, will ensure an attractive project.

* **Ease of Use.** With the use of an on-off switch, a front panel-mounted LED indicator, and a panel-numbered voltage indicator, the project will have a high degree of functionality and be easy to use.

Project Responsibility

Because this is to be a learning experience for the electronics student, he or she will be responsible for the design (which, in this case, is being supplied), drawings, experimenting, prototyping, and testing, troubleshooting, and final documentation phases of the Variable Power Supply project. Due dates for each phase will be assigned by the electronics instructor.

Project Theory of Operation

The Variable Power Supply changes the 120 V ac from the wall outlet to a varying dc of 1.2 to 15 V. The ac line cord couples, via switch S_1 and internal fuse F_1, the 120 V ac to the primary of transformer T_1. The transformer steps down the 120 V ac to approximately 15.5 V ac. Diodes D_1–D_4 form a full-wave bridge to change (rectify) the ac to pulsating direct current (pdc). Light-emitting diode D_5 is used as a power "on" indicator; resistor R_1 limits current to the LED. Capacitor C_1 filters the pdc to smooth dc. Capacitor C_2 provides high-frequency bypassing. The variable voltage regulator U_1 sets its output voltage by a control loop consisting of resistor R_2 and potentiometer R_3. The output voltage can be varied over the 1.2- to 15-V range by adjusting the potentiometer. Finally, capacitor C_3 acts to swamp any ringing effect developed by the control loop and thereby ensures stability of the voltage regulator's output.

Electronics instructor's approval _Bob Jones_ Date _4-11_

4

Experiment Results Document
Variable Power Supply
by
Austin Babayan

Conclusions

The Variable Power Supply project performs as it was designed to perform, under breadboard conditions. Output voltage is adjustable from 1.2 to 17.2 V; however, it should be noted that at the breadboarding stage, the Variable Power Supply was tested under a no-load condition. This is because at this time voltage regulator U_1 was not equipped with a heat sink.

Test Results

1. Output resistance test OK. Measured 290 to 5.2 k ohms.

2. Project was plugged into the wall outlet and switch S_1 closed.

3. LED D_5 did not light. Close examination revealed that the LED had been installed backward. When the LED was reversed, it lit as it should.

4. A voltmeter was placed across the output terminals. While potentiometer R_3 was being adjusted through its range the output voltage reading recorded was 1.2 to 17.2 V.

5. The project was left on for 1 hour, after which output voltage readings were again taken. They were identical with those taken in step 4.

Test Results Document

Name of Project _____ *VARIABLE POWER SUPPLY* _____

Person(s) Performing Tests _____ *AUSTIN BABAYAN* _____

Preliminary Test Results

1. Is all wiring complete and correct?   Init. *A.B.*

                                                             Date *6-1*

2. Are all soldering connections good?   Init. *A.B.*

                                                             Date *6-1*

3. Are all components in the correct location?

                                                             Init. *A.B.*

                                                             Date *6-1*

4. Are all components installed in the correct direction?

                                                             Init. *A.B.*

                                                             Date *6-1*

Operational Test Results

1. Any catastrophic failure?   Yes _____   No __*X*__   Init. __*A.B.*__

Explain: _____

_____

_____

2. Any minor problems?   Yes _____   No __*X*__   Init. __*A.B.*__

Explain: _____

_____

_____

6

3. Calibration, if necessary.   Yes _X_   No _____   Init. _A.B._

CALIBRATION OF PANEL FACE FOR VOLTAGE READINGS. USED TRANSISTOR RADIO AS LOAD.

4. Is troubleshooting required? Yes _____   No _X_   Init. _A.B._

What is anticipated? _____

_____

_____

5. Is project operation satisfactory at this time?

         Yes _X_   No _____   Init. _A.B._   Date _6-1_

Troubleshooting Results

1. Define the problem.

NONE

_____

_____

2. Steps taken to correct the problem.

N/A

_____

_____

3. Is project functioning satisfactorily?

         Yes _✓_   No _____   Init. _A.B._   Date _6-1_

Performance Test Results

1. User test performed. *FELLOW STUDENT GIVEN PROJECT AND ASKED TO OPERATE IT UNDER VARIOUS LOAD CONDITIONS.*

7

2. User test results. *POWER SUPPLY PERFORMED WELL*
UNDER ALL LOAD CONDITIONS. HEAT SINK GOT WARM,
BUT NOT EXCESSIVELY HOT.

Concluding Comments

VARIABLE POWER SUPPLY OPERATES AS EXPECTED UNDER
VARIOUS LOAD CONDITIONS.

8

Summary and Recommendations Document
Variable Power Supply
by
Austin Babayan

Our goal was to design and fabricate a 1.2- to 15-V Variable Power Supply
prototype project. The goal has been met and the power supply meets all
preliminary project objectives. The initial design has proven valid, and no
unanticipated design or fabrication problems developed during prototype
construction. It is the recommendation of this technician that full-scale
production of the Variable Power Supply be given serious consideration.

Name ___*AUSTIN BABAYAN*___

Signature ___*Austin Babayan*___

Date ___*6-15*___

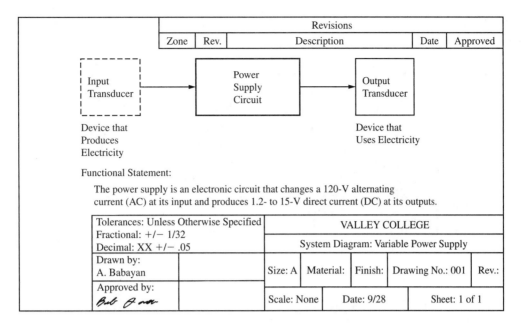

		Revisions		
Zone	Rev.	Description	Date	Approved

Functional Statement:

The power supply is an electronic circuit that changes a 120-V alternating current (AC) at its input and produces 1.2- to 15-V direct current (DC) at its outputs.

Tolerances: Unless Otherwise Specified Fractional: +/− 1/32 Decimal: XX +/− .05		VALLEY COLLEGE				
		System Diagram: Variable Power Supply				
Drawn by: A. Babayan		Size: A	Material:	Finish:	Drawing No.: 001	Rev.:
Approved by: *Bob Garza*		Scale: None	Date: 9/28		Sheet: 1 of 1	

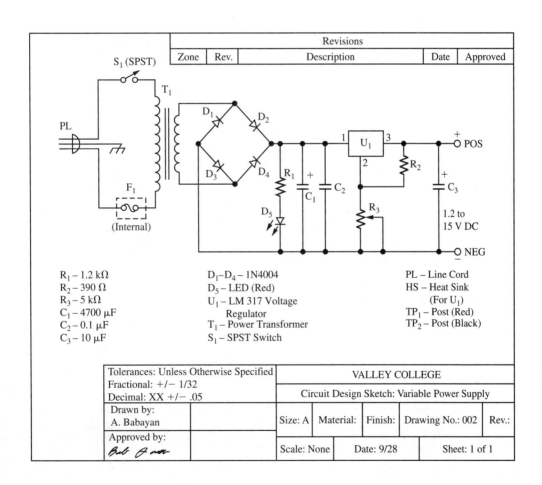

		Revisions		
Zone	Rev.	Description	Date	Approved

R_1 – 1.2 kΩ
R_2 – 390 Ω
R_3 – 5 kΩ
C_1 – 4700 μF
C_2 – 0.1 μF
C_3 – 10 μF

D_1–D_4 – 1N4004
D_5 – LED (Red)
U_1 – LM 317 Voltage
  Regulator
T_1 – Power Transformer
S_1 – SPST Switch

PL – Line Cord
HS – Heat Sink
  (For U_1)
TP_1 – Post (Red)
TP_2 – Post (Black)

Tolerances: Unless Otherwise Specified Fractional: +/− 1/32 Decimal: XX +/− .05		VALLEY COLLEGE				
		Circuit Design Sketch: Variable Power Supply				
Drawn by: A. Babayan		Size: A	Material:	Finish:	Drawing No.: 002	Rev.:
Approved by: *Bob Garza*		Scale: None	Date: 9/28		Sheet: 1 of 1	

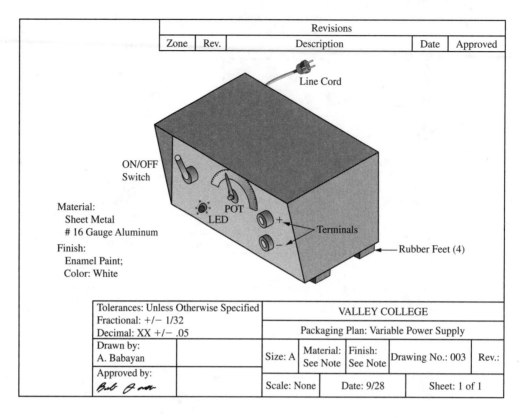

		Revisions		
Zone	Rev.	Description	Date	Approved

Line Cord

ON/OFF
Switch

Material:
   Sheet Metal
   # 16 Gauge Aluminum
Finish:
   Enamel Paint;
   Color: White

POT
LED
+
−
Terminals
Rubber Feet (4)

Tolerances: Unless Otherwise Specified Fractional: +/− 1/32 Decimal: XX +/− .05		VALLEY COLLEGE				
		Packaging Plan: Variable Power Supply				
Drawn by: A. Babayan		Size: A	Material: See Note	Finish: See Note	Drawing No.: 003	Rev.:
Approved by:		Scale: None	Date: 9/28		Sheet: 1 of 1	

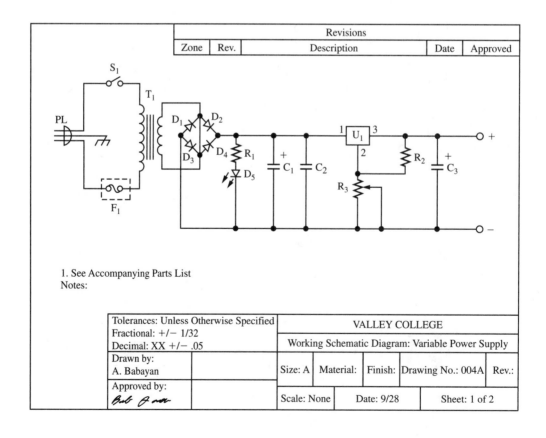

		Revisions		
Zone	Rev.	Description	Date	Approved

1. See Accompanying Parts List
Notes:

Tolerances: Unless Otherwise Specified Fractional: +/− 1/32 Decimal: XX +/− .05		VALLEY COLLEGE				
		Working Schematic Diagram: Variable Power Supply				
Drawn by: A. Babayan		Size: A	Material:	Finish:	Drawing No.: 004A	Rev.:
Approved by:		Scale: None	Date: 9/28		Sheet: 1 of 2	

		Revisions			
Zone	Rev.	Description		Date	Approved

Qty.	Symbol	Description	Part No.
1	R_1	Resistor, 1.2 Ω, ¼ W	
1	R_2	Resistor, 390 Ω, ¼ W	
1	R_3	Potentiometer, 5 kΩ	
1	C_1	Capacitor, 4700 μF	
1	C_2	Capacitor, 0.1 μF	
1	C_3	Capacitor, 10 μF	
4	D_1–D_4	Diode, 1N4004	
1	D_5	Diode, LED-Red	
1	U_1	Voltage Regulator, LM 317	
1	T_1	Transformer, 120/12 V AC	
1	S_1	Switch, S.P.S.T.	
1	PL	AC Line Cord	
1		Heat Sink (LM 317)	
1		Terminal Post-Red	
1		Terminal Post-Black	

Tolerances: Unless Otherwise Specified		VALLEY COLLEGE					
Fractional: +/− 1/32 Decimal: XX +/− .05		Schematic Drawing Parts List: Variable Power Supply					
Drawn by: A. Babayan		Size: A	Material:	Finish:	Drawing No.: 004B		Rev.:
Approved by: *Bob Jones*		Scale: None		Date: 9/28		Sheet: 2 of 2	

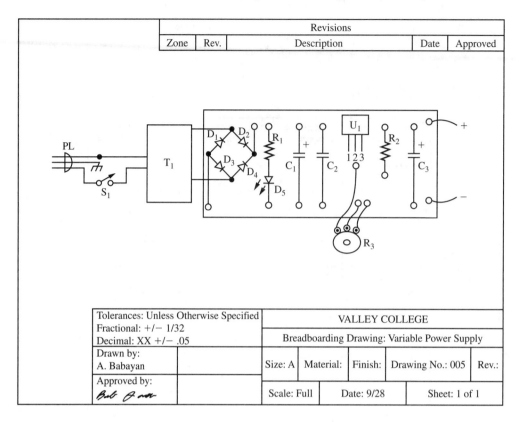

		Revisions		
Zone	Rev.	Description	Date	Approved

Tolerances: Unless Otherwise Specified Fractional: +/− 1/32 Decimal: XX +/− .05		VALLEY COLLEGE			
		Breadboarding Drawing: Variable Power Supply			
Drawn by: A. Babayan		Size: A	Material:	Finish:	Drawing No.: 005	Rev.:
Approved by: *Bob Jones*		Scale: Full	Date: 9/28		Sheet: 1 of 1

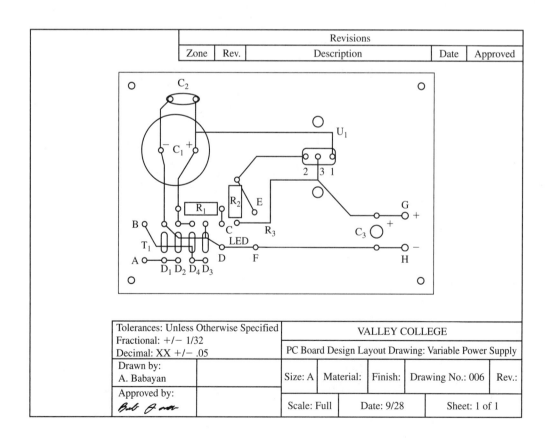

		Revisions		
Zone	Rev.	Description	Date	Approved

Tolerances: Unless Otherwise Specified Fractional: +/− 1/32 Decimal: XX +/− .05		VALLEY COLLEGE			
		PC Board Design Layout Drawing: Variable Power Supply			
Drawn by: A. Babayan		Size: A	Material:	Finish:	Drawing No.: 006	Rev.:
Approved by: *Bob Jones*		Scale: Full	Date: 9/28		Sheet: 1 of 1

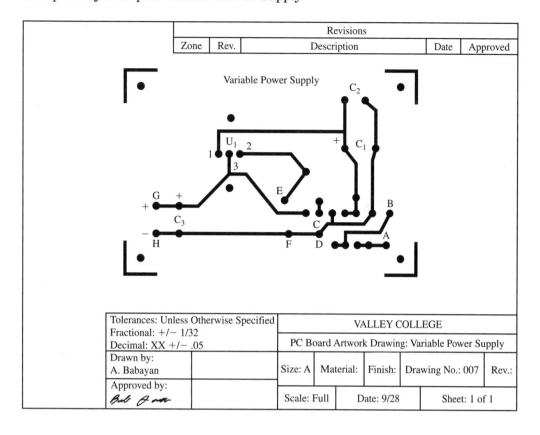

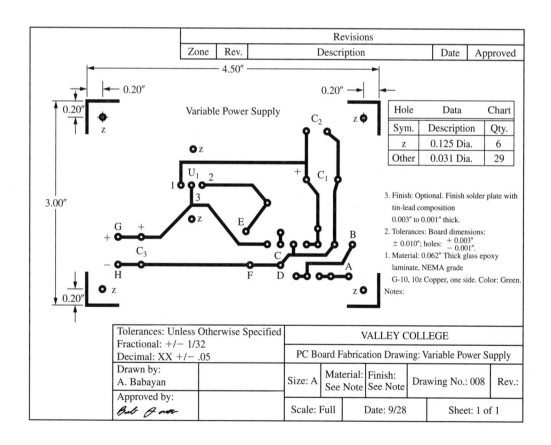

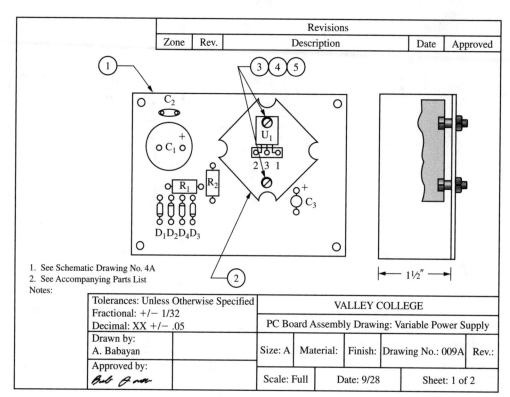

	Revisions			
Zone	Rev.	Description	Date	Approved

1. See Schematic Drawing No. 4A
2. See Accompanying Parts List
Notes:

Tolerances: Unless Otherwise Specified		VALLEY COLLEGE				
Fractional: +/− 1/32						
Decimal: XX +/− .05		PC Board Assembly Drawing: Variable Power Supply				
Drawn by: A. Babayan		Size: A	Material:	Finish:	Drawing No.: 009A	Rev.:
Approved by: *Bob Brown*		Scale: Full	Date: 9/28		Sheet: 1 of 2	

	Revisions				
Zone	Rev.	Description		Date	Approved

Item No.	Ref. Des.	Description	Qty.
1	1	PC Board	1
2	R_1	Resistor, 1.2 kΩ, $\frac{1}{4}$ W	1
3	R_2	Resistor, 390 Ω, $\frac{1}{4}$ W	1
4	C_1	Capacitor, 4700 μF	1
5	C_2	Capacitor, 0.1 μF	1
6	C_3	Capacitor, 10 μF	1
7	D_1–D_4	Diode, 1N4004	4
8	U_1	Voltage Reg. LM 317	1
9	2	Heat Sink	1
10	3	Spacer, $\frac{1}{8}''$	2
11	4	Screw, 6–32 $\times$ $\frac{1}{2}''$	2
12	5	Nut, 6–32	2

Tolerances: Unless Otherwise Specified Fractional: +/− 1/32 Decimal: XX +/− .05		VALLEY COLLEGE				
		PC Board Assembly Drawing Parts List: Variable Power Supply				
Drawn by: A. Babayan		Size: A	Material:	Finish:	Drawing No.: 009B	Rev.:
Approved by: *Bob Jnot*		Scale: None	Date: 9/28		Sheet: 2 of 2	

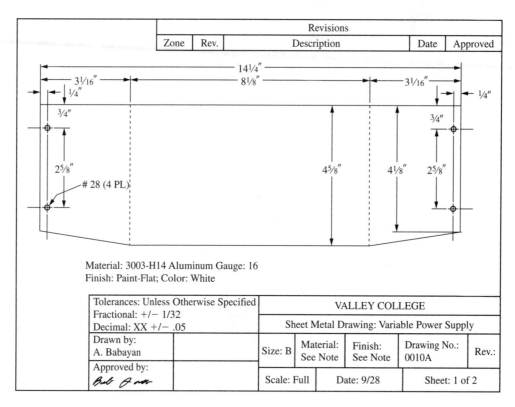

Material: 3003-H14 Aluminum Gauge: 16
Finish: Paint-Flat; Color: White

Revisions				
Zone	Rev.	Description	Date	Approved

Tolerances: Unless Otherwise Specified		VALLEY COLLEGE				
Fractional: +/− 1/32						
Decimal: XX +/− .05		Sheet Metal Drawing: Variable Power Supply				
Drawn by: A. Babayan		Size: B	Material: See Note	Finish: See Note	Drawing No.: 0010A	Rev.:
Approved by: *Bob Jones*		Scale: Full	Date: 9/28		Sheet: 1 of 2	

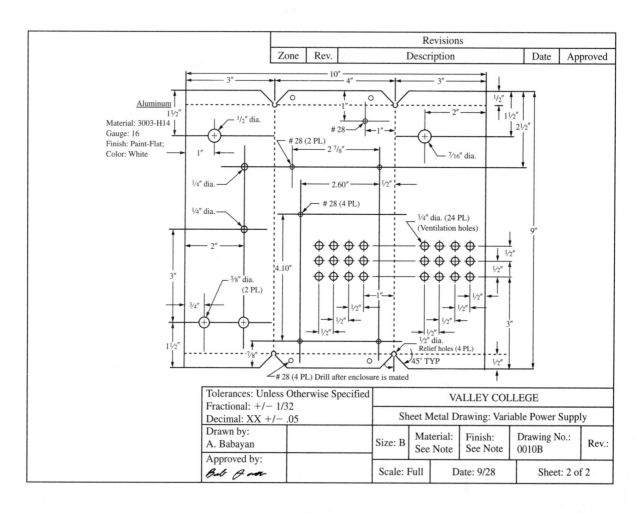

Revisions				
Zone	Rev.	Description	Date	Approved

Tolerances: Unless Otherwise Specified		VALLEY COLLEGE				
Fractional: +/− 1/32						
Decimal: XX +/− .05		Sheet Metal Drawing: Variable Power Supply				
Drawn by: A. Babayan		Size: B	Material: See Note	Finish: See Note	Drawing No.: 0010B	Rev.:
Approved by: *Bob Jones*		Scale: Full	Date: 9/28		Sheet: 2 of 2	

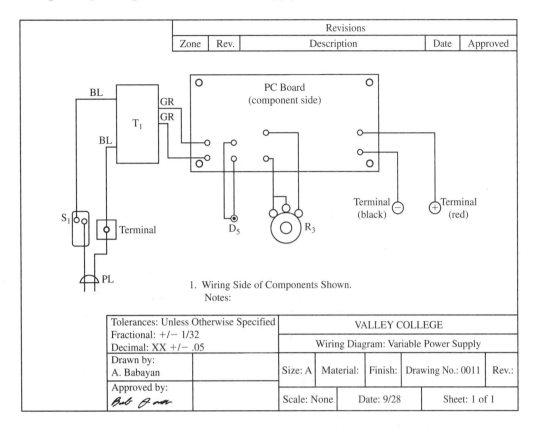

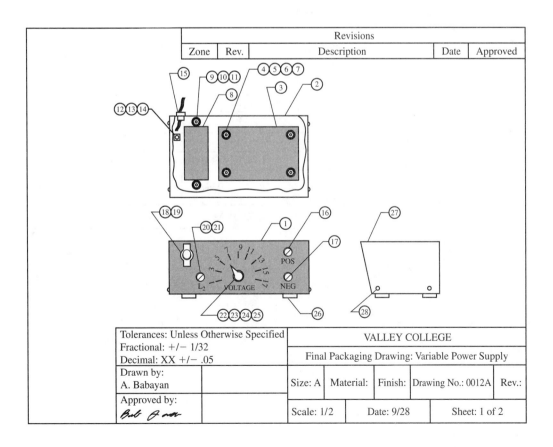

	Revisions				
Zone	Rev.	Description		Date	Approved

PARTS LIST
Final Assembly

Item No.	Qty.	Part No.	Indenture				Description	Comments
1	1		X				Variable Power Supply	
2	1			X			Chassis (Bottom)	
3	1				X		PC Board Subassembly	
4	4					X	Screw, $4-40 \times 3/4''$	
5	4					X	Nut, $4-40$	
6	4					X	Washer, $1/4''$	
7	4					X	Spacer, $1/4''$	
8	1				X		Transformer, T_1	
9	2					X	Screw, $4-40 \times 1/2''$	
10	2					X	Nut, $4-40$	
11	2					X	Washer, $1/4''$	
12	1				X		Terminal Strip (2–LUG)	
13	1					X	Screw, $4-40 \times 1/2''$	
14	1					X	Nut, $4-40$	
15	1				X		Strain Relief	
16	1				X		STD Terminal-Red	
17	1				X		STD Terminal-Black	
18	1				X		Switch, S_1	
19	1					X	ON/OFF Plate	
20	1				X		LED, D_5	
21	1					X	Grommet, $1/4''$	
22	1				X		Potentiometer, R_3	
23	1					X	Washer (Match POT)	
24	1					X	Nut (Match POT)	
25	1					X	Knob, Pointer	
26	4				X		Rubber Feet $3/4'' \times 3/4''$	Self-adhesive
27	1			X			Enclosure (Top)	
28	4					X	Screw SM, $3/8''$	Self-tapping

Tolerances: Unless Otherwise Specified Fractional: $+/- 1/32$ Decimal: XX $+/- .05$		VALLEY COLLEGE				
		Final Packaging Drawing Parts List: Variable Power Supply				
Drawn by: A. Babayan		Size: A	Material:	Finish:	Drawing No.: 0012B	Rev.:
Approved by: *Bob G. nor*		Scale: None	Date: 9/28		Sheet: 2 of 2	

16

Exercise Project Report/3-Channel Color Organ

Project Report

3-Channel Color Organ

by

Larry Hyde

June 15

Contents

Written Documentation

* Concepts and Requirements Document

* Experiment Results Document

* Test Results Document

* Summary and Recommendations Document

Graphics Documentation

* Design Drawings

* Working Schematic Drawing

* Breadboard Drawing

* Printed Circuit Board Design Layout Drawing

* Printed Circuit Board Artwork Drawing

* Printed Circuit Board Fabrication Drawing

* Printed Circuit Board Assembly Drawing

* Sheet Metal Drawing

* Wiring Diagram

* Final Packaging Drawing

2

Concepts and Requirements Document
3-Channel Color Organ
by
Larry Hyde

Project Goals

This Concepts and Requirements Document proposes that students of Electronics 150 design and build a working prototype for a 3-Channel Color Organ that may later be mass-produced. The color organ is a project that flashes three strings of Christmas lights to the sound of music produced by a stereo system or FM radio. Each string of lights responds to a range of frequencies: bass, midrange, and treble. When the lights are placed behind a piece of diffused plastic, the display can be quite dramatic.

Commercial units, although generally available, are expensive, easily costing more than $50. Furthermore, undertaking the design and construction of such a project, rather than purchasing the item, will provide electronics students with a valuable electronics design and fabrication experience.

Project Objectives

The project is to have the following objectives: inexpensive to produce (less than $20), reliable, safe, attractive, and have good amplitude and frequency sensitivity. These objectives will be met in the following ways:

* **Cost.** By using noncritical, readily available hardware and electronic components, bought in bulk quantities, the cost will be kept low.

* **Reliability.** By using solid-state active components and printed circuit board construction techniques, a high degree of reliability will be obtained.

* **Safety.** The use of a fuse in the ac line and a completely sealed metal enclosure will assure safe operation.

* **Appearance.** A metal enclosure, either painted or covered with self-adhesive vinyl paper, will ensure an attractive project.

* Sensitivity. The color organ will control overall amplitude sensitivity with the use of a limiting resistor and a bypass switch. The frequency sensitivity of individual channels will be controlled by filter networks consisting of fixed resistors and capacitors, as well as a potentiometer.

Project Responsibility

Because this is to be a learning experience for the electronics student, he or she will be responsible for the design (which, in this case, is being supplied), drawings, experimenting, prototyping, and testing, troubleshooting, and final documentation phases of the 3-Channel Color Organ project. Due dates for each phase will be assigned by the electronics instructor.

Project Theory of Operation

The 3-Channel Color Organ gives a dramatic effect by flashing three strings of lights (each a different color) to the sound of music. Each string (with up to 200 W of light) responds to a particular range of audio tones: one for the bass, another for the midrange, and a third for the treble. Any audio source will provide the input signal, tapped directly off the speaker leads.

Resistor R_1 limits current to the indicator bulb I_1. The audio input signal, connected via jack J_1, is current-limited by resistor R_2. This resistor can be bypassed by slide switch S_2, applying the full audio input signal to transformer T_1. When the audio source provides a high input signal, S_2 must be open; when a low-level signal is present, S_2 is closed. T_1 is used as a step-up transformer to provide sufficient audio levels to trigger SCRs 1, 2, and 3. Potentiometers R_3, R_4, and R_5 are used as voltage dividers, allowing for variable adjustment of the SCR trigger levels. Resistor R_6 and capacitor C_1 form a low-pass filter for frequencies below 500 Hz for SCR_1 gate drive. Components R_7, R_8, C_2, and C_3 form a band-pass filter for frequencies between 500 and 3000 Hz for SCR_2 gate drive. Finally, capacitor C_4 and resistor R_9 form a high-pass filter for frequencies above 3000 Hz for SCR_3 gate drive.

Electronics instructor's approval *Bob Jones*   Date_**4-11**___

4

Experiment Results Document
3-Channel Color Organ
by
Larry Hyde

Conclusions

The 3-Channel Color Organ project performs as it was designed to perform, under breadboard conditions. All three channels responded with "dancing" lights to an audio input. Both an FM radio and a tape player were used as the signal source.

Test Results

1. Project was plugged into wall socket and switch S_1 closed. An FM radio was used as the signal source. Each channel was connected with fifteen 10-W Christmas bulbs, for a total of 150 W per channel.

2. Channels 2 and 3 responded as they should. The lights for channel 1 did not come on.

3. Close inspection revealed that SCR_1 was installed backward. When it was correctly installed, the channel worked as it should.

4. Switch S_2 was closed to check for operation at low volume. The circuit worked as it should.

5. The FM radio signal source was replaced with a tape player. Tests were run again and the circuit performed as it should.

6. The project was left on for 1 hour, after which time the circuit continued to work fine.

Test Results Document

Name of Project   *3-CHANNEL COLOR ORGAN*

Person(s) Performing Tests   *LARRY HYDE*

Preliminary Test Results

1. Is all wiring complete and correct?   Init.   *LH*

                                                    Date   *6-1*

2. Are all soldering connections good?   Init.   *LH*

                                                    Date   *6-1*

3. Are all components in the correct location?

                                                    Init.   *LH*

                                                    Date   *6-1*

4. Are all components installed in the correct direction?

                                                    Init.   *LH*

                                                    Date   *6-1*

Operational Test Results

1. Any catastrophic failure?   Yes _____   No ✓   Init.   *LH*

Explain: _____

_____

_____

2. Any minor problems?   Yes _____   No ✓   Init.   *LH*

Explain: _____

_____

_____

6

3. Calibration, if necessary.   Yes _____   No __✓___   Init. _LH_

_____

_____

4. Is troubleshooting required? Yes __✓___   No _____   Init. _LH_

What is anticipated? _____

_____

_____

5. Is project operation satisfactory at this time?

         Yes __✓___   No _____ Init. _LH_   Date _6-1_

Troubleshooting Results

1. Define the problem.

   _NONE_____

_____

_____

2. Steps taken to correct the problem.

   _N/A_____

_____

_____

3. Is project functioning satisfactorily?

         Yes __✓___   No _____ Init. _LH_   Date _6-1_

Performance Test Results

1. User test performed. _FELLOW STUDENT GIVEN PROJECT TO_

USE WITH STEREO SYSTEM.

2. User test results. _ALL CHANNELS WORKED WITH GOOD_

SENSITIVITY ON FM AND TAPE.

7

Concluding Comments

3 CHANNEL COLOR ORGAN OPERATES AS INTENDED WITH GOOD
FREQUENCY RESPONSE AT WIDE VOLUME RANGE.

8

Summary and Recommendations Document
3-Channel Color Organ
by
Larry Hyde

Our goal was to design and fabricate a 3-channel color organ that would cause three strings of colored lights to dance to an audio input signal. The goal has been met and the color organ meets all preliminary project objectives. The initial design has proven valid, and no unanticipated design or fabrication problems developed during prototype construction. It is the recommendation of this technician that full-scale production of the 3-channel color organ be given serious consideration.

Name *LARRY HYDE*

Signature *Larry Hyde*

Date *6-15*

9

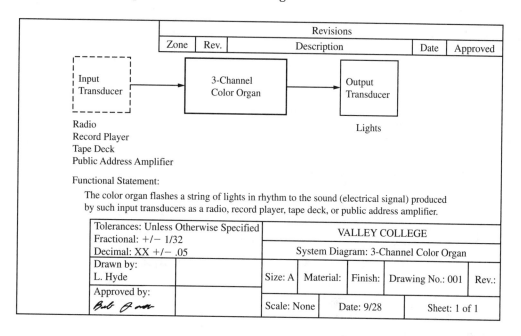

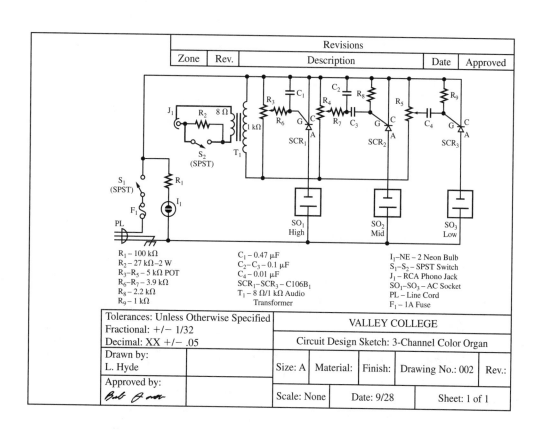

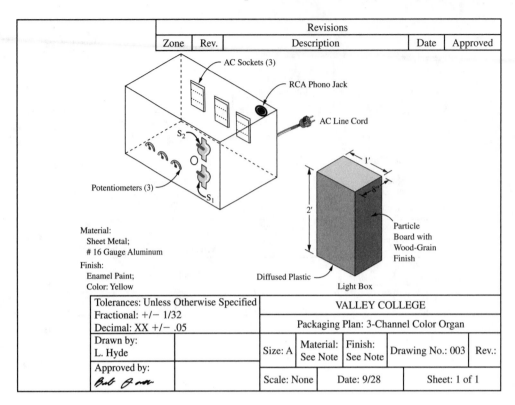

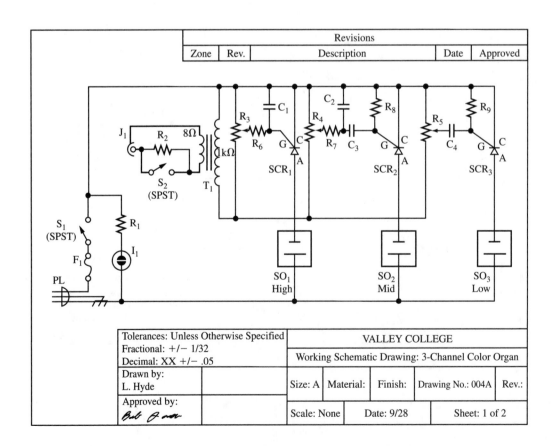

		Revisions			
Zone	Rev.	Description		Date	Approved

Qty.	Symbol	Description	Part No.
1	R_1	Resistor, 100 kΩ, $\frac{1}{4}$ W	
1	R_2	Resistor, 27 Ω, 2W	
3	R_3–R_5	Potentiometer, 5 kΩ	
2	R_6, R_7	Resistor, 3.9 kΩ, $\frac{1}{4}$ W	
1	R_8	Resistor, 2.2 kΩ, $\frac{1}{4}$ W	
1	R_9	Resistor, 1 kΩ, $\frac{1}{4}$ W	
1	C_1	Capacitor, 0.47 μF	
2	C_2, C_3	Capacitor, 0.1 μF	
1	C_4	Capacitor, 0.01 μF	
3	SCR_1–SCR_3	SCR, C106B1	
1	T_1	Transformer, Audio 8 Ω/1 kΩ	
1	I_1	Neon bulb, NE_2	
2	S_1, S_2	Switch, SPST	
1	J_1	RCA Phono jack	
3	SO_1–SO_3	AC Socket	
1	PL	Line cord	
1	F_1	Fuse, 1A	

Tolerances: Unless Otherwise Specified Fractional: +/− 1/32 Decimal: XX +/− .05		VALLEY COLLEGE				
		Schematic Drawing Parts List: 3-Channel Color Organ				
Drawn by: L. Hyde		Size: A	Material:	Finish:	Drawing No.: 004B	Rev.:
Approved by: *Bub J nor*		Scale: None	Date: 9/28		Sheet: 2 of 2	

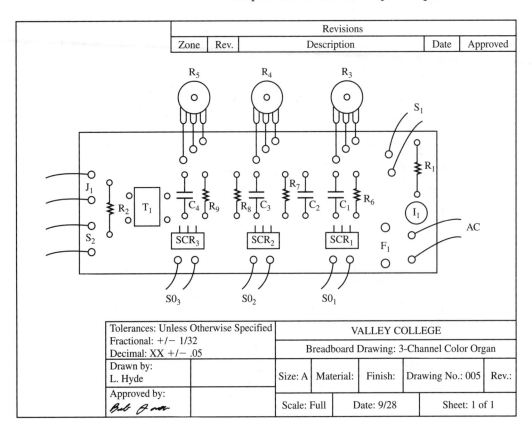

Revisions					
Zone	Rev.	Description		Date	Approved

Tolerances: Unless Otherwise Specified		VALLEY COLLEGE				
Fractional: +/− 1/32 Decimal: XX +/− .05		Breadboard Drawing: 3-Channel Color Organ				
Drawn by: L. Hyde		Size: A	Material:	Finish:	Drawing No.: 005	Rev.:
Approved by: *Bob Garner*		Scale: Full	Date: 9/28		Sheet: 1 of 1	

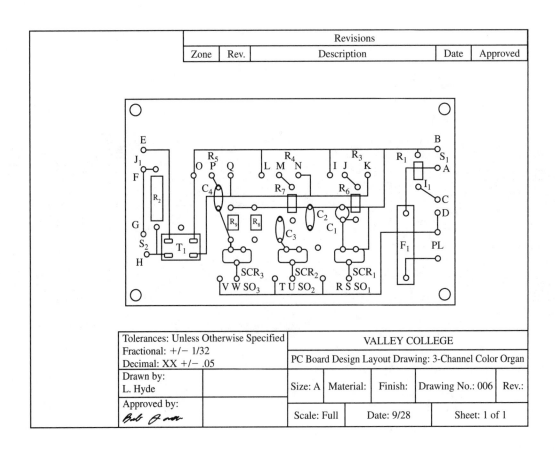

Revisions					
Zone	Rev.	Description		Date	Approved

Tolerances: Unless Otherwise Specified		VALLEY COLLEGE				
Fractional: +/− 1/32 Decimal: XX +/− .05		PC Board Design Layout Drawing: 3-Channel Color Organ				
Drawn by: L. Hyde		Size: A	Material:	Finish:	Drawing No.: 006	Rev.:
Approved by: *Bob Garner*		Scale: Full	Date: 9/28		Sheet: 1 of 1	

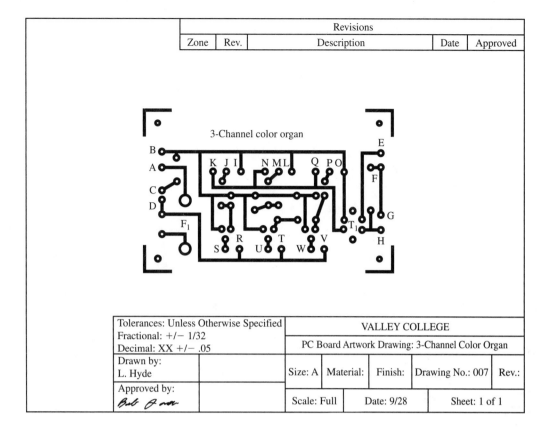

Revisions				
Zone	Rev.	Description	Date	Approved

3-Channel color organ

Tolerances: Unless Otherwise Specified Fractional: +/− 1/32 Decimal: XX +/− .05		VALLEY COLLEGE				
		PC Board Artwork Drawing: 3-Channel Color Organ				
Drawn by: L. Hyde		Size: A	Material:	Finish:	Drawing No.: 007	Rev.:
Approved by:		Scale: Full	Date: 9/28		Sheet: 1 of 1	

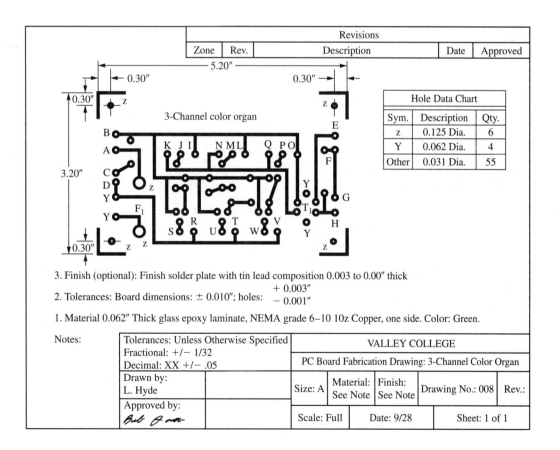

Revisions				
Zone	Rev.	Description	Date	Approved

3-Channel color organ

Hole Data Chart		
Sym.	Description	Qty.
z	0.125 Dia.	6
Y	0.062 Dia.	4
Other	0.031 Dia.	55

3. Finish (optional): Finish solder plate with tin lead composition 0.003 to 0.00″ thick

2. Tolerances: Board dimensions: ± 0.010″; holes: $^{+\,0.003''}_{-\,0.001''}$

1. Material 0.062″ Thick glass epoxy laminate, NEMA grade 6–10 10z Copper, one side. Color: Green.

Notes:

Tolerances: Unless Otherwise Specified Fractional: +/− 1/32 Decimal: XX +/− .05		VALLEY COLLEGE			
		PC Board Fabrication Drawing: 3-Channel Color Organ			
Drawn by: L. Hyde		Size: A	Material: See Note	Finish: See Note	Drawing No.: 008	Rev.:
Approved by:		Scale: Full	Date: 9/28		Sheet: 1 of 1

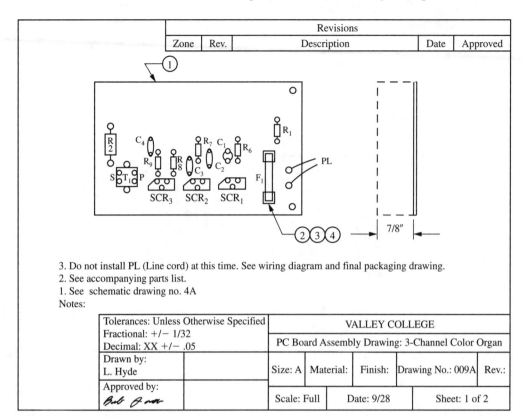

3. Do not install PL (Line cord) at this time. See wiring diagram and final packaging drawing.
2. See accompanying parts list.
1. See schematic drawing no. 4A
Notes:

Tolerances: Unless Otherwise Specified Fractional: +/− 1/32 Decimal: XX +/− .05		VALLEY COLLEGE					
		PC Board Assembly Drawing: 3-Channel Color Organ					
Drawn by: L. Hyde		Size: A	Material:	Finish:	Drawing No.: 009A		Rev.:
Approved by: *Bob Jron*		Scale: Full		Date: 9/28		Sheet: 1 of 2	

		Revisions		
Zone	Rev.	Description	Date	Approved

Item No.	Ref. Des.	Description	Qty.
1	1	PC Board	1
2	2	Fuse Clip	2
3	3	Screw, 4–40X ¼″	2
4	4	Nut, 4–40	2
5	F_1	Fuse , 1A	1
6	R_1	Resistor, 100 kΩ, $\frac{1}{4}$ W	1
7	R_2	Resistor, 27 kΩ, $\frac{2}{4}$ W	1
8	R_6–R_7	Resistor, 3.9 kΩ, $\frac{1}{4}$ W	2
9	R_8	Resistor, 2.2 kΩ, $\frac{1}{4}$ W	1
10	R_9	Resistor, 1 kΩ $\frac{1}{4}$ W	1
11	C_1	Capacitor, 0.47 μF	1
12	C_2, C_3	Capacitor, 0.1 μF	2
13	C_4	Capacitor, 0.01 μF	1
14	T_1	Transformer, Audio 8 Ω / 1 kΩ	1
15	SCR_1– SCR_3	SCR, C106B$_1$	3
16	PL	Line Cord	1

Tolerances: Unless Otherwise Specified Fractional: +/− 1/32 Decimal: XX +/− .05		**VALLEY COLLEGE**				
		PC Board Assembly Drawing Parts List: 3-Channel Color Organ				
Drawn by: L. Hyde		Size: A	Material:	Finish:	Drawing No.: 009B	Rev.:
Approved by: *Bob Brown*		Scale: None	Date: 9/28		Sheet: 2 of 2	

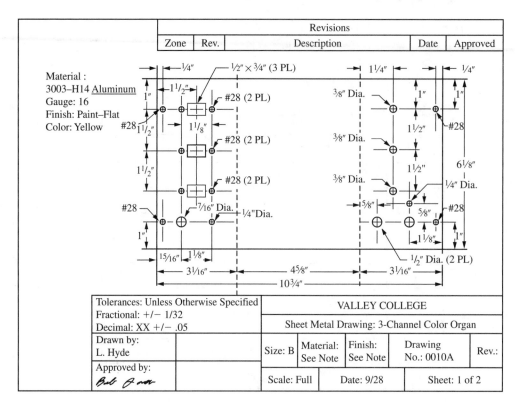

Material :
3003–H14 <u>Aluminum</u>
Gauge: 16
Finish: Paint–Flat
Color: Yellow

		Revisions		
Zone	Rev.	Description	Date	Approved

Tolerances: Unless Otherwise Specified		VALLEY COLLEGE			
Fractional: +/− 1/32					
Decimal: XX +/− .05		Sheet Metal Drawing: 3-Channel Color Organ			
Drawn by: L. Hyde		Size: B	Material: See Note	Finish: See Note	Drawing No.: 0010A	Rev.:
Approved by: Bob O*****		Scale: Full	Date: 9/28		Sheet: 1 of 2

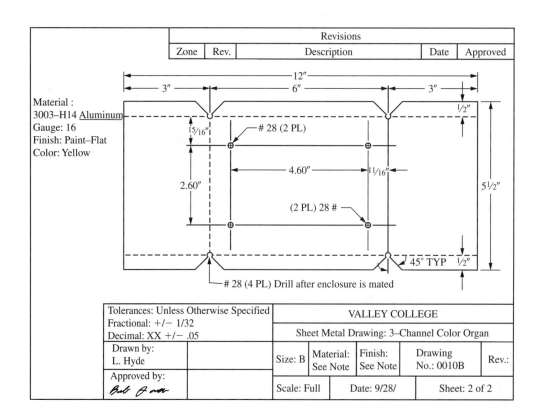

Material :
3003–H14 <u>Aluminum</u>
Gauge: 16
Finish: Paint–Flat
Color: Yellow

		Revisions		
Zone	Rev.	Description	Date	Approved

Tolerances: Unless Otherwise Specified		VALLEY COLLEGE			
Fractional: +/− 1/32					
Decimal: XX +/− .05		Sheet Metal Drawing: 3–Channel Color Organ			
Drawn by: L. Hyde		Size: B	Material: See Note	Finish: See Note	Drawing No.: 0010B	Rev.:
Approved by: Bob O*****		Scale: Full	Date: 9/28/		Sheet: 2 of 2

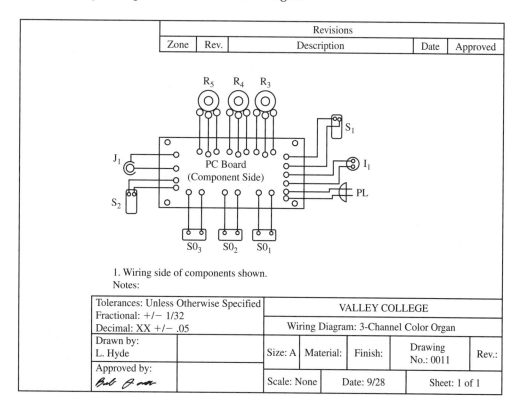

Revisions				
Zone	Rev.	Description	Date	Approved

1. Wiring side of components shown.

Notes:

Tolerances: Unless Otherwise Specified Fractional: +/− 1/32 Decimal: XX +/− .05		VALLEY COLLEGE				
		Wiring Diagram: 3-Channel Color Organ				
Drawn by: L. Hyde		Size: A	Material:	Finish:	Drawing No.: 0011	Rev.:
Approved by: *Bob Garner*		Scale: None		Date: 9/28		Sheet: 1 of 1

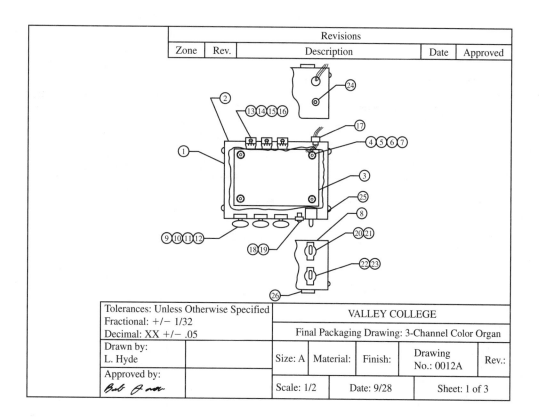

Revisions				
Zone	Rev.	Description	Date	Approved

Tolerances: Unless Otherwise Specified Fractional: +/− 1/32 Decimal: XX +/− .05		VALLEY COLLEGE				
		Final Packaging Drawing: 3-Channel Color Organ				
Drawn by: L. Hyde		Size: A	Material:	Finish:	Drawing No.: 0012A	Rev.:
Approved by: *Bob Garner*		Scale: 1/2		Date: 9/28		Sheet: 1 of 3

Revisions					
Zone	Rev.	Description		Date	Approved

PARTS LIST
Final Assembly

Item No.	Qty.	Part	Indent				Description	Comments
1	1		X				3-Channel Color Organ	
2	1			X			Chassis (Bottom)	
3	1				X		PC Board Subassembly	
4	4					X	Screw, $4-40 \times \frac{3}{4}''$	
5	4					X	Nut, $4-40$	
6	4					X	Washer, $\frac{1}{4}''$	
7	4					X	Spacer, $\frac{1}{4}''$	
8	1			X			Enclosure (TOP)	
9	3				X		Potentiometer, R_3, R_4, R_5	
10	3					X	Washer (Match POT)	
11	3					X	Nut (Match POT)	
12	3					X	Knob (Pointer)	
13	3				X		AC Sockets, SO_1, SO_2, SO_3	
14	6					X	Screw, $4-40 \times \frac{1}{2}''$	
15	6					X	Nut, $4-40$	
16	6					X	Washer, $\frac{1}{4}''$	
17	1				X		Strain Relief	
18	1				X		I1, Lamp Neon	
19	1					X	Grommet, $\frac{1}{4}''$	
20	1				X		Switch, S_1	
21	1					X	ON/OFF Plate	
22	1				X		Switch, S_2	
23	1					X	ON/OFF Plate	
24	1				X		RCA Phono Jack	
25	4					X	Screw SM, $\frac{3}{8}$	Self-tapping
26	4					X	Rubber Feet $\frac{3}{4}'' \times \frac{3}{4}''$	Self-adhesive

Tolerances: Unless Otherwise Specified Fractional: $+/-$ 1/32 Decimal: XX $+/-$ .05		VALLEY COLLEGE					
		Final Packaging Drawing Parts List: 3-Channel Color Organ					
Drawn by: L. Hyde		Size: A	Material:	Finish:	Drawing No.: 0012B		Rev.:
Approved by: *Bob Grove*		Scale: None		Date: 9/28		Sheet: 2 of 3	

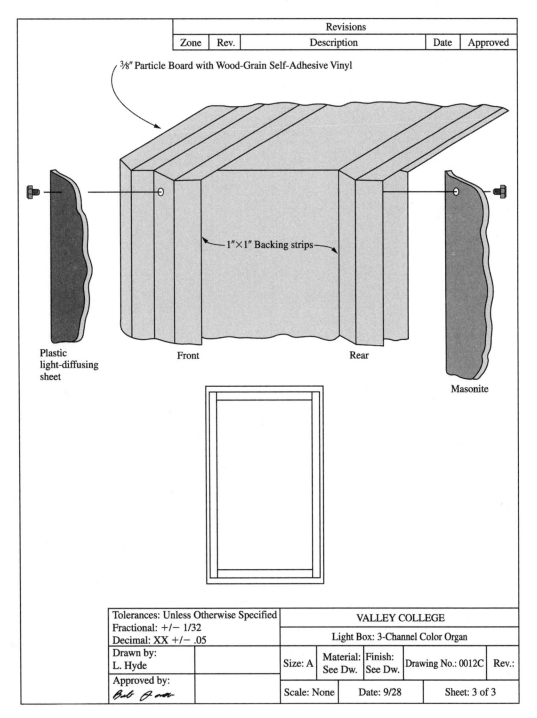

⅜″ Particle Board with Wood-Grain Self-Adhesive Vinyl

1″×1″ Backing strips

Plastic
light-diffusing
sheet

Front

Rear

Masonite

Revisions					
Zone	Rev.	Description		Date	Approved

Tolerances: Unless Otherwise Specified Fractional: +/− 1/32 Decimal: XX +/− .05		VALLEY COLLEGE				
		Light Box: 3-Channel Color Organ				
Drawn by: L. Hyde		Size: A	Material: See Dw.	Finish: See Dw.	Drawing No.: 0012C	Rev.:
Approved by: Bob Jones		Scale: None	Date: 9/28		Sheet: 3 of 3	

17 Concepts and Requirements Documents and Three-Drawing Set/Elective Projects

Concepts and Requirements Document

Adjustable Dual Power Supply

by

(student's name)

Project Goals

This Concepts and Requirements Document proposes that a working prototype be built of an adjustable dual power supply, for which there may be a market. If it is determined that such a market exists, the project could be mass-produced.

Today's electronic circuits often require a variety of voltages and two polarities. TTL circuits demand 5 V, CMOS circuits use up to 15 V, and many op-amps require 9-, 12-, or 15-V positive and negative supplies. Clearly, an inexpensive dual power supply, providing both positive and negative voltages, independently adjustable from 1.2 V to 15 V dc, would be a valuable addition to any electronics technician's workbench. The circuit described here will meet that need.

Project Objectives

The project is to have the following objectives: inexpensive to produce, reliable, safe, and easy to use. These objectives will be met in the following ways:

*　**Cost.** By using noncritical, readily available hardware and electronic components, bought in bulk, the cost will be kept low.

*　**Reliability.** By using quality components and printed circuit board construction techniques, a high degree of reliability will be obtained.

*　**Safety.** Although potentially dangerous ac and dc voltages are present, safe operation is assured by enclosing the project in a suitable case.

*　**Ease of Use.** Circuit operation is straightforward: a toggle switch turns the supply on or off, and two potentiometers adjust the output voltages.

Project Responsibility

Because this is to be a learning experience for the electronics student, he or she will be responsible for the design (which, in this case, is being supplied), drawings, experimenting, prototyping, and testing, troubleshooting, and final documentation phases of the Adjustable Dual Power Supply. Due dates for each phase will be assigned by the electronics instructor.

1

Project Theory of Operation

When switch S1 is closed, the 120-V ac line power is delivered to the primary of power transformer T1. The secondary of the transformer supplies approximately 12.6 V to rectifiers D1 and D2. The pulsating dc emerging from the rectifiers is fed to filter capacitor C1 in the positive supply and filter capacitor C2 in the negative supply. The reference point for the two supply sections is established by connecting both grounds to one end of the transformer's secondary winding (binding post BP2).

Filtered dc voltages are now fed to adjustable voltage regulator IC1 in the positive supply and adjustable voltage regulator IC2 in the negative supply. Potentiometer R3 adjusts for the positive supply, while potentiometer R4 does the same for the negative supply. From the outputs of both regulators, the final voltages are fed to binding posts BP1 (positive) and BP3 (negative), where they are made available for external use. LED1 serves as an on/off indicator.

Electronics instructor's approval _____ **Date** _____

2

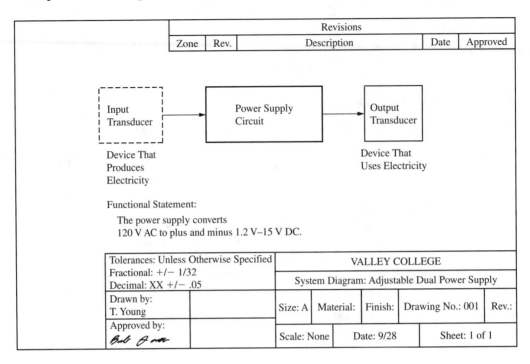

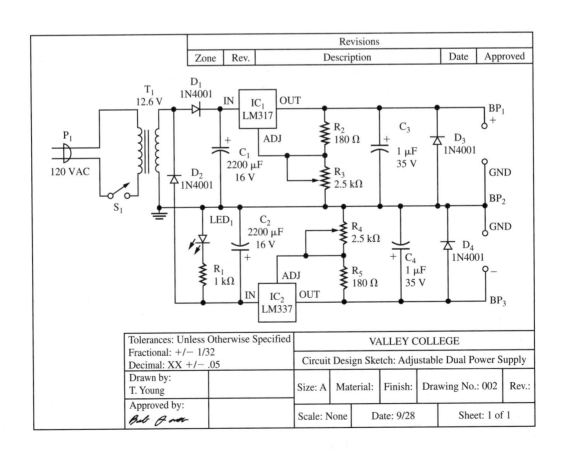

Revisions					
Zone	Rev.	Description		Date	Approved

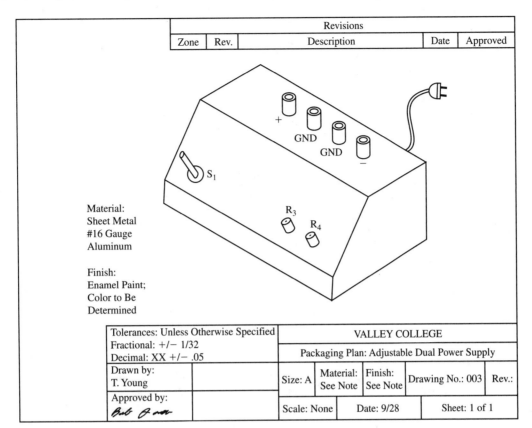

Material:
Sheet Metal
#16 Gauge
Aluminum

Finish:
Enamel Paint;
Color to Be
Determined

Tolerances: Unless Otherwise Specified		VALLEY COLLEGE					
Fractional: +/− 1/32							
Decimal: XX +/− .05		Packaging Plan: Adjustable Dual Power Supply					
Drawn by:			Material:	Finish:			
T. Young		Size: A	See Note	See Note	Drawing No.: 003		Rev.:
Approved by:							
Bob Jones		Scale: None	Date: 9/28			Sheet: 1 of 1	

Concepts and Requirements Document

Burglar Alarm

by

(student's name)

Project Goals

This Concepts and Requirements Document proposes that a working prototype be built of a burglar alarm, for which there may be a commercial market. If it is determined that such a market exists, the project could then be mass-produced.

There is considerable demand for a simple, inexpensive burglar alarm that will protect windows and doors in the home or automobile. The proposed device will do these things and, in addition, will operate from a 9-V battery and can be built for less than $10.

Project Objectives

The project is to have the following objectives: very inexpensive to produce (less than $10), reliable, easy to install, be powered by a dc source, monitor many locations simultaneously, and, once triggered, remain on even if the pre-alarm condition is reestablished. These objectives will be met in the following ways:

*   **Cost.** The project parts count can be held down by incorporating a simple silicon-controlled rectifier design, and avoiding complex disarming and delay circuits. The cost will be kept low by having few parts and by using noncritical, readily available hardware and electronic components, bought in bulk quantities.

*   **Reliability.** Use of an extremely dependable solid-state active component, the SCR, and printed circuit board construction techniques will ensure a high degree of reliability.

*   **Ease of Installation.** Installation is completed by simply stringing wires, in series or parallel, to the various locations to be monitored.

1

* **DC Power.** With SCR design, the circuit will operate on 5 to 24 V dc. The circuit may be powered using standard batteries because standby current drain is extremely low (less than 10 mA).

* **Monitoring of Many Locations.** With the use of both normally open (N/O) and normally closed (N/C) contacts, many spots can be monitored by simply connecting successive locations in series or parallel.

* **Latching Condition.** With the use of an SCR, the alarm will latch on once it is triggered. The only way to turn off the alarm is to reset it with a simple disarming switch, which is under the user's control.

Project Responsibility

Because this is to be a learning experience for the electronics student, he or she will be responsible for the design (which, in this case, is being supplied), drawings, experimenting, prototyping, and testing, troubleshooting, and final documentation phases of the Burglar Alarm Project. Due dates for each phase will be assigned by the electronics instructor.

Project Theory of Operation

Silicon-controlled rectifier SCR1 is the basis for the latching switch alarm circuit. The SCR will conduct current from its cathode to its anode when, and only when, it is triggered with a positive voltage on its gate. Even when the gate voltage is removed, the SCR will continue to conduct. Only when the SCR current is interrupted, by opening switch S1, will the SCR turn off. While the SCR conducts, the buzzer, of course, will sound.

When a normally closed (N/C) switch opens, resistor R1 supplies current to the gate of the SCR via diode D1, thus triggering the SCR on. When a normally open (N/O) switch closes, resistor R2 supplies current to the gate of the SCR and triggers it on. Diode D1 prevents the N/C switches from shorting the gate of the SCR to ground when an N/O switch is used. Capacitor C1 and resistor R3 help eliminate transients from falsely triggering the SCR. Diode D2 will clamp any inductive transients generated by the buzzer. Resistor R4 ensures that continuous current flow will be available in the event that an electromechanical or make/break type of buzzer is used. Finally, switch S1 acts as an on/off disarm and reset switch.

As shown in drawing No. 2, when N/C connections are used, the alarm switches (contact points to be monitored) are connected in series. When N/O connections are used, the alarm switches should be placed in parallel.

Electronics instructor's approval _____ **Date** _____

2

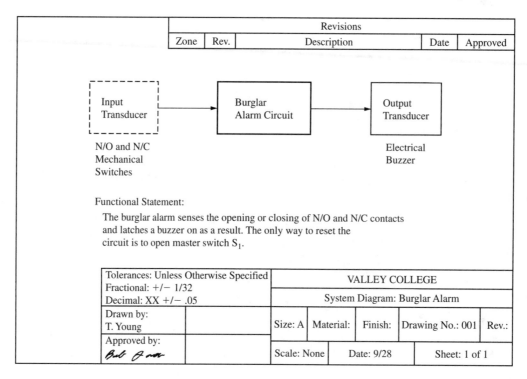

	Revisions			
Zone	Rev.	Description	Date	Approved

Functional Statement:

The burglar alarm senses the opening or closing of N/O and N/C contacts and latches a buzzer on as a result. The only way to reset the circuit is to open master switch S_1.

Tolerances: Unless Otherwise Specified		VALLEY COLLEGE				
Fractional: +/− 1/32						
Decimal: XX +/− .05		System Diagram: Burglar Alarm				
Drawn by: T. Young		Size: A	Material:	Finish:	Drawing No.: 001	Rev.:
Approved by: *Bob Grant*		Scale: None	Date: 9/28		Sheet: 1 of 1	

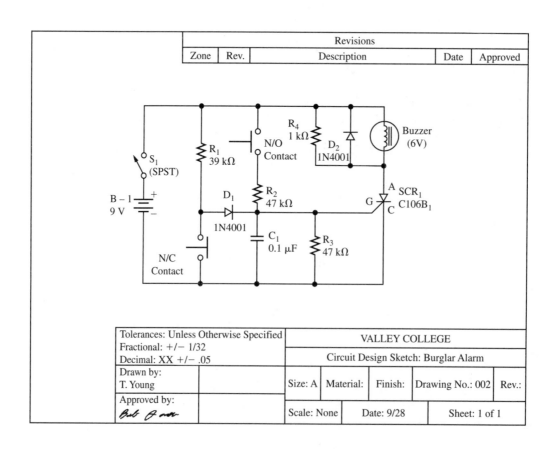

	Revisions			
Zone	Rev.	Description	Date	Approved

Tolerances: Unless Otherwise Specified		VALLEY COLLEGE				
Fractional: +/− 1/32						
Decimal: XX +/− .05		Circuit Design Sketch: Burglar Alarm				
Drawn by: T. Young		Size: A	Material:	Finish:	Drawing No.: 002	Rev.:
Approved by: *Bob Grant*		Scale: None	Date: 9/28		Sheet: 1 of 1	

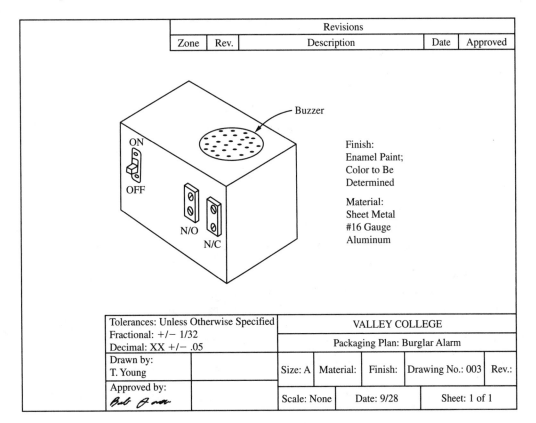

Revisions					
Zone	Rev.	Description		Date	Approved

Buzzer

Finish:
Enamel Paint;
Color to Be
Determined

Material:
Sheet Metal
#16 Gauge
Aluminum

Tolerances: Unless Otherwise Specified Fractional: +/− 1/32 Decimal: XX +/− .05		VALLEY COLLEGE				
		Packaging Plan: Burglar Alarm				
Drawn by: T. Young		Size: A	Material:	Finish:	Drawing No.: 003	Rev.:
Approved by: *Bob Brown*		Scale: None	Date: 9/28		Sheet: 1 of 1	

Concepts and Requirements Document

Clap-on/Clap-off Switch

by

(student's name)

Project Goals

This Concepts and Requirements Document proposes that a working prototype be built of a clap-on/clap-off switch, for which there may be a market. If it is determined that such a market exists, the project could be mass-produced.

A clap-on/clap-off switch, operated from a 9-V battery, is ideal for the remote control of practically any electrical or electronic device, with a clap of the hands. With the first clap, the device turns on and remains on. With a second clap, it shuts off. Since the output load for the circuit we propose to build is a relay, virtually any device, be it a TV, stereo, radio, alarm, light, or motor, can be controlled by the clap-on/clap-off circuit.

Project Objectives

The project is to have the following objectives: inexpensive to produce, reliable, safe, and easy to use. These objectives will be met in the following ways:

* **Cost.** Use of noncritical, readily available hardware and electronic components, bought in bulk, will keep the cost low.

* **Reliability.** Use of integrated circuits, a quality relay, and printed circuit board construction techniques will ensure a high degree of reliability.

* **Safety.** Shielding the relay contacts within a metal or plastic box will assure human isolation from the 120-V line.

* **Ease of Use.** Operation couldn't be simpler. Once two internal potentiometers are adjusted and set, and the relay contacts are connected to the device to be controlled, operation is simply a matter of clapping the hands.

Project Responsibility

Because this is to be a learning experience for the electronics student, he or she will be responsible for the design (which, in this case, is being supplied), drawing, experimenting, prototyping, and testing, troubleshooting, and final documentation phases of the Clap-on/Clap-off Switch. Due dates for each phase will be assigned by the electronics instructor.

Project Theory of Operation

When sound waves from a clap hit the microphone, they are converted into an electrical signal. Only a specific range of frequencies are allowed to pass the band-pass filter made up of IC1A, an op-amp. The frequencies that get through and have sufficient signal strength trigger the voltage comparator circuit, IC1B. A pulse from the voltage comparator is then sent to IC2A, a monostable multivibrator.

On the first clap, the monostable multivibrator, or timer, outputs a high pulse, having received a trigger from the voltage comparator circuit. That pulse in turn triggers the flip-flop, IC2B, which turns on transistor Q1. With the next clap, the voltage comparator circuit triggers the timer again. The flip-flop now resets. Q1 turns off. Since the transistor acts as a switch to control the relay, the relay is on when the transistor is on, and when the transistor is off, so is the relay. The relay turns whatever is connected to its contacts on and off.

Potentiometer P1 controls the circuit's sensitivity to the loudness level of the clap. Potentiometer P2, connected as a rheostat, controls an RC time delay that prevents the circuit from turning on and off too quickly. Thus, clapping the hands a few times will just turn on the device, not turn it on and off in quick succession.

Electronics instructor's approval _____ **Date** _____

2

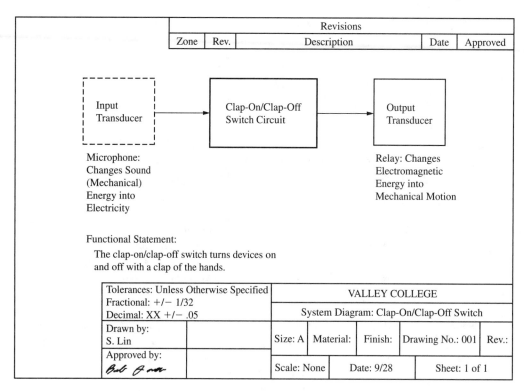

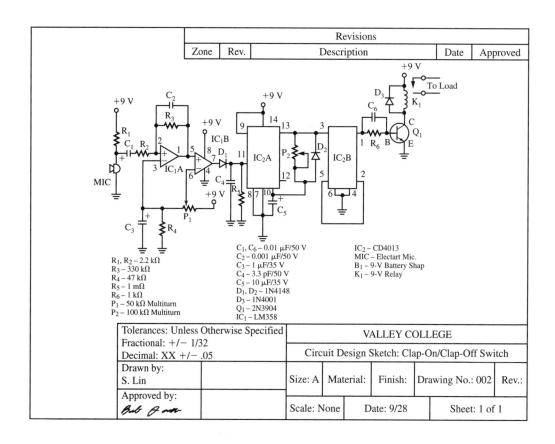

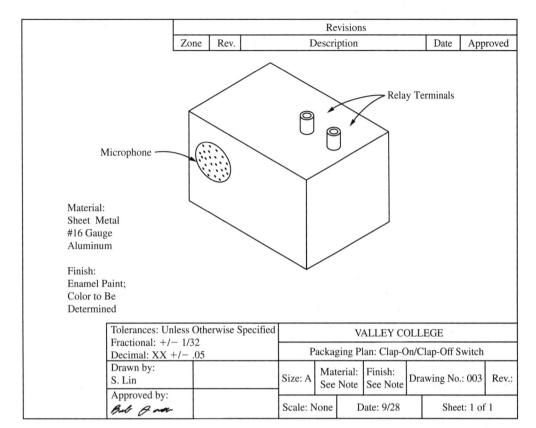

Revisions					
Zone	Rev.	Description		Date	Approved

Relay Terminals

Microphone

Material:
Sheet Metal
#16 Gauge
Aluminum

Finish:
Enamel Paint;
Color to Be
Determined

Tolerances: Unless Otherwise Specified		VALLEY COLLEGE				
Fractional: +/− 1/32 Decimal: XX +/− .05		Packaging Plan: Clap-On/Clap-Off Switch				
Drawn by: S. Lin		Size: A	Material: See Note	Finish: See Note	Drawing No.: 003	Rev.:
Approved by: *Bob Brown*		Scale: None	Date: 9/28		Sheet: 1 of 1	

Concepts and Requirements Document

Events Timer

by

(student's name)

Project Goals

This Concepts and Requirements Document proposes that a working prototype be built of an events timer, for which there may be a market. If it is determined that such a market exists, the project could be mass-produced.

Many events, from basketball games to speech contests, require a time delay followed by a short audio burst. The so-called 2-minute warning is an example.

This circuit provides such a delay. Consisting of three 555 ICs, the circuit is activated by pressing a normally open push-button switch. Approximately 2 minutes later, a 500-Hz tone is sounded for 2 seconds. Substitution of different capacitor and/or resistor values allows the 2-minute time delay, 2-second tone burst, and 500-Hz tone to be altered.

Project Objectives

The project is to have the following objectives: inexpensive to produce, reliable, safe, and easy to use. These objectives will be met in the following ways:

* **Cost.** Use of noncritical, readily available hardware and electronic components, bought in bulk, will keep the cost low.

* **Reliability.** Use of integrated circuits, a transistor, and printed circuit board construction techniques will ensure a high degree of reliability.

* **Safety.** The circuit operates from a 9-V transistor battery and therefore is quite safe.

* **Ease of Use.** Operation couldn't be simpler. One simply presses a button to start the timing sequence.

1

Project Responsibility

Because this is to be a learning experience for the electronics student, he or she will be responsible for the design (which, in this case, is being supplied), drawing, experimenting, prototyping, and testing, troubleshooting, and final documentation phases of the Events Timer Project. Due dates for each phase will be assigned by the electronics instructor.

Project Theory of Operation

Referring to the schematic drawing, note the three 555 ICs. IC1 and IC2 are configured as monostable multivibrators (timers), whereas IC3 works as an astable multivibrator (clock).

The moment S1 is pressed, pin 2 of IC1 goes low, and pin 3 outputs a high pulse for 2 minutes. When its timing cycle is complete, the output of monostable IC1 drops low, triggering monostable IC2, through capacitor C3. IC2's output in turn goes high for 2 seconds. As it does it releases astable multivibrator IC3 from its reset condition, and a 500-Hz tone blares from the speaker.

The initial delay can be altered by changing the values of R1 and C1. Increasing either increases the delay; decreasing either decreases the delay. The on time of the tone is adjusted with R_4 and C4. Increasing either increases the on time; decreasing either decreases the on time. The output pitch is varied by changing R6, R7, and C6. In this case, increasing either decreases the pitch; decreasing either increases the output tone.

Electronics instructor's approval _____ **Date** _____

2

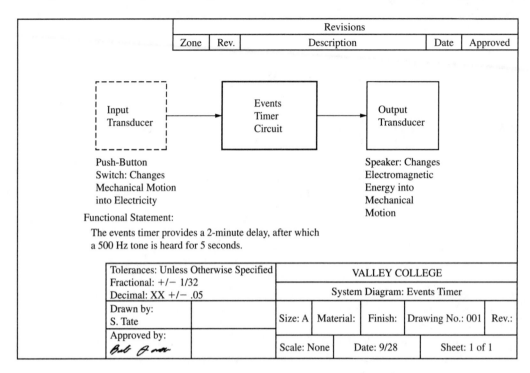

Revisions				
Zone	Rev.	Description	Date	Approved

Input Transducer

Events Timer Circuit

Output Transducer

Push-Button Switch: Changes Mechanical Motion into Electricity

Speaker: Changes Electromagnetic Energy into Mechanical Motion

Functional Statement:

The events timer provides a 2-minute delay, after which a 500 Hz tone is heard for 5 seconds.

Tolerances: Unless Otherwise Specified Fractional: +/− 1/32 Decimal: XX +/− .05		VALLEY COLLEGE				
		System Diagram: Events Timer				
Drawn by: S. Tate		Size: A	Material:	Finish:	Drawing No.: 001	Rev.:
Approved by: *Bob J nor*		Scale: None	Date: 9/28		Sheet: 1 of 1	

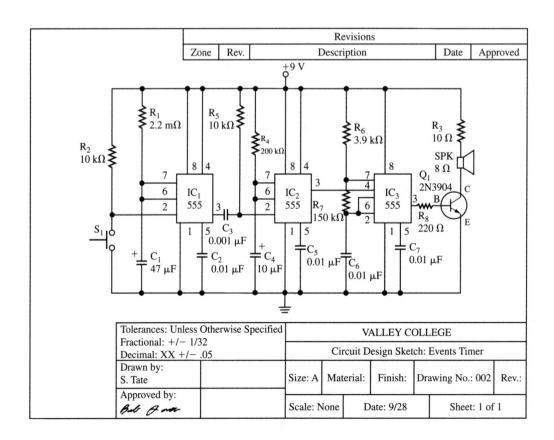

Revisions				
Zone	Rev.	Description	Date	Approved

Tolerances: Unless Otherwise Specified Fractional: +/− 1/32 Decimal: XX +/− .05		VALLEY COLLEGE				
		Circuit Design Sketch: Events Timer				
Drawn by: S. Tate		Size: A	Material:	Finish:	Drawing No.: 002	Rev.:
Approved by: *Bob J nor*		Scale: None	Date: 9/28		Sheet: 1 of 1	

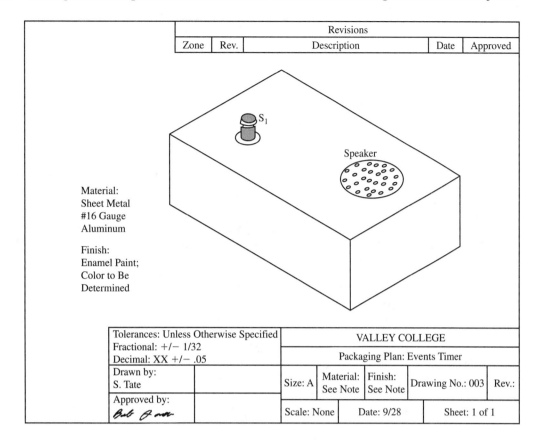

Revisions					
Zone	Rev.	Description		Date	Approved

Material:
Sheet Metal
#16 Gauge
Aluminum

Finish:
Enamel Paint;
Color to Be
Determined

S₁ (labeled on knob)

Speaker

Tolerances: Unless Otherwise Specified		VALLEY COLLEGE				
Fractional: +/− 1/32 Decimal: XX +/− .05		Packaging Plan: Events Timer				
Drawn by: S. Tate		Size: A	Material: See Note	Finish: See Note	Drawing No.: 003	Rev.:
Approved by: *Bob Gross*		Scale: None	Date: 9/28		Sheet: 1 of 1	

Concepts and Requirements Document

LED Attention-Getter

by

(student's name)

Project Goals

Strong student interest has been shown in a "Knight Rider" type of attention-getting display that can be placed in the rear window of an automobile, in a window of a home, or in the display window of a store. Applications are numerous, limited only by one's imagination.

This Concepts and Requirements Document proposes that a working prototype be built of an all-digital LED attention-getter, for which there may be a commercial market. If it is determined that such a market exists, the project could then be mass-produced.

The proposed project would consist of 16 red LEDs arranged in a horizontal row. At a rate determined by a resistance/capacitance (RC) network, the LEDs would "race" back and forth, with only one LED on at any given time. The circuit would require only 5 V dc and thus could be powered by a 12-V car battery (with appropriate voltage regulation to supply 5 V to the circuit), or for home or office use, by a 6-V lantern battery or 5-V dc power supply.

Project Objectives

The project is to have the following objectives: inexpensive to produce, reliable, safe, and easy to use. These objectives will be met in the following ways:

*   **Cost.** Use of noncritical, readily available hardware and electronic components, bought in bulk quantities, will keep the cost low.

1

* **Reliability.** Use of integrated circuits and printed circuit board construction techniques will ensure a high degree of reliability.

* **Safety.** The use of a low-voltage dc power source assures safe operation.

* **Ease of Use.** To use, one just connects dc power and turns on an SPST switch.

Project Responsibility

Because this is to be a learning experience for the electronics student, he or she will be responsible for the design (which, in this case, is being supplied), drawing, experimenting, prototyping, and testing, troubleshooting, and final documentation phases of the LED Attention-Getter Project. Due dates for each phase will be assigned by the electronics instructor.

Project Theory of Operation

This all-digital project consists of a 555 IC acting as an astable multivibrator (clock), a TTL 74193 up/down counter, a TTL 74154 1-of-16 data distributor, and four NAND gates in a TTL 7400 IC. Operation is as follows:

* All the outputs of the 74154 except one are high at any given time.

* Assume output O (pin 1) of the 74154 is low. This assumption means that LED 0 (there are 16 LEDs, but they are counted 0–15) is lit. Also, pin 13 of IC 7400D is low. Its output, pin 11, is high (construct a NAND gate truth table to check it out). With the output of IC 7400D high, both inputs to IC 7400C are high, and its output is low, placing a low on input pin 12 of IC 7400D.

* With the outputs of ICs 7400C and 7400D low and high, respectively, inputs to IC 7400A (pin 1) and IC 7400B (pin 5) are low and high, respectively.

* As square-wave pulses from the 555 IC appear on input pins 2 and 4 of ICs 7400A and 7400B, respectively, their output voltages (pins 3 and 6, respectively) are sent to IC 74193, the up/down counter. Output of pin 4 of IC 74193 will remain high (check the truth table for the NAND gate) while the signal arriving on pin 5 of IC 74193 will alternate from high to low and so on.

* With the down pin (4) or IC 74193 held high and the up pin (5) alternating from high to low, to high, and so on, the IC 74193 will count up.

2

*   On the next count, LED number 1 (the second LED) turns on; its cathode goes low because pin 2 of IC 74154 is now low. Even though pin 1 on the IC 74154 is now high, the outputs to both NAND gates C and D remain as before (low and high, respectively).

*   IC 74193 continues to count up, and the LEDs connected to IC 74154 "advance," lighting in sequence.

*   When pin 17 of IC 74154 goes low (LED 15 turns on), pin 9 of IC 7400C goes low. The output, pin 8, goes high, while the output of IC 7400D is low.

*   With pin 5 of IC 7400B now held low, its output is high, as is the up input (pin 5) to the 74193 IC. The low input (pin 4) of the 74193 IC alternates between high and low with pulses from the 555 IC, via the 7400D IC. Now the 74193 counts down, and the LEDs reverse direction.

*   When LED 0 lights again, the whole process repeats.

*   The values of resistors R1 and R2 and capacitor C1 determine the pulse rate of the 555 clock. Increasing the value of R1 or C1 slows the pulse rate. Reducing the value of R1 or C1 increases the pulse rate.

Electronics instructor's approval _____ **Date** _____

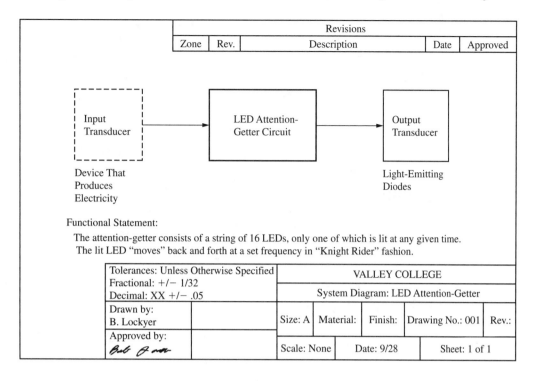

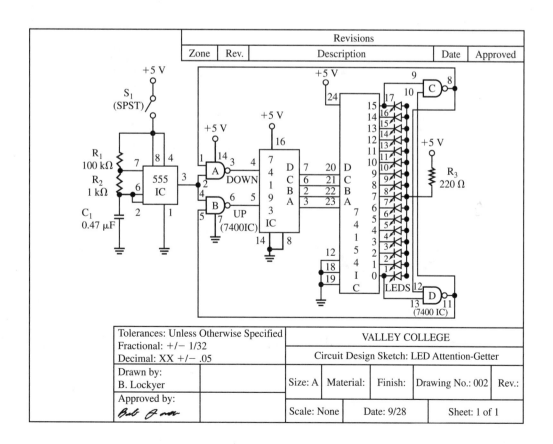

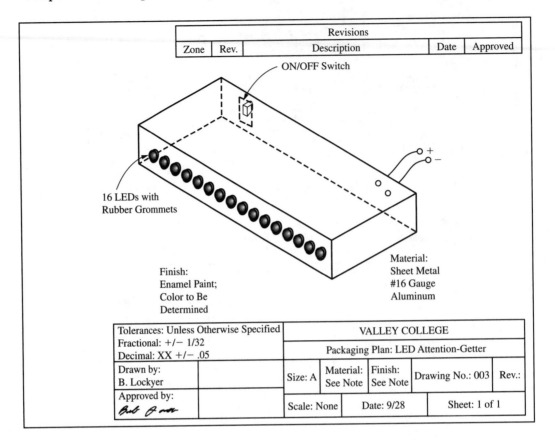

		Revisions		
Zone	Rev.	Description	Date	Approved

ON/OFF Switch

16 LEDs with
Rubber Grommets

+
−

Finish:
Enamel Paint;
Color to Be
Determined

Material:
Sheet Metal
#16 Gauge
Aluminum

Tolerances: Unless Otherwise Specified		VALLEY COLLEGE				
Fractional: +/− 1/32 Decimal: XX +/− .05		Packaging Plan: LED Attention-Getter				
Drawn by: B. Lockyer		Size: A	Material: See Note	Finish: See Note	Drawing No.: 003	Rev.:
Approved by: *Bob Jones*		Scale: None	Date: 9/28		Sheet: 1 of 1	

Concepts and Requirements Document

Light-Controlled Switch

by

(student's name)

Project Goals

This Concepts and Requirements Document proposes that a working prototype be built of a light-controlled switch, for which there may be a commercial market. If it is determined that such a market exists, the project could then be mass-produced.

The proposed project is an electronic switch that is controlled (turned on and off) by the amount of light shining on a photocell. When light strikes the photocell, the switch opens (turns off); when the photocell is in darkness, the switch closes (turns on). In series with the switch is placed any ac appliance rated at up to 300 W. Table and floor lamps, ac radios, coffeemakers, electronic counters, and the like are common devices used with the light-controlled switch.

In a typical application, a lamp is plugged into the light-controlled switch, which, in turn, is plugged into the ac wall outlet. The unit is then placed near a window such that daylight bathes the photocell. At night, when the photocell is in darkness, the switch closes, turning the lamp on. As daytime approaches, the photocell turns off the switch, and the switch turns off the lamp. Used in this way, the light-controlled switch gives unwanted guests the impression that someone is at home.

Project Objectives

The project is to have the following objectives: inexpensive to produce, reliable, completely safe, adjustable to various light levels, and easy to use. These objectives will be met in the following ways:

1

*   Cost. Use of noncritical, readily available hardware and electronic components, bought in bulk quantities, will keep the cost low.

*   Reliability. Use of an extremely dependable solid-state active component, the triac, and printed circuit board construction techniques will ensure a high degree of reliability.

*   Safety. Fusing of the ac line and completely enclosing the project in a metal or plastic enclosure will assure safe operation.

*   Sensitivity Adjustment. A potentiometer adjustment is used to ensure switch triggering at various light sensitivities.

*   Ease of Use. All the user need do to operate the light-controlled switch is to plug an appliance into the device, plug the switch into the ac wall outlet, and make a simple light sensitivity adjustment with a potentiometer.

Project Responsibility

Because this is to be a learning experience for the electronics student, he or she will be responsible for the design (which, in this case, is being supplied), drawing, experimenting, prototyping, and testing, troubleshooting, and final documentation phases of the Light-Controlled Switch Project. Due dates for each phase will be assigned by the electronics instructor.

Project Theory of Operation

Four components—resistor R1, potentiometer R2, photocell R3, and the load plugged into ac socket SO1—form a voltage divider across the ac line voltage. The resistance of R3 rises as the light level decreases; it falls as the light level increases. Thus, the voltage level at the junction of R2, R3, C1, and D1 is a function of the amount of light shining on R3. R2, along with R1, allows adjustment of this junction voltage for different light levels. Diac D1, a bilateral trigger diode, will switch on in either direction. When capacitor C1, which is across R3, reaches a potential of approximately 0.32 V, D1 switches on. The capacitor then discharges its voltage into the gate of triac Q1, turning it on. Because both Q1 and D1 are bilateral devices, the output to ac socket SO1 will be a near-perfect ac sine wave. R4 acts to eliminate false triggering of Q1 due to ac line transients.

Electronics instructor's approval _____ **Date** _____

2

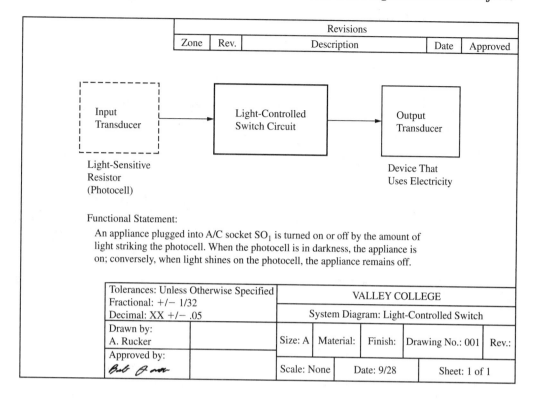

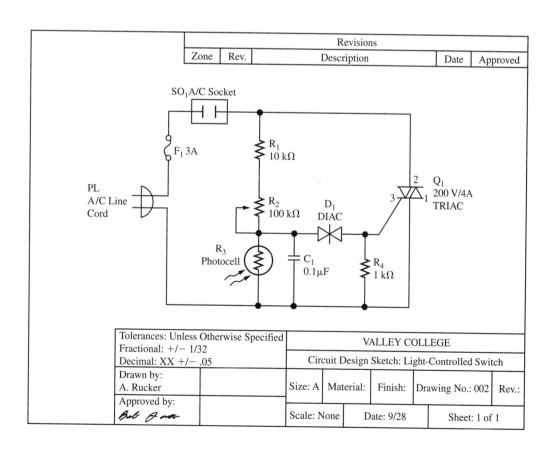

	Revisions				
Zone	Rev.	Description		Date	Approved

Material:
Sheet Metal
#16 Gauge
Aluminum

Finish:
Enamel Paint;
Color to Be
Determined

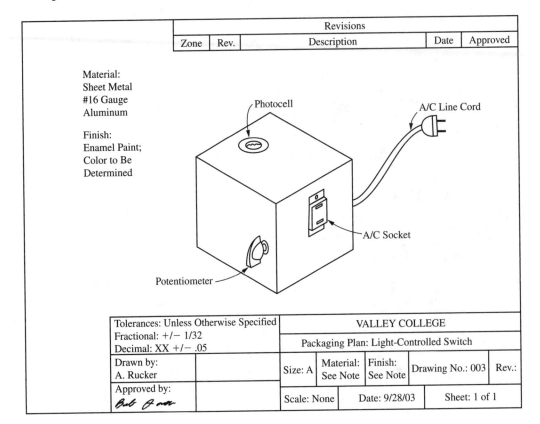

Photocell

A/C Line Cord

A/C Socket

Potentiometer

Tolerances: Unless Otherwise Specified		VALLEY COLLEGE				
Fractional: +/− 1/32 Decimal: XX +/− .05		Packaging Plan: Light-Controlled Switch				
Drawn by: A. Rucker		Size: A	Material: See Note	Finish: See Note	Drawing No.: 003	Rev.:
Approved by: *Bob Graner*		Scale: None	Date: 9/28/03		Sheet: 1 of 1	

Concepts and Requirements Document

Mosquito Repeller

by

(student's name)

Project Goals

This Concepts and Requirements Document proposes that a working prototype be built of a mosquito repeller, for which there may be application in the commercial market. If it is determined that such a market exists, the project could be mass-produced.

Those pesty little mosquitoes—how can we let something so small bother us so much? Yet, they do: on picnics, at the beach, on camping trips, or in our own backyards. Although all manner of chemical solutions, in the form of sprays and ointments, exist, these approaches are uncomfortable and, in some cases, environmentally questionable. An electronics solution is needed, yet nothing reliable exists commercially. This mosquito repeller could do the trick.

The circuit uses two oscillators, working together, to produce a pulsating, near-ultrasonic, quick up-and-down frequency sweep, which, according to the latest research on the subject, drives mosquitoes away. Power is derived from a 9-V battery.

Project Objectives

The project is to have the following objectives: inexpensive to produce, reliable, safe, and easy to use. These objectives will be met in the following ways:

*   **Cost**. Use of noncritical, readily available hardware and electronic components, bought in bulk, will keep the cost low.

*   **Reliability**. Use of two integrated circuits will ensure a high degree of reliability.

*   **Safety**. Use of a 9-V battery to power the circuit will assure safety.

*   **Ease of Use**. To operate, one simply connects a 9-V battery, turns on a switch, and hopes for the best. What could be easier?

Project Responsibility

Because this is to be a learning experience for the electronics student, he or she will be responsible for the design (which, in this case, is being supplied), drawing, experimenting, prototyping, and testing, troubleshooting, and final documentation phases of the Mosquito Repeller Project. Due dates for each phase will be assigned by the electronics instructor.

1

Project Theory of Operation

The mosquito repeller consists of two oscillators working together to produce the desired near-ultrasonic tone necessary to drive mosquitoes away. Let's see how.

The Gated Oscillator

*   An oscillator is a circuit that produces a self-generating signal. In this project we use two oscillators, one of which is configured around a CMOS 4001 quad two-input NOR Gate IC. Only two of the IC's four gates are used, and they are configured as NOT gates. As shown in the schematic drawing, the basic oscillator can be represented by two NOT gates with two feedback components: a resistor and a capacitor. The resistor and capacitor values determine the output frequency of the oscillator. In our circuit the oscillator's feedback resistor is R3 and its feedback capacitor is C2. The circuit's output frequency is approximately 10 Hz.

*   The Mosquito Repeller Project uses a second oscillator, a 555 IC timer, to drive a speaker. The 555 IC's output (pin 3) alternately goes high, then low, then high, and so on (see the schematic drawing). When the output is high the speaker gets the positive and negative voltage it needs, it conducts current through its voice coil, and the cone moves. When the 555 IC's output is low both ends of the speaker are at the same voltage, no current can flow through its voice coil, and the cone remains motionless. The alternating high, low signal at the IC's output turns the speaker on, then off, then on, and so on. We, and, more importantly, the mosquitoes, hear a tone.

*   The frequency of the 555 oscillator is determined primarily by an RC (resistance/capacitance) subcircuit made up, in our case, of R1, R2, and C1. Given the values of R1, R2, and C1, the tone will be a high-pitch whine.

*   Note that the output of the first oscillator is connected to pin 5 of the second oscillator (see the schematic drawing). Pin 5 is known as the voltage-controlled oscillator (VCO) input. As the voltage on pin 5 changes so does the output frequency of IC1. The effect is to produce a high-pitched but pulsing tone that, ideally, drives mosquitoes away.

Electronics instructor's approval _____ **Date** _____

2

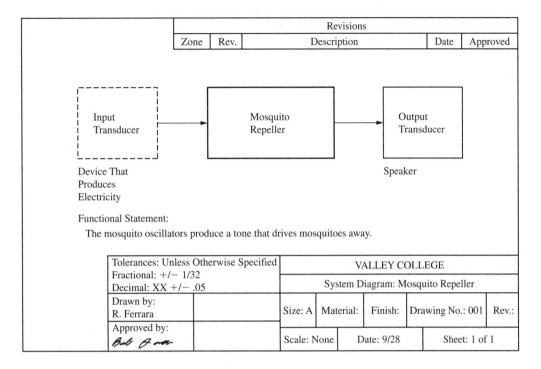

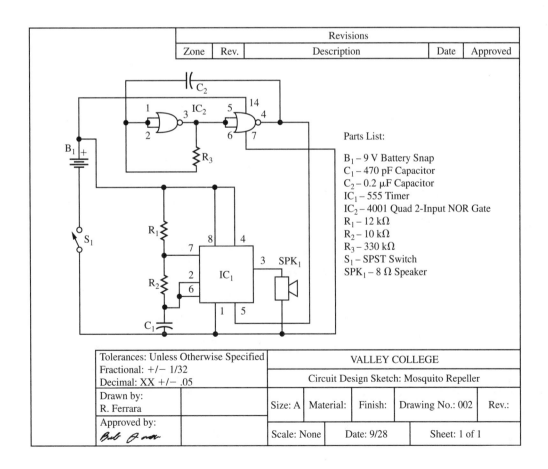

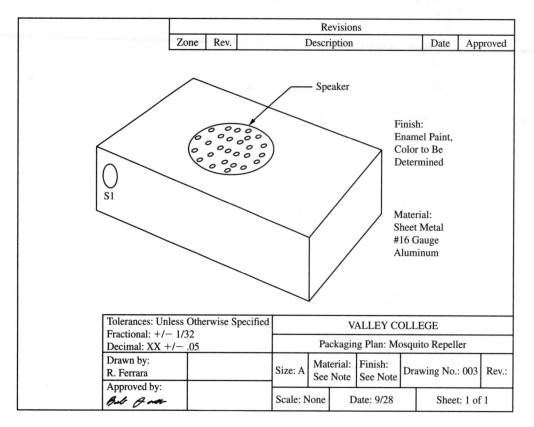

Revisions				
Zone	Rev.	Description	Date	Approved

Speaker

Finish:
Enamel Paint,
Color to Be
Determined

Material:
Sheet Metal
#16 Gauge
Aluminum

S1

Tolerances: Unless Otherwise Specified Fractional: +/− 1/32 Decimal: XX +/− .05		VALLEY COLLEGE				
		Packaging Plan: Mosquito Repeller				
Drawn by: R. Ferrara		Size: A	Material: See Note	Finish: See Note	Drawing No.: 003	Rev.:
Approved by: *Bob Brown*		Scale: None	Date: 9/28		Sheet: 1 of 1	

Concepts and Requirements Document

Push-Button Combination Lock

by

(student's name)

Project Goals

This Concepts and Requirements Document proposes that a working prototype be built of an electronic combination lock, for which there may be a commercial market. If it is determined that such a market exists, the project could be mass-produced.

There is considerable demand for a simple, inexpensive, yet nearly foolproof electronic combination lock that can be used to turn battery- and ac-powered equipment on or off. The circuit presented here requires pressing 4 of 10 push-button switches in the correct sequence. If, at any time, the remaining 6 "dummy" push-button switches are pressed, the combination lock immediately resets. When the valid 4-button sequence is entered, a 6-V relay is activated, the contacts of which can switch on or off an electrical or electronic device. The push-button combination lock is to be powered by two 9-V transistor batteries placed in parallel. Alternatively, a 6–9-V power supply may be used.

Project Objectives

The project is to have the following objectives: inexpensive to produce (less than $20), highly reliable, easy to use, a high degree of code security, and the ability to switch high-current-demand loads. These objectives will be met in the following ways:

*   **Cost.** Incorporation of a simple flip-flop circuit design that uses just two 4013 CMOS D flip-flops will hold project costs to a minimum. Even though high-quality push-button switches, requiring hard wiring, are necessary, the cost per unit will not exceed $20 when mass-produced.

*   **Reliability.** The use of quality push-button switches, a durable relay, CMOS flip-flop ICs, and partial construction on a printed circuit board, will ensure a high degree of reliability.

*   **Ease of Use.** Operation of the project is straightforward: one just presses the correct 4 push buttons in the correct sequence. The circuit is reset by pressing any one of 6 dummy switches.

1

*   **Code Security.** With the use of 10 identical-looking push-button switches, 4 of which must be pressed in the exact sequence while avoiding contact with the 6 remaining unknown dummy switches, the odds of obtaining the correct combination by guessing are greater than 1000:1.

*   **Ability to Handle High-Current Loads.** Choice of a relay with suitably rated normally open and normally closed contacts (at least 10 A), and use of parallel multiple contacts when necessary, will allow heavy-current-carrying devices, such as machinery and appliances, to be switched on or off.

Project Responsibility

Because this is to be a learning experience for the electronics student, he or she will be responsible for the design (which, in this case, is being supplied), drawings, experimenting, prototyping, and testing, troubleshooting, and final documentation phases of the Push-Button Combination Lock Project. Due dates for each phase will be assigned by the electronics instructor.

Project Theory of Operation

The push-button combination lock consists of two 4013 dual D flip-flops, 10 normally open push-button switches, a single on/off SPST key switch, a PNP transistor, a 6-V relay, a diode, and 11 resistors. With the on/off switch closed, switches S1–S4 are pressed in *sequence,* transistor Q1 turns on, and the relay is activated. Should one of the dummy switches, S5–S10, be pressed, the combination lock immediately resets and the 4-bit combination (S1–S4) must be reentered to activate the relay. Once the lock has been activated, pressing any of the dummy switches will deactivate it again.

The project works in the following way:

1. On/off key switch S11 is closed. Any dummy switch is pressed, causing all flip-flops to set and output Q4 to go high. A high on the base of transistor Q1 turns it off, and the relay is deactivated.

2. S1 is pressed. As a result, output Q1 goes low. (Remember that D1 of IC1a is tied low.)

3. When S2 is pressed, the low on output Q1 is transferred to D2 and output Q2 goes low. Pressing S3 brings output Q3 low and pressing S4 brings output Q4 low, which places a low on the base of transistor Q1. The transistor turns on and so does the relay.

4. Pressing any dummy switch sets all flip-flops, placing output Q4 high and deactivating the relay.

One power line to the load (electrical appliance, machinery, etc.) is wired in series with the relay contacts. If the device to be controlled is to turn on when the combination lock is activated, the normally open relay contacts are used. If the device is to turn off when the combination lock is energized, the normally closed relay contacts are used.

Electronics instructor's approval _____ **Date** _____

3

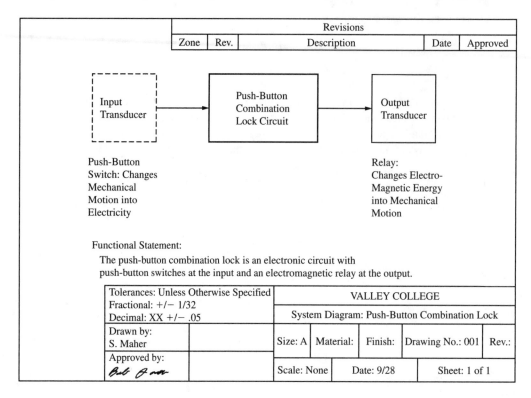

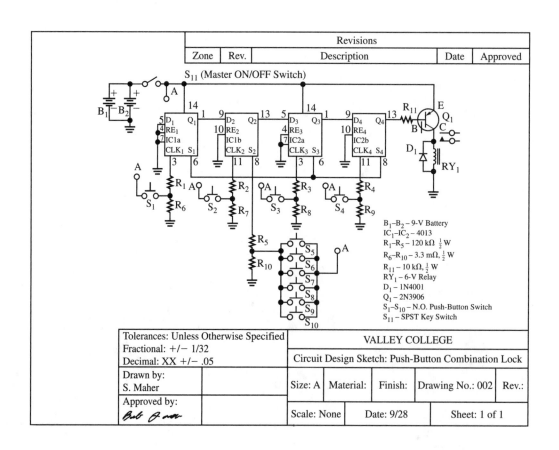

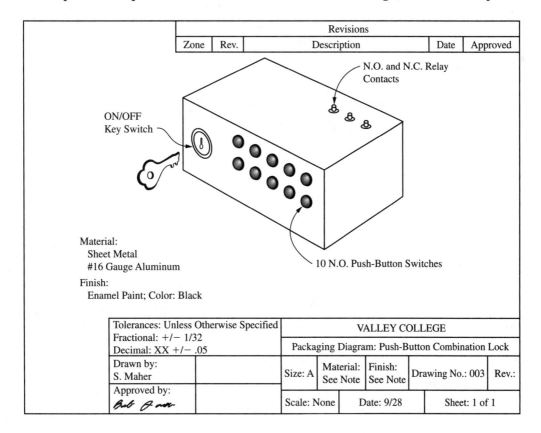

Revisions				
Zone	Rev.	Description	Date	Approved

N.O. and N.C. Relay Contacts

ON/OFF Key Switch

Material:
  Sheet Metal
  #16 Gauge Aluminum
Finish:
  Enamel Paint; Color: Black

10 N.O. Push-Button Switches

Tolerances: Unless Otherwise Specified		VALLEY COLLEGE				
Fractional: +/− 1/32						
Decimal: XX +/− .05		Packaging Diagram: Push-Button Combination Lock				
Drawn by: S. Maher		Size: A	Material: See Note	Finish: See Note	Drawing No.: 003	Rev.:
Approved by: *Bob Grant*		Scale: None	Date: 9/28		Sheet: 1 of 1	

Concepts and Requirements Document

Sound-Activated Switch

by

(student's name)

Project Goals

This Concepts and Requirements Document proposes that a working prototype be built of a sound-activated switch, for which there may be a commercial market. If it is determined that such a market exists, the project could then be mass-produced.

Any ac device requiring 600 W or less is plugged into the sound-activated switch. The switch, in turn, is plugged into the ac wall outlet. With each sharp sound (handclapping, for example), the device will go on or off. This clap-on/clap-off project can be used to turn on and off many household appliances, such as televisions, radios, lamps, computers, electronic clocks, and coffeemakers. Regardless of where a person is in the room, a quick clap of the hands is all it takes to turn on or off any of these, or similar, devices.

Project Objectives

The project is to have the following objectives: inexpensive to produce, reliable, completely safe, and easy to use. These objectives will be met in the following ways:

* **Cost.** Use of noncritical, readily available hardware and electronic components, bought in bulk quantities, will keep the cost low.

* **Reliability.** Use of mostly solid-state active components and printed circuit board construction techniques will ensure a high degree of reliability.

* **Safety.** Completely enclosing the project in a metal enclosure will assure safe operation.

1

*   **Ease of Use.** Because the user is required to adjust only one simple control (a sensitivity potentiometer), ease of operation is assured. Once the sensitivity setting for a particular device is obtained, no further adjustment need be made.

Project Responsibility

Because this is to be a learning experience for the electronics student, he or she will be responsible for the design (which, in this case, is being supplied), drawing, experimenting, prototyping, and testing, troubleshooting, and final documentation phases of the Sound-Activated Switch Project. Due dates for each phase will be assigned by the electronics instructor.

Project Theory of Operation

With a handclap, the sound-activated switch will turn on or off any ac appliance plugged into it. Diodes D1 and D2 provide half-wave rectification of the ac power for two power supplies. The first power supply is current-limited by resistor R1 in series with the amplifier (Q1–Q3) and logic circuits (U1). Capacitor C1 filters the supply voltage, while zener diode D5 acts as a 12-V regulator. The second power supply is current-limited by resistor R2 and filtered by capacitor C6.

Transistors Q1, Q2, and Q3, with their associated components, make up a high-gain ac amplifier. Resistor R9 provides dc stabilization through the dc feedback signal applied to the emitter of Q1. Amplifier gain, or sensitivity, is controlled by the position of potentiometer R10. The first half integrated circuit U1, a dual "D" flip-flop, is configured so that its output will change state (low to high) with the first positive input transition on pin 11. The output will reset after a preset time delay, determined by the RC circuit consisting of R11 and C5. The second half of U1 will change states with each positive input transition of pin 3. When its output is high, Q4 turns on. Current now finds a path through relay RLY1 to ground. Thus, ac power is applied to ac socket SO1, and the appliance turns on. When the second half of U1 receives another positive input transition, its output goes low. Transistor Q4 now turns off, the relay is deactivated, and power is no longer available at SO1. The appliance shuts off.

Electronics instructor's approval _____ **Date** _____

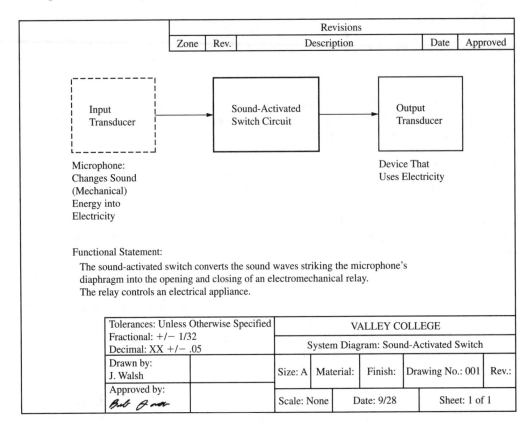

Revisions					
Zone	Rev.	Description		Date	Approved

Input Transducer

Sound-Activated Switch Circuit

Output Transducer

Microphone: Changes Sound (Mechanical) Energy into Electricity

Device That Uses Electricity

Functional Statement:

The sound-activated switch converts the sound waves striking the microphone's diaphragm into the opening and closing of an electromechanical relay.
The relay controls an electrical appliance.

Tolerances: Unless Otherwise Specified
Fractional: +/− 1/32
Decimal: XX +/− .05

Drawn by: J. Walsh

Approved by:

VALLEY COLLEGE

System Diagram: Sound-Activated Switch

Size: A | Material: | Finish: | Drawing No.: 001 | Rev.:

Scale: None | Date: 9/28 | Sheet: 1 of 1

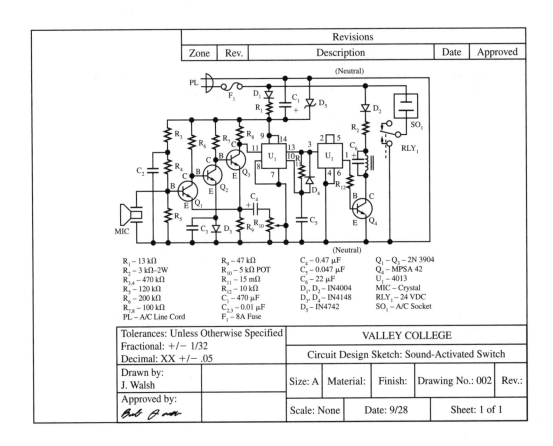

Revisions					
Zone	Rev.	Description		Date	Approved

R_1 – 13 kΩ
R_2 – 3 kΩ–2W
$R_{3,4}$ – 470 kΩ
R_5 – 120 kΩ
R_6 – 200 kΩ
$R_{7,8}$ – 100 kΩ
PL – A/C Line Cord

R_9 – 47 kΩ
R_{10} – 5 kΩ POT
R_{11} – 15 mΩ
R_{12} – 10 kΩ
C_1 – 470 μF
$C_{2,3}$ – 0.01 μF
F_1 – 8A Fuse

C_4 – 0.47 μF
C_5 – 0.047 μF
C_6 – 22 μF
D_1, D_2 – IN4004
D_3, D_4 – IN4148
D_5 – IN4742

$Q_1 – Q_3$ – 2N 3904
Q_4 – MPSA 42
U_1 – 4013
MIC – Crystal
RLY_1 – 24 VDC
SO_1 – A/C Socket

Tolerances: Unless Otherwise Specified
Fractional: +/− 1/32
Decimal: XX +/− .05

Drawn by: J. Walsh

Approved by:

VALLEY COLLEGE

Circuit Design Sketch: Sound-Activated Switch

Size: A | Material: | Finish: | Drawing No.: 002 | Rev.:

Scale: None | Date: 9/28 | Sheet: 1 of 1

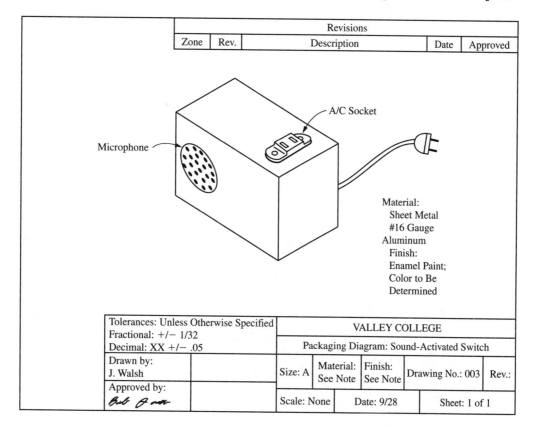

Revisions					
Zone	Rev.	Description		Date	Approved

A/C Socket

Microphone

Material:
Sheet Metal
#16 Gauge
Aluminum
Finish:
Enamel Paint;
Color to Be
Determined

Tolerances: Unless Otherwise Specified Fractional: +/− 1/32 Decimal: XX +/− .05		VALLEY COLLEGE				
		Packaging Diagram: Sound-Activated Switch				
Drawn by: J. Walsh		Size: A	Material: See Note	Finish: See Note	Drawing No.: 003	Rev.:
Approved by: *Bob Green*		Scale: None	Date: 9/28		Sheet: 1 of 1	

Concepts and Requirements Document

10-Note Tunable Electronic Organ

by

(student's name)

Project Goals

This Concepts and Requirements Document proposes that a working prototype be built of a 10-note tunable electronic organ, for which there may be application in the commercial toy market. If it is determined that such a market exists, the project could be mass-produced.

Young children love to push buttons and make noise as a result. This project will allow them to do just that, 10 times over, and create pleasing musical tones as a result. Although noise-producing toys do exist, the 10-note tunable electronic organ will allow the user to adjust each tone output to his or her liking.

The 10-note tunable electronic organ consists of a CMOS 4001 quad 2-Input NOR gate IC, a Darlington amplifier, 10 push buttons, and 10 small, adjustable PC board-mounted potentiometers. It can be powered by a standard 9-V battery.

Project Objectives

The project is to have the following objectives: inexpensive to produce, reliable, safe, and easy to use. These objectives will be met in the following ways:

*   **Cost**. Use of noncritical, readily available hardware and electronic components, bought in bulk, will keep the cost low.

*   **Reliability**. Use of an integrated circuit, two transistors, and quality potentiometers and switches will ensure a high degree of reliability.

*   **Safety**. Use of a 9-V battery to power the circuit will assure safety.

*   **Ease of Use**. Although the product will come preadjusted, the user is free to alter the settings on the various potentiometers to produce new sounds. Other than performing that action, all one does is press a push-button switch.

Project Responsibility

Because this is to be a learning experience for the electronics student, he or she will be responsible for the design (which, in this case, is being supplied), drawing, experimenting, prototyping, and testing, troubleshooting, and final documentation phases of the 10-Note Tunable Electronic Organ Project. Due dates for each phase will be assigned by the electronics instructor.

1

Project Theory of Operation

The 10-Note Tunable Electronic Organ plays a different tone with the pressing of each of
10 push-button switches. The circuit consists of three subcircuits: (1) an oscillator, (2) the
10 tone controls, and (3) an amplifier. Let's look briefly at each subcircuit.

*   An oscillator is a circuit that produces a self-generating signal. Here, we use a CMOS 4001
quad, two-input NOR gate IC to generate the tone. Only two of the IC's four gates are used, and
they are configured as NOT gates. The resistor and capacitor values determine the output
frequency of the oscillator. If the value of either the resistor or the capacitor increases, the
circuit's output tone goes down. If the value of either component decreases, the circuit's output
tone goes up.

In our circuit, the oscillator's feedback resistor is R2, and its feedback capacitor is C1. If
these were the only two external components in the oscillator circuit, the output tone would
remain at a fixed value.

*   The tone-control subcircuit places a potentiometer in series with feedback resistor R2.
When a given push button, S1–S10, is pressed, its associated potentiometer, P1–P10, along with
fixed resistor R2, is placed in the feedback circuit. Because each potentiometer is adjusted to a
different value, when a given push button is pressed, a different resistive sum, consisting of R2
and an associated potentiometer, becomes part of the feedback circuit. Thus, a different tone is
heard with each button press.

*   Our amplifier subcircuit consists of two 2N3906 PNP transistors connected in what is
known as a *Darlington pair.* As such, the two-transistor arrangement produces a multiplication
of current gain.

Electronics instructor's approval _____ **Date** _____

2

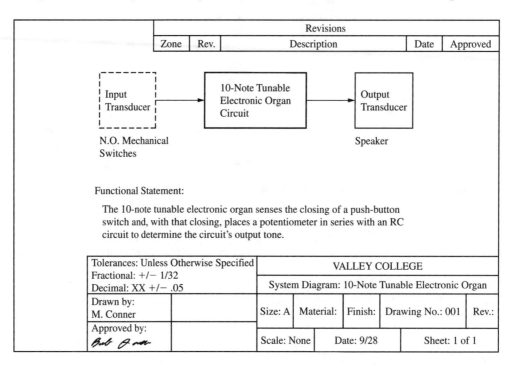

Revisions				
Zone	Rev.	Description	Date	Approved

Input Transducer

N.O. Mechanical Switches

10-Note Tunable Electronic Organ Circuit

Output Transducer

Speaker

Functional Statement:

The 10-note tunable electronic organ senses the closing of a push-button switch and, with that closing, places a potentiometer in series with an RC circuit to determine the circuit's output tone.

Tolerances: Unless Otherwise Specified Fractional: +/− 1/32 Decimal: XX +/− .05		VALLEY COLLEGE				
		System Diagram: 10-Note Tunable Electronic Organ				
Drawn by: M. Conner		Size: A	Material:	Finish:	Drawing No.: 001	Rev.:
Approved by: *Bob Gross*		Scale: None	Date: 9/28		Sheet: 1 of 1	

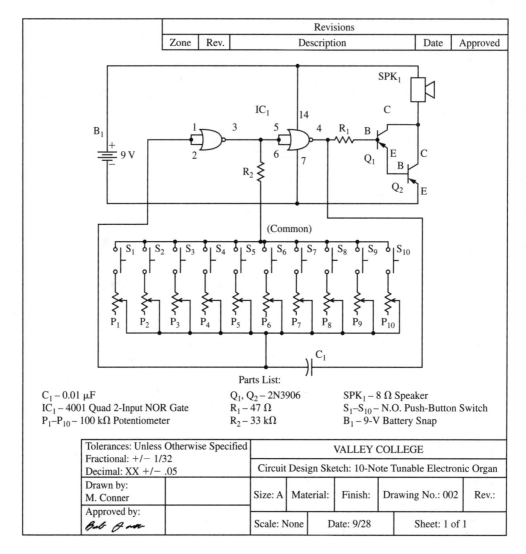

Revisions				
Zone	Rev.	Description	Date	Approved

Parts List:

$C_1 - 0.01\ \mu F$

$IC_1 - 4001$ Quad 2-Input NOR Gate

$P_1 - P_{10} - 100\ k\Omega$ Potentiometer

$Q_1, Q_2 - 2N3906$

$R_1 - 47\ \Omega$

$R_2 - 33\ k\Omega$

$SPK_1 - 8\ \Omega$ Speaker

$S_1 - S_{10} - $ N.O. Push-Button Switch

$B_1 - 9\text{-V}$ Battery Snap

Tolerances: Unless Otherwise Specified Fractional: +/− 1/32 Decimal: XX +/− .05		VALLEY COLLEGE				
		Circuit Design Sketch: 10-Note Tunable Electronic Organ				
Drawn by: M. Conner		Size: A	Material:	Finish:	Drawing No.: 002	Rev.:
Approved by: *Bob Gross*		Scale: None	Date: 9/28		Sheet: 1 of 1	

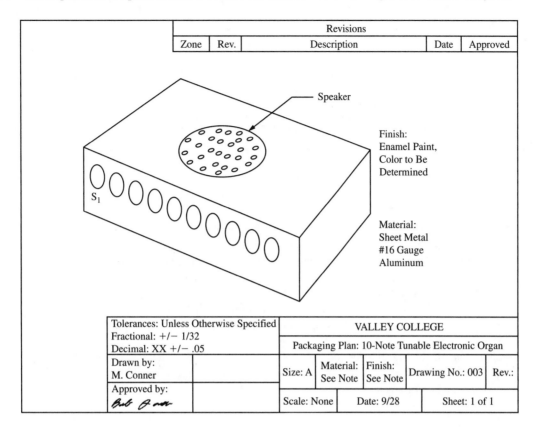

Revisions					
Zone	Rev.	Description		Date	Approved

Speaker

Finish:
Enamel Paint,
Color to Be
Determined

Material:
Sheet Metal
#16 Gauge
Aluminum

S_1

Tolerances: Unless Otherwise Specified		VALLEY COLLEGE				
Fractional: +/− 1/32						
Decimal: XX +/− .05		Packaging Plan: 10-Note Tunable Electronic Organ				
Drawn by: M. Conner		Size: A	Material: See Note	Finish: See Note	Drawing No.: 003	Rev.:
Approved by: *Bob Jones*		Scale: None	Date: 9/28		Sheet: 1 of 1	

Concepts and Requirements Document

2-W Stereo Amplifier

by

(student's name)

Project Goals

Many music lovers would like to give their Walkman, or other small stereo players, a boost. Such devices are fine when used with earphones, as intended, but for them to drive speakers, a stereo amplifier must be added to their outputs. What is needed is a relatively simple, portable, easy-to-use stereo amplifier, rated, say, at 2 W per channel, that can simply be plugged into the headphone jack of a Walkman or similar device.

This Concepts and Requirements Document proposes that a working prototype be built of an inexpensive, truly portable 2-W stereo amplifier. If it is then determined that a commercial market exists for the device, it could be mass-produced.

The amplifier project, operated from four AA cells, uses two identical LM386 IC amplifier chips, one for each stereo channel. Separate volume adjustment for each channel is provided via a potentiometer.

Project Objectives

The project is to have the following objectives: inexpensive to produce, reliable, safe, and easy to use. These objectives will be met in the following ways:

*   **Cost.** Use of noncritical, readily available hardware and electronic components, bought in bulk quantities, will keep the cost low.

*   **Reliability.** Use of solid-state active components and printed circuit board construction techniques will ensure a high degree of reliability.

*   **Safety.** The circuit is powered from four 1.5-V AA cells connected in series to provide 6 V. Thus, safe operation is assured.

*   **Ease of Use.** To use, one just connects the plug input of the stereo amplifier to the output of the audio source. Two speakers are either wired into the amplifier or connected via plugs and jacks.

1

Project Responsibility

Because this is to be a learning experience for the electronics student, he or she will be responsible for the design (which, in this case, is being supplied), drawing, experimenting, prototyping, and testing, troubleshooting, and final documentation phases of the 2-W Stereo Amplifier Project. Due dates for each phase will be assigned by the electronics instructor.

Project Theory of Operation

There are, essentially, two identical circuits built into the stereo amplifier, one for each channel. Operation is as follows:

*   IC1 (an LM386 audio amplifier, developed in 1975 by National Semiconductor) amplifies an input signal via potentiometer P1, acting as a voltage divider, and coupling capacitor C5. Although this IC requires fairly large electrolytic capacitors in support, the result is a quite stable circuit.

*   The amplified output signal is taken off capacitor C2.

*   Operation of both channels is the same.

*   The LED, with its current-limiting resistor R5, is simply an on/off indicator. Switch S1 turns the circuit on.

   To get the best results from the stereo amplifier, adjust the volume control of your Walkman, or similar device, so that the sound is loud, yet not distorted, when the variable potentiometers are in their middle position.

2

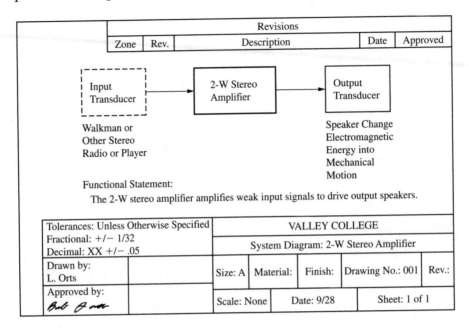

		Revisions		
Zone	Rev.	Description	Date	Approved

Input Transducer

2-W Stereo Amplifier

Output Transducer

Walkman or
Other Stereo
Radio or Player

Speaker Change
Electromagnetic
Energy into
Mechanical
Motion

Functional Statement:
The 2-W stereo amplifier amplifies weak input signals to drive output speakers.

Tolerances: Unless Otherwise Specified		VALLEY COLLEGE			
Fractional: +/− 1/32 Decimal: XX +/− .05		System Diagram: 2-W Stereo Amplifier			
Drawn by: L. Orts		Size: A	Material:	Finish:	Drawing No.: 001	Rev.:
Approved by:		Scale: None	Date: 9/28		Sheet: 1 of 1

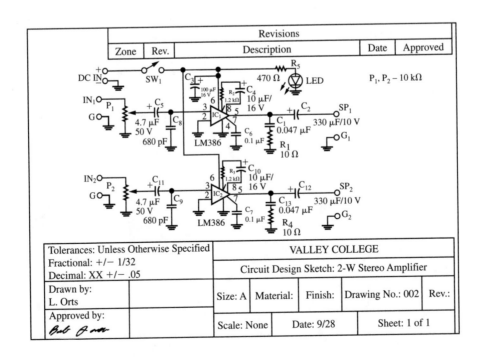

		Revisions		
Zone	Rev.	Description	Date	Approved

Tolerances: Unless Otherwise Specified		VALLEY COLLEGE			
Fractional: +/− 1/32 Decimal: XX +/− .05		Circuit Design Sketch: 2-W Stereo Amplifier			
Drawn by: L. Orts		Size: A	Material:	Finish:	Drawing No.: 002	Rev.:
Approved by:		Scale: None	Date: 9/28		Sheet: 1 of 1

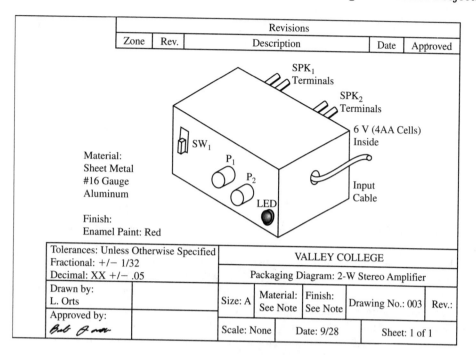

Revisions					
Zone	Rev.	Description		Date	Approved

Material:
Sheet Metal
#16 Gauge
Aluminum

Finish:
Enamel Paint: Red

Tolerances: Unless Otherwise Specified Fractional: +/− 1/32 Decimal: XX +/− .05		VALLEY COLLEGE				
		Packaging Diagram: 2-W Stereo Amplifier				
Drawn by: L. Orts		Size: A	Material: See Note	Finish: See Note	Drawing No.: 003	Rev.:
Approved by: *Bob Brown*		Scale: None	Date: 9/28		Sheet: 1 of 1	

Concepts and Requirements Document

Variable Strobe Light

by

(student's name)

Project Goals

This Concepts and Requirements Document proposes that a working prototype be built of a variable strobe light, for which there may be a commercial market. If it is determined that such a market exists, the project could then be mass-produced.

Strobe lights—lamps that produce a brilliant white flash of light for extremely short periods of time—have been used for years as attention-getters, to create special effects, and as safety beacons. Commercial units on the market range in price from $50 to $1000 and more. There is considerable demand for a low-cost strobe that can be used primarily for special effects at parties and as attention-getters for window displays and sales promotions. The proposed variable strobe light is ideal for these and similar purposes.

Project Objectives

The project is to have the following objectives: inexpensive to produce (less than $10), reliable, safe, having a variable flash rate, and attractive in appearance. These objectives will be met in the following ways:

*   **Cost.** Use of noncritical, readily available hardware and electronic components, bought in bulk quantities, will keep the cost low.

*   **Reliability.** Use of an extremely dependable solid-state active component, the SCR, and printed circuit board construction techniques will ensure a high degree of reliability.

*   **Safety.** Use of a fused ac line, a polarized plug, and a wooden or metal enclosure for the project will assure safe operation.

1

* **Variable Flash.** The use of a potentiometer in the RC timing circuit will produce a range of flash rates, from one every 2 seconds to approximately eight per second.

* **Appearance.** A wooden or metal enclosure, either painted or covered with self-adhesive vinyl, will ensure an attractive project.

Project Responsibility

Because this is to be a learning experience for the electronics student, he or she will be responsible for the design (which, in this case, is being supplied), drawing, experimenting, prototyping, and testing, troubleshooting, and final documentation phases of the Variable Strobe Light Project. Due dates for each phase will be assigned by the electronics instructor.

Project Theory of Operation

Capacitors C1, C2, and C3 with diodes D1–D4 form a half-wave voltage doubler circuit. This circuit produces a stored charge on C3 of more than 300 V dc. Components R1, R4, C4, and I1 form an RC timing circuit, with I1 being used as a trigger for the gate of SCR1. When SCR1 turns on, a low-voltage pulse is sent to the primary of trigger transformer T1 and is stepped up to approximately 4000 V on the secondary. The high voltage applied to the trigger lead of FT ionizes the xenon gas. When this happens, the energy stored in C3 discharges through the flashtube, creating a bright flash.

Electronics instructor's approval _____ **Date** _____

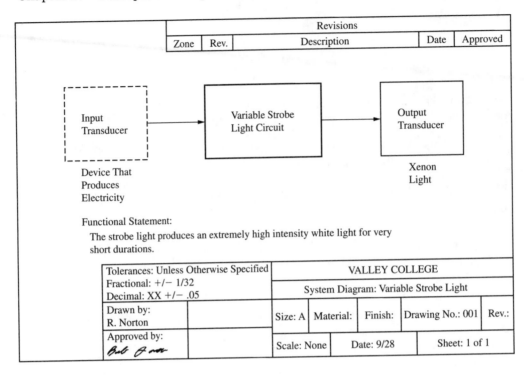

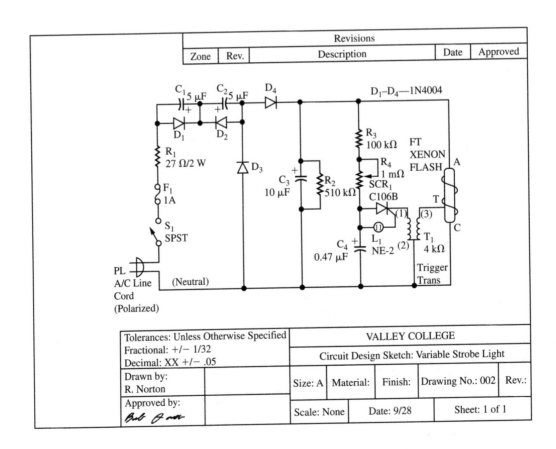

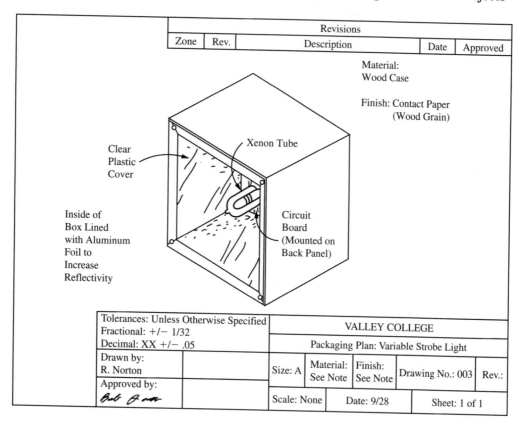

		Revisions		
Zone	Rev.	Description	Date	Approved

Material:
Wood Case

Finish: Contact Paper
(Wood Grain)

Xenon Tube

Clear
Plastic
Cover

Inside of
Box Lined
with Aluminum
Foil to
Increase
Reflectivity

Circuit
Board
(Mounted on
Back Panel)

Tolerances: Unless Otherwise Specified
Fractional: +/− 1/32
Decimal: XX +/− .05

VALLEY COLLEGE

Packaging Plan: Variable Strobe Light

Drawn by:
R. Norton

Approved by:
Bob Brown

Size: A | Material: See Note | Finish: See Note | Drawing No.: 003 | Rev.:

Scale: None | Date: 9/28 | Sheet: 1 of 1

Appendices

Name _____
Date _____
Course No. _____

ELECTRONICS SAFETY TEST

Directions: Choose the *best* answer and place the appropriate letter in the space provided at the left of the question.

General Safety

_____ 1. Tools and machines that are not operating properly should be reported immediately to the instructor. Failure to do so may result in
    a. less accurate work.
    b. excessive wear on the equipment.
    c. injury to the operator or fellow students.

_____ 2. In order to provide assistance in case of fire, each student should know the
    a. phone number of the local fire department.
    b. location of the appropriate fire extinguisher.
    c. cause of the fire.

_____ 3. You should observe all safety zones. Crowding someone who is operating a machine might cause the person to
    a. injure him- or herself.
    b. do his or her work incorrectly.
    c. damage the machine.

_____ 4. When engaged in any activity where eye hazards exist, you should wear suitable
    a. sunglasses.
    b. clothing.
    c. eye protection.

_____ 5. If you have the slightest doubt about a safe method of doing a job, you should
    a. ask the instructor for help.
    b. read a book on safety.
    c. proceed with caution.

_____ 6. In case of an accident or injury, immediately notify the
    a. custodian.
    b. instructor.
    c. head of the school.

_____ 7. Chewing or eating solder could be hazardous to your health because
    a. solder contains toxic substances.
    b. solder doesn't taste good.
    c. it is wrong to waste solder.

_____ 8. A cathode ray tube (CRT) is dangerous because it may
    a. catch fire.
    b. implode.
    c. scratch.

Avoiding Electrical Shock

_____ **9.** To discharge a circuit or component after the power has been turned off, you should use a
   a. short piece of wire.
   b. screwdriver with an insulated handle.
   c. resistor of the appropriate value.

_____ **10.** Using too heavy a fuse (overfusing) should be avoided because
   a. special tools are required to insert heavy fuses.
   b. a heavy fuse may cause the wiring to overheat.
   c. heavy fuses are expensive.

_____ **11.** When changing the components in electronic equipment, you should
   a. remove the ac plug from the wall outlet.
   b. turn off the switch on the equipment.
   c. change the component with the utmost caution.

_____ **12.** A person using portable electric equipment should stand on a dry surface (which is a poor conductor of electricity) to
   a. reduce the possibility of electrical shock.
   b. allow a good electron flow in the motor.
   c. get a good grip on the tool.

_____ **13.** If you get hold of a hot wire and can't let go, you should
   a. reach for someone's hand.
   b. yell for help.
   c. squeeze very tightly.

_____ **14.** You should not flip switches and turn controls on electrical equipment in a playful manner because it may
   a. cause the equipment to explode.
   b. cause damage or injury.
   c. anger the instructor.

_____ **15.** When testing for live wires, a suitable device to use is
   a. a screwdriver.
   b. wires.
   c. a test lamp or meter.

_____ **16.** All worn or defective electrical cords should be replaced because they
   a. can cause a short circuit and personal injury.
   b. can break or cause an open circuit.
   c. are cheap and easy to replace.

Handling Toxic and Hazardous Materials

_____ **17.** If any part of your body or clothing comes in contact with acid, wash immediately with
   a. paint thinner.
   b. rubbing alcohol.
   c. cool water.

_____ **18.** You should wear eye protection when placing in or removing copper-clad panels (PC boards) from acid or etching solution because
   a. acid or etching solution in the eye may cause permanent blindness.
   b. fumes from the acid or etching solution may cause the eyes to water.
   c. odors from the acid or etching solution may be strong.

_____ **19.** When using tongs to place work in or remove it from an acid or etching solution, you should be certain you are
   a. holding the project securely before placing it over the acid or etching solution.
   b. being assisted by another student.
   c. using tongs with special grips.

_____ **20.** The work area where etching takes place must be thoroughly cleaned and dried when etching is completed because
   a. someone may accidentally touch the work area or other wet items.
   b. corrosion may affect any wet items.
   c. this will ensure a clean laboratory and promote better workmanship.

_____ **21.** When using a soldering iron, you should avoid fumes from the soldering flux because they
   a. will corrode the work.
   b. will oxidize the soldering iron tip.
   c. are harmful to the lungs.

_____ **22.** Many paints should never be used in a poorly ventilated area because of the danger of
   a. suffocation.
   b. spillage.
   c. waste.

_____ **23.** An injury resulting from handling a broken cathode ray or fluorescent tube should be reported to the instructor at once because these tubes
   a. often contain a poisonous gas.
   b. usually have coatings of a poisonous material.
   c. frequently have wires that may cause skin infection.

_____ **24.** All oily cloths should be placed
   a. in the wastebasket.
   b. in a closed metal container.
   c. in an easily disposable wooden box.

Safe Use of Hand Tools

_____ **25.** Sheet metal has rough edges that can cause cuts. To prevent injury
   a. avoid grasping rough edges.
   b. remove rough edges with a file.
   c. smooth rough edges by hammering them.

_____ **26.** A file without a handle is dangerous to use because
   a. if it sticks or catches on the work, the tang may puncture the wrist or hand.
   b. it may break more easily than a file with a handle.
   c. when pressure is applied to it, the teeth may cut the hand.

_____ **27.** You should not carry tools in your pockets because you may
   a. be injured or cause an injury to another person.
   b. tear your clothing.
   c. walk out of the laboratory with the tools.

_____ **28.** You should pass tools to fellow students with the handles turned
   a. toward you.
   b. toward the other person.
   c. toward the ground.

_____ **29.** Screws, wires, and small electronic components are not to be
   a. put in your mouth.
   b. held in your hand.
   c. put on the workbench.

_____ **30.** Cut away from your body when using a
   a. sharp tool.
   b. dull tool.
   c. soldering iron.

_____ **31.** The right kind of shoes to wear in the laboratory are
   a. sandals.
   b. leather shoes.
   c. thongs.

_____ **32.** Components and tools must be kept off the laboratory floor because
   a. they may cause a person to slip or fall.
   b. they can easily become lost.
   c. cluttered floors are unsightly.

Safe Use of Power Tools

_____ **33.** You should not leave a machine until it has completely stopped running. If a machine that is running is left unattended, there is a danger of
   a. excessive machine wear and vibration.
   b. extensive damage to the machine.
   c. injury to the next operator who may not realize the machine is still running.

_____ **34.** Power or machine tools with the guards removed should
   a. not be used.
   b. be replaced.
   c. be used with more caution.

_____ **35.** Never touch a piece of stock that is turning in a power machine because of the danger of
   a. injuring yourself.
   b. damaging the machine.
   c. damaging your work.

_____ **36.** If you are operating a drill press, be sure that small pieces of metal to be drilled are held securely by
   a. your hand.
   b. a vise or suitable clamp.
   c. a fellow student.

_____ **37.** When drilling sheet metal, hold it in place with a clamp and back it up with a piece of
   a. wood.
   b. rubber.
   c. iron.

_____ **38.** It is dangerous to wear long sleeves or loose clothing when operating power equipment because
   a. such clothing is uncomfortable when operating machines.
   b. the clothing may catch on the work and damage your part.
   c. such clothing may catch and pull you into the machine, causing injury.

_____ **39.** You should always remove the chuck key from the power machine before turning on the power because
   a. the chuck key may fly out and get lost.
   b. the chuck key may fly out and cause injury.
   c. the chuck key may fly out and break.

_____ **40.** When you are through using a soldering iron
   a. cool it with water.
   b. place it back in its holder and unplug it.
   c. leave it plugged in.

Project Safety

_____ **41.** MOS-type integrated circuits are packaged in antistatic materials to
   a. prevent a voltage buildup between their pins.
   b. avoid moisture damage.
   c. prevent them from being crushed.

_____ **42.** When handling MOS components
   a. don't squeeze them too hard.
   b. avoid touching the pins.
   c. wear rubber gloves.

_____ **43.** Never install or remove integrated circuits while power is applied because
   a. the sudden voltage jolt may damage them.
   b. doing so may ruin other components.
   c. the power supply may be damaged.

_____ **44.** A soldering iron with a grounded tip should be used
   a. because it reduces the danger of electrical shock.
   b. because it is cheaper.
   c. to prevent static charge transfer to sensitive components while they are being soldered.

_____ **45.** Increasing a room's humidity
   a. increases the risk from static electricity.
   b. reduces the risk from static electricity.
   c. has no effect on static electricity.

_____ **46.** An antistatic wrist strap
   a. brings the user to ground potential.
   b. prevents electrical shock.
   c. makes installing and removing ICs easier.

_____ **47.** Static electricity is a
   a. stationary charge.
   b. moving charge.
   c. nonconductive charge.

_____ **48.** The most common way to create a static charge is through
   a. friction.
   b. heat.
   c. low conductor paths.

_____ **49.** Dissimilar materials rubbed together will
   a. develop a static charge on either material.
   b. get hot.
   c. wear out.

_____ **50.** The best way to avoid damage from the "static electricity monster" is to
   a. ground all equipment.
   b. ground yourself.
   c. prevent static buildup in the first place.

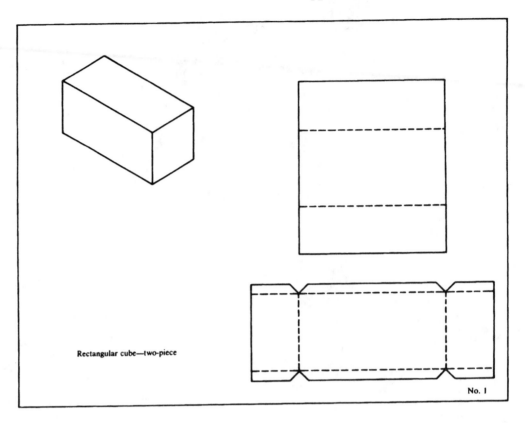

Rectangular cube—two-piece

No. 1

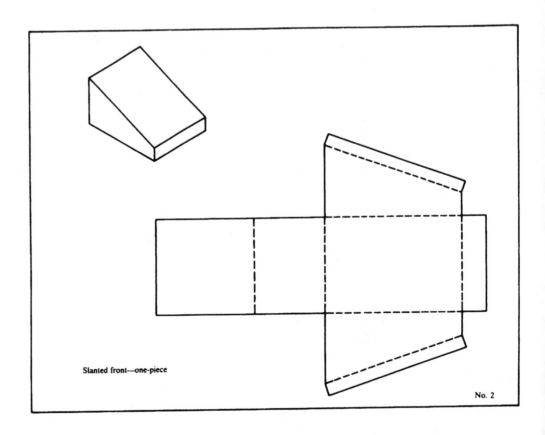

Slanted front—one-piece

No. 2

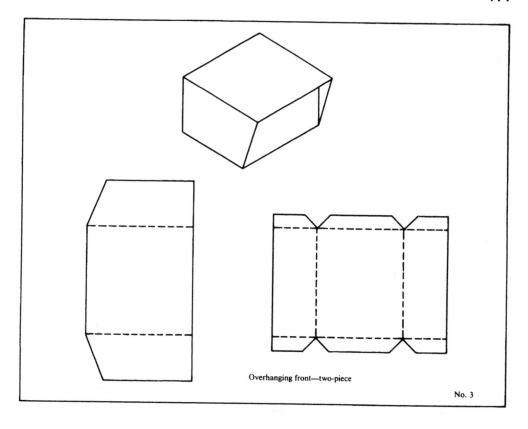

Overhanging front—two-piece

No. 3

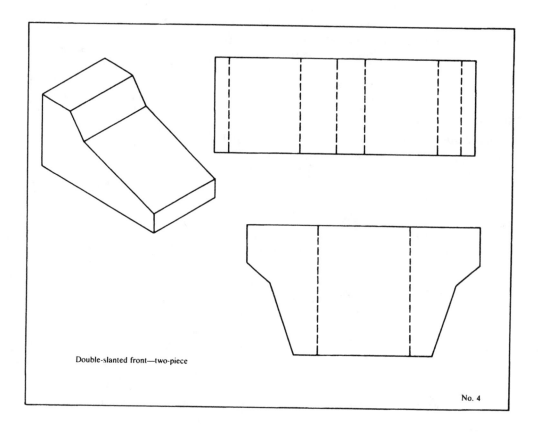

Double-slanted front—two-piece

No. 4

For increased accuracy in determining output voltage settings, you may want to add a panel-type voltmeter to your Variable Power Supply Project. To do so, you can take one of two approaches. The easiest way to go is simply to purchase a 0–20 (or 0–25)-V dc panel meter and connect it across the power supply output terminals, as shown in Figure C–1.

A second approach allows you to use any dc panel meter and "customize" it to measure 0–20 V. This method requires five simple steps:

1. You must determine the sensitivity of your meter; that is, its full-scale current deflection.
2. You must determine the internal resistance of the meter.
3. You must determine the value of the meter's multiplier resistance to be added in series with the meter movement.
4. You must redraw the meter scale to indicate 0–20 V.
5. You must install the meter.

Here is how to perform each step:

Step 1

To determine the full-scale current deflection of your panel meter (assuming it is not stated on the meter face), place it in series with a DMM or VOM set to measure current, a 10-k ohm potentiometer, and a 5-V power supply, as shown in Figure C–2. Adjust the potentiometer until the panel meter shows full-scale deflection. The DMM or VOM will now indicate current flow in the circuit and thus the current sensitivity of your panel meter. (Recall, current in a series circuit is the same everywhere.) Let's assume that the current is 1 mA, which is a typical value.

Step 2

To determine the internal resistance of your panel meter (again, assuming it is not stated), simply measure the resistance across the terminals with an

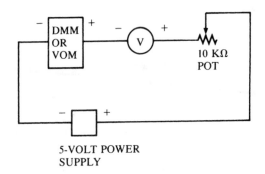

FIGURE C–2
Determining meter sensitivity

ohmmeter. Let's assume it is 1000 ohms; again, this is a typical value.

Step 3

To determine the value of the multiplier resistor required to be placed in series with the panel meter, first determine the total resistance of the meter circuit (multiplier resistor plus internal resistance of panel meter). The Variable Power Supply Project produces an output voltage of 1.2 to 15 V. To give yourself an additional margin, you'll want a panel meter that will indicate 0–20 V. To determine total resistance, use Ohm's law, where $R = V/I$. In your case, you want a voltage of 20 V and an I of 1 mA. The calculation is as follows:

$$R = V/I = 20/0.001 = 20,000 \text{ ohms}$$

The total resistance is 20,000 ohms. Since the meter in this example has an internal resistance (R_i) of 1000 ohms, a 19,000-ohm multiplier resistor (R_m) is required ($20,000 - 1000 = 19,000$). The 19-k ohm resistor (1% tolerance preferred) is connected in series with the panel meter, as shown in Figure C–3.

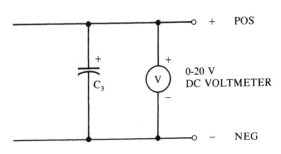

FIGURE C–1
Adding a dc voltmeter to the Variable Power Supply Project

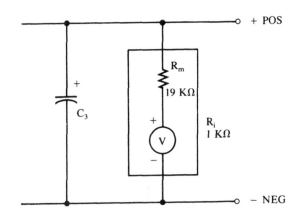

FIGURE C–3
Adding a multiplier resistor

Step 4

The simplest way to draw a new meter scale to indicate 0–20 V is to remove the existing scale, turn it over, and spray paint the reverse side a semigloss white. When it is dry, use a drafting compass and appropriate rub-on decals to create your new scale. To determine the location of various marks and voltage indications, you will need to parallel-connect your DMM or VOM set to measure voltage. See Figure C–4. As you slowly increase the voltage output of your power supply, your DMM or VOM gives you an accurate voltage reading. Using a pencil, mark various voltage points on the panel meter scale.

Step 5

Finally, install your panel meter in an appropriate location on the power supply cabinet. The electrical hook-up is as shown in Figure C–4. Be sure to observe correct meter polarity.

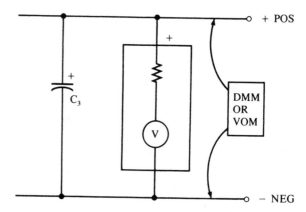

FIGURE C–4
Calibrating the panel meter

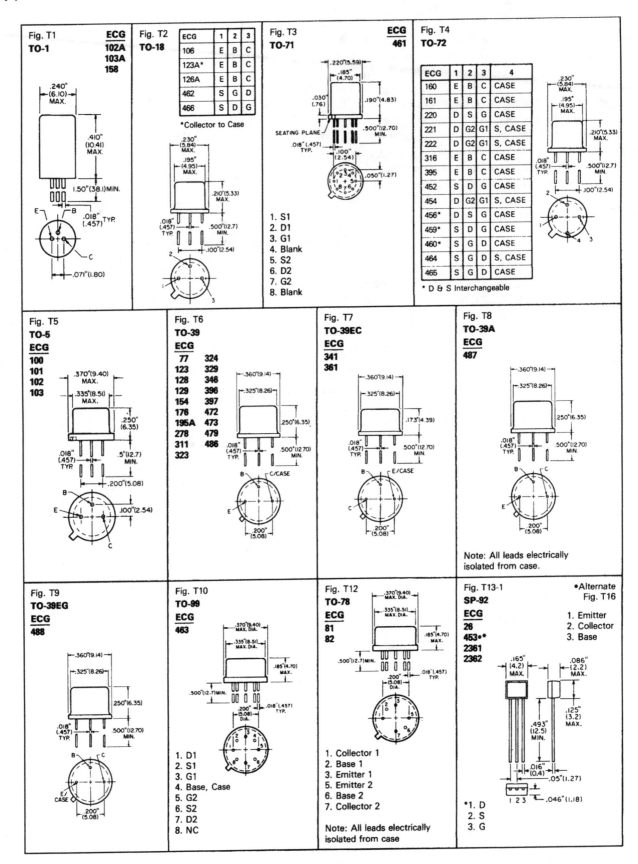

Fig. T1 **TO-1** **ECG** 102A 103A 158

.240" (6.10) MAX.
.410" (10.41) MAX.
1.50" (38.1) MIN.
.018" (.457) TYP.
E B C
.071" (1.80)

Fig. T2 **TO-18**

ECG	1	2	3
106	E	B	C
123A*	E	B	C
126A	E	B	C
462	S	G	D
466	S	D	G

*Collector to Case

.230" (5.84) MAX.
.195" (4.95) MAX.
.210" (5.33) MAX.
.018" (.457) TYP.
.500" (12.7) MIN.
.100" (2.54)

Fig. T3 **TO-71** **ECG** 461

.220" (5.59)
.185" (4.70)
.030" (.76)
.190" (4.83)
SEATING PLANE
.018" (.457) TYP.
.500" (12.70) MIN.
.100" (2.54)
.050" (1.27)

1. S1
2. D1
3. G1
4. Blank
5. S2
6. D2
7. G2
8. Blank

Fig. T4 **TO-72**

ECG	1	2	3	4
160	E	B	C	CASE
161	E	B	C	CASE
220	D	S	G	CASE
221	D	G2	G1	S, CASE
222	D	G2	G1	S, CASE
316	E	B	C	CASE
395	E	B	C	CASE
452	S	D	G	CASE
454	D	G2	G1	S, CASE
456*	D	S	G	CASE
459*	S	D	G	CASE
460*	S	G	D	CASE
464	S	G	D	S, CASE
465	S	G	D	CASE

* D & S Interchangeable

.230" (5.84) MAX.
.195" (4.95) MAX.
.210" (5.33) MAX.
.018" (.457) TYP.
.500" (12.7) MIN.
.100" (2.54)

Fig. T5 **TO-5** **ECG** 100 101 102 103

.370" (9.40) MAX.
.335" (8.51) MAX.
.250" (6.35)
.018" (.457) TYP.
.5" (12.7) MIN.
.200" (5.08)
.100" (2.54)
B E C

Fig. T6 **TO-39** **ECG**

77	324
123	329
128	346
129	396
154	397
176	472
195A	473
278	479
311	486
323	

.360" (9.14)
.325" (8.26)
.250" (6.35)
.018" (.457) TYP.
.500" (12.70) MIN.
B C/CASE E
.200" (5.08)

Fig. T7 **TO-39EC** **ECG** 341 361

.360" (9.14)
.325" (8.26)
.173" (4.39)
.018" (.457) TYP.
.500" (12.70) MIN.
B E/CASE C
.200" (5.08)

Fig. T8 **TO-39A** **ECG** 487

.360" (9.14)
.325" (8.26)
.250" (6.35)
.018" (.457) TYP.
.500" (12.70) MIN.
B C E
.200" (5.08)

Note: All leads electrically isolated from case.

Fig. T9 **TO-39EG** **ECG** 488

.360" (9.14)
.325" (8.26)
.250" (6.35)
.018" (.457) TYP.
.500" (12.70) MIN.
B C E/CASE
.200" (5.08)

Fig. T10 **TO-99** **ECG** 463

.370" (9.40) MAX. DIA.
.335" (8.51) MAX. DIA.
.185" (4.70) MAX.
.500" (12.7) MIN.
.200" (5.08) DIA.
.018" (.457) TYP.

1. D1
2. S1
3. G1
4. Base, Case
5. G2
6. S2
7. D2
8. NC

Fig. T12 **TO-78** **ECG** 81 82

.370" (9.40) MAX. DIA.
.335" (8.51) MAX. DIA.
.185" (4.70) MAX.
.500" (12.7) MIN.
.200" (5.08) DIA.
.018" (.457) TYP.

1. Collector 1
2. Base 1
3. Emitter 1
5. Emitter 2
6. Base 2
7. Collector 2

Note: All leads electrically isolated from case

Fig. T13-1 **SP-92** **ECG** 26 453•* 2361 2362

•Alternate Fig. T16

1. Emitter
2. Collector
3. Base

.165" (4.2) MAX.
.086" (2.2) MAX.
.125" (3.2) MAX.
.493" (12.5) MIN.
.016" (0.4)
.05" (1.27)
.046" (1.18)
1 2 3

*1. D
2. S
3. G

Drawings Courtesy Philips ECG

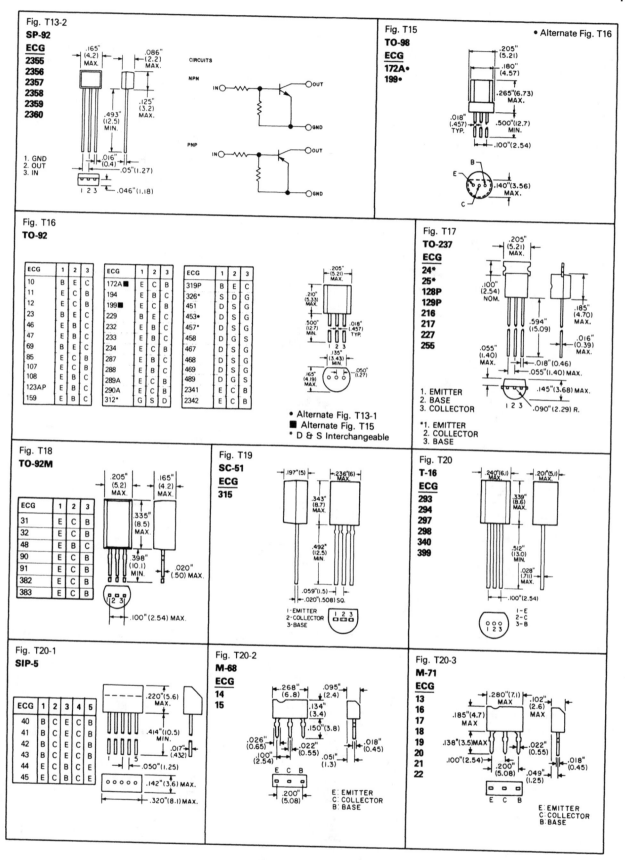

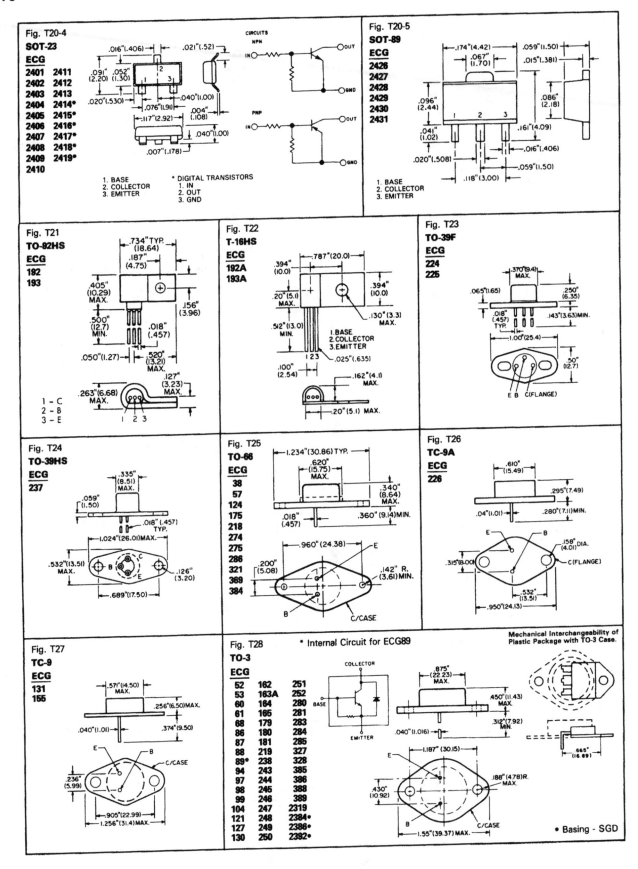

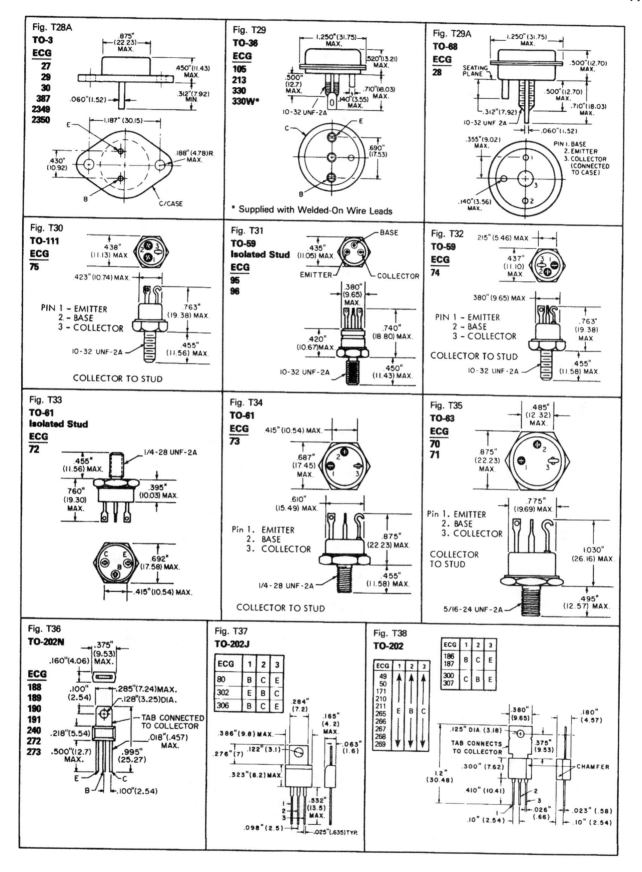

Fig. T28A
TO-3
ECG
27
29
30
387
2349
2350

.875"
(22.23)
MAX.
.450"(11.43) MAX.
.312"(7.92) MIN.
.060"(1.52)
1.187" (30.15)
.430" (10.92)
.188" (4.78)R. MAX.
E
B
C/CASE

Fig. T29
TO-36
ECG
105
213
330
330W*

1.250"(31.75) MAX.
.520"(13.21) MAX.
.500" (12.7) MAX.
.710"(18.03) MAX.
.140"(3.55) MAX.
10-32 UNF-2A
C
E
B
.690" (17.53)

* Supplied with Welded-On Wire Leads

Fig. T29A
TO-68
ECG
28

1.250"(31.75) MAX.
SEATING PLANE
.500" (12.70) MAX.
.500" (12.70) MAX.
.710"(18.03) MAX.
.312"(7.92)
10-32 UNF 2A
.060"(1.52)
.355"(9.02) MAX.
PIN I. BASE
2. EMITTER
3. COLLECTOR (CONNECTED TO CASE)
.140"(3.56) MAX.

Fig. T30
TO-111
ECG
75

.438" (11.13) MAX
.423" (10.74) MAX.
PIN 1 – EMITTER
2 – BASE
3 – COLLECTOR
10-32 UNF-2A
.763" (19.38) MAX.
.455" (11.56) MAX.
COLLECTOR TO STUD

Fig. T31
TO-59
Isolated Stud
ECG
95
96

BASE
.435" (11.05) MAX.
EMITTER
COLLECTOR
.380" (9.65) MAX.
.420" (10.67)MAX.
10-32 UNF-2A
.740" (18.80) MAX.
.450" (11.43) MAX.

Fig. T32
TO-59
ECG
74

.215" (5.46) MAX
.437" (11.10) MAX
.380" (9.65) MAX
PIN 1 – EMITTER
2 – BASE
3 – COLLECTOR
COLLECTOR TO STUD
10-32 UNF-2A
.763" (19.38) MAX
.455" (11.58) MAX.

Fig. T33
TO-61
Isolated Stud
ECG
72

1/4-28 UNF-2A
.455" (11.56) MAX.
760" (19.30) MAX.
.395" (10.03) MAX.
C   E
B
.692" (17.58) MAX.
.415" (10.54) MAX.

Fig. T34
TO-61
ECG
73

.415" (10.54) MAX.
.687" (17.45) MAX.
2   1
1   3
.610" (15.49) MAX.
Pin 1. EMITTER
2. BASE
3. COLLECTOR
1/4-28 UNF-2A
.875" (22.23) MAX.
.455" (11.58) MAX.
COLLECTOR TO STUD

Fig. T35
TO-63
ECG
70
71

.485" (12.32) MAX.
.875" (22.23) MAX.
1   2
3
Pin 1. EMITTER
2. BASE
3. COLLECTOR
COLLECTOR TO STUD
5/16-24 UNF-2A
.775" (19.69) MAX.
1.030" (26.16) MAX.
.495" (12.57) MAX.

Fig. T36
TO-202N
ECG
188
189
190
191
240
272
273

.375" (9.53) MAX.
.160"(4.06) MAX.
.100" (2.54)
.285"(7.24)MAX.
.128"(3.25)DIA.
.218"(5.54)
TAB CONNECTED TO COLLECTOR
.018"(.457) MAX.
.500"(12.7) MAX.
.995" (25.27)
E   C
B
.100"(2.54)

Fig. T37
TO-202J

ECG	1	2	3
80	B	C	E
302	E	B	C
306	B	C	E

.284" (7.2)
.386"(9.8)MAX.
.276"(7)
.122"(3.1)
.323"(8.2)MAX.
.165" (4.2) MAX.
.063" (1.6)
1
2
3
.532" (13.5) MAX.
.098" (2.5)
.025"(.635)TYP.

Fig. T38
TO-202

ECG	1	2	3
49			
50			
171			
210			
211			
265	E	B	C
266			
267			
268			
269			

ECG	1	2	3
186	B	C	E
187			
300	C	B	E
307			

.380" (9.65)
.125" DIA. (3.18)
TAB CONNECTS TO COLLECTOR
.300" (7.62)
1.2" (30.48)
.410" (10.41)
.10" (2.54)
.180" (4.57)
.375" (9.53)
CHAMFER
.023" (.58)
.10" (2.54)
2
3
1
.026" (.66)

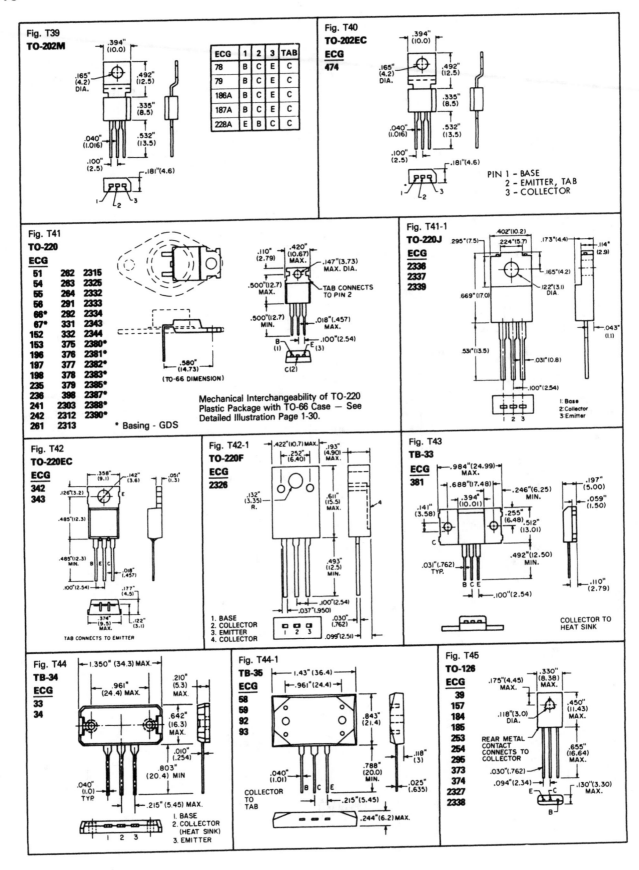

Fig. T39
TO-202M

ECG	1	2	3	TAB
78	B	C	E	C
79	B	C	E	C
186A	B	C	E	C
187A	B	C	E	C
228A	E	B	C	C

Fig. T40
TO-202EC
ECG
474

PIN 1 – BASE
2 – EMITTER, TAB
3 – COLLECTOR

Fig. T41
TO-220
ECG

51	262	2315
54	263	2325
55	264	2332
56	291	2333
66*	292	2334
67*	331	2343
152	332	2344
153	375	2380*
196	376	2381*
197	377	2382*
198	378	2383*
235	379	2385*
236	398	2387*
241	2303	2388*
242	2312	2390*
261	2313	

* Basing - GDS

TAB CONNECTS TO PIN 2

(TO-66 DIMENSION)

Mechanical Interchangeability of TO-220
Plastic Package with TO-66 Case — See
Detailed Illustration Page 1-30.

Fig. T41-1
TO-220J
ECG
2336
2337
2339

1: Base
2: Collector
3: Emitter

Fig. T42
TO-220EC
ECG
342
343

TAB CONNECTS TO EMITTER

Fig. T42-1
TO-220F
ECG
2326

1. BASE
2. COLLECTOR
3. EMITTER
4. COLLECTOR

Fig. T43
TB-33
ECG
381

COLLECTOR TO
HEAT SINK

Fig. T44
TB-34
ECG
33
34

1. BASE
2. COLLECTOR
(HEAT SINK)
3. EMITTER

Fig. T44-1
TB-35
ECG
58
59
92
93

COLLECTOR
TO
TAB

Fig. T45
TO-126
ECG
39
157
184
185
253
254
295
373
374
2327
2338

REAR METAL
CONTACT
CONNECTS TO
COLLECTOR

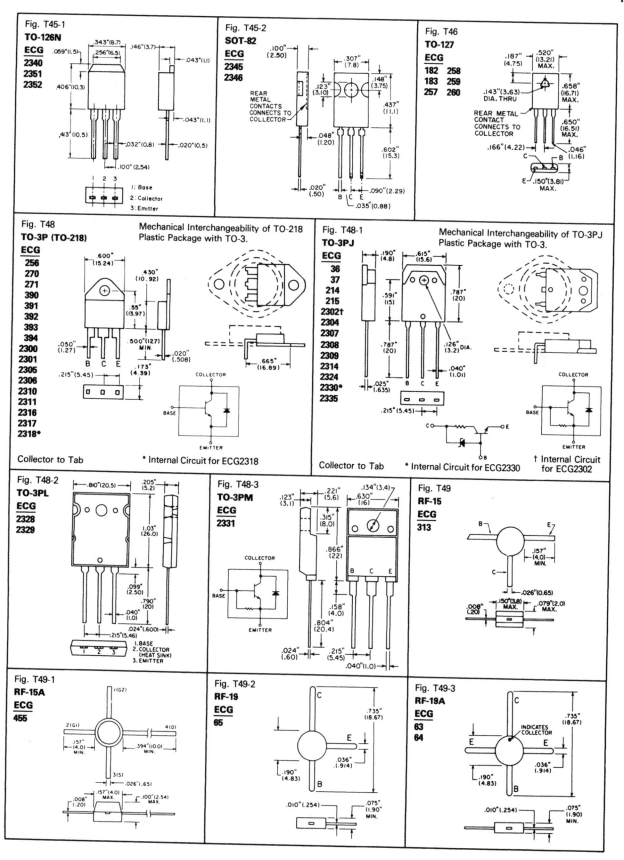

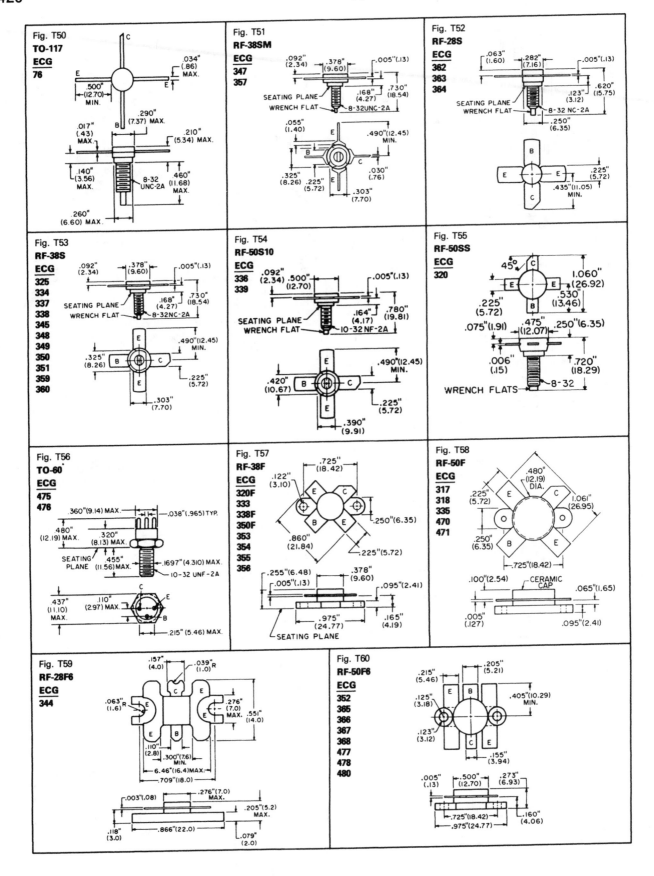

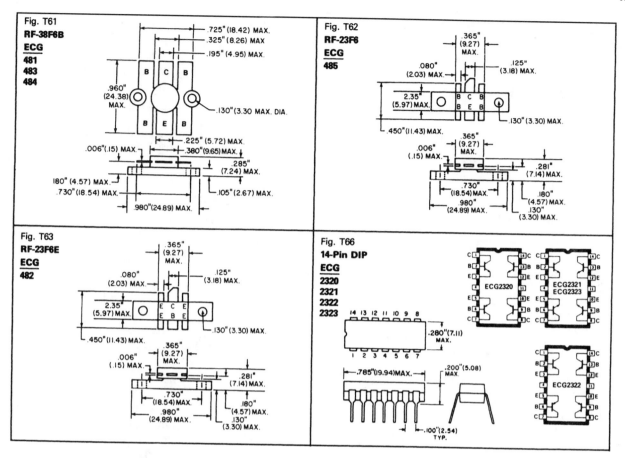

555 DUAL LED FLASHER PROJECT (SMT PROJECT)

Project Description

The 555 Dual LED Flasher Project operates to flash two LEDs alternately at a 1-Hz rate. It is powered from a 9-V battery.

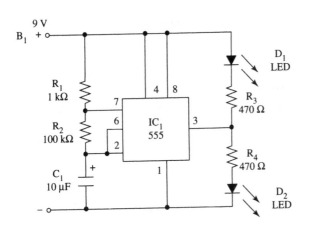

Parts List

1   SMT 555 timer (IC_1)
2   SMT LED (any color) (D_1, D_2)
1   SMT 10-μF capacitor (C_1)
2   SMT 470-Ω resistor (R_3, R_4)
1   SMT 1-kΩ resistor (R_1)
1   SMT 100-kΩ resistor (R_2)
1   9-V battery snap (B_1)

FLASHER PROJECT (SMT PROJECT)

Project Description

The Flasher Project alternately flashes two LEDs at a 0.5-Hz rate. It is powered from a 9-V battery.

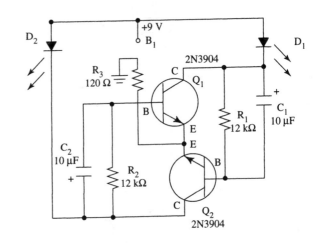

Parts List

2   SMT 2N3904 transistor (Q_1, Q_2)
2   SMT LED (any color) (D_1, D_2)
2   SMT 10-μF capacitor (C_1, C_2)
1   SMT 120-Ω resistor (R_3)
2   SMT 12-kΩ resistor (R_1, R_2)
1   9-V battery snap (B_1)

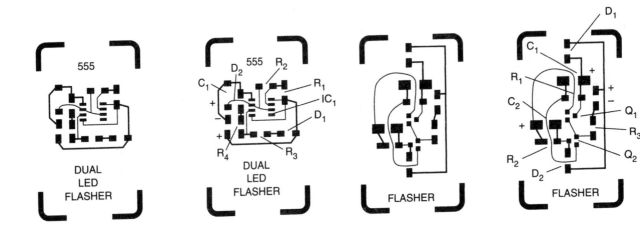

TONE BURST PROJECT (SMT PROJECT)

Project Description

The Tone Burst Project puts out a 500-Hz tone at a 1-Hz pulse rate. It is powered from a 9-V battery.

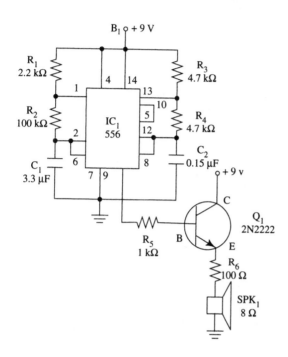

PRACTICE BOARD

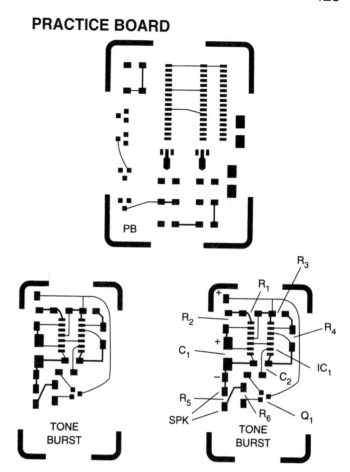

Parts List

1 SMT 556 timer (IC_1)
1 SMT 2N2222 transistor (Q_1)
1 SMT 0.15-μF capacitor (C_2)
1 SMT 3.3-μF capacitor (C_1)
1 SMT 100-Ω resistor (R_6)
1 SMT 1-kΩ resistor (R_5)
1 SMT 2.2-kΩ resistor (R_1)
2 SMT 4.7-kΩ resistor (R_3, R_4)
1 SMT 100-kΩ resistor (R_2)
1 8-Ω speaker (SPK_1)
1 9-V battery snap (B_1)

SURFACE MOUNT TOOLS AND EQUIPMENT

Although there are many companies that supply equipment and tools for the electronics industry, the following have particularly outstanding offerings with regard to surface mount technology fabrication and rework.

☐ **AccuGraphics Plus**
P.O. Box 508
San Fernando, CA 91341
(800) 223-1396
http://www.accugraphicsplus.net

☐ **Hexagon Electric Company**
P.O. Box 36
161 West Clay Avenue
Roselle Park, NJ 07204-1946
Phone: (908) 245-6200
Fax: (908) 245-6176
http://www.hexagonelectric.com

☐ **Hosfelt Electronics, Inc.**
2700 Sunset Boulevard
Steubenville, OH 43952-1158
Phone: (800) 524-6464
Fax: (614) 264-5414
http://www.hosfelt.com

☐ **Knight Electronics**
10940 Alder Circle
Dallas, TX 75238
Phone: (800) 323-2439
Fax: (214) 340-5870
http://www.knightonline.com

☐ **OK Industries, Inc.**
4 Executive Plaza
Yonkers, NY 10701
Phone: (914) 969-6800
Fax: (914) 969-6650
http://www.okindustries.com

☐ **Pace, Inc.**
9893 Brewers Court
Laurel, MD 20723-1990
Phone: (301) 490-9860
Fax: (301) 498-3252
http://www.paceworldwide.com

SURFACE MOUNT COMPONENTS

Although most distributors of electronic components offer a line of SMCs, the following ones have a particularly wide selection.

☐ **Capital Advanced Technologies, Inc.**
309-A Village Drive
Carol Stream, IL 60188
Phone: (630) 690-1696
Fax: (630) 690-2498
http://www.capitaladvanced.com

☐ **Chaney Electronics, Inc.**
P.O. Box 4116
Scottsdale, AZ 85261
Phone: (800) 227-7312
http://www.chaneyelectronics.com

☐ **Digi-Key, Corp.**
701 Brooks Avenue South
Thief River Falls, MN 56701-0677
Phone: (800) 344-4539
Fax: (218) 681-3380
http://www.digikey.com

☐ **Electronic Goldmine**
P.O. Box 5408
Scottsdale, AZ 85261
Phone: (602) 451-7454
Fax: (602) 451-9495
http://www.goldmine-elec.com

☐ **Jameco**
1355 Shoreway Road
Belmont, CA 94002
Phone: (800) 831-4242
Fax: (800) 237-6948
http://www.jameco.com

☐ **Surface Mount Center**
1580 Oakland Road, C115
San Jose, CA 95131
Phone: (408) 453-2023
Fax: (408) 453-5699
http://www.surfacemountcenter.com

BOOKS TO CONSIDER

- Blechman, Fred. *Simple, Low-Cost Electronics Projects.* Eagle Rock, Va.: LLH Technology.
- Bergsman, Paul. *Controlling the World with Your PC.* Solana Beach, Calif.: HighText Publications.
- Brittain, Robert. *Barron's A Pocket Guide to Correct Punctuation.* Woodbury, N.Y.: Barron's Educational Series.
- Edwards, Scott. *Programming and Customizing the BASIC Stamp Computer.* New York: McGraw-Hill.
- Kamichik, Steven. *IC Design Projects.* Indianapolis, Ind.: Prompt Publications.
- Lines, David. *Power Supplies: Projects for the Hobbyist and Technician,* 2d ed. Indianapolis, Ind.: Prompt Publications.
- McMurrey, David A. *Power Tools for Technical Communications.* New York: Harcourt College Publishers.
- Mims, Forrest. *The Forrest Mims Circuit Scrapbook,* Vol. II. Eagle Rock, Va.: LLH Technology.
- Strunk, Jr., William, and White E.B. *The Elements of Style.* New York: Macmillan.
- Zinsser, William. *On Writing Well.* New York: Harper Perennial.

MAGAZINES TO GET

- *Nuts and Volts Magazine*
  An excellent project magazine that's also loaded with theory you can understand.
- *Tech Directions*
  Primarily sent to technology educators, it is worth a student's effort to obtain a copy.

There are two, five-band resistor color codes in common use: the *Precision Five-Band Resistor Color Code* and the *E.I.A. Five-Band Resistor Color Code*. Here is how both are read:

Precision Five-Band Resistor Color Code

In the precision five-band resistor color code, the first three colored bands represent significant numbers, the fourth colored band is the multiplier, and the fifth colored band is the resistor's tolerance (see Chapter 3).

The tolerance band reads as follows:

Red	plus/minus	2%
Brown	plus/minus	1%
Green	plus/minus	0.5%
Blue	plus/minus	0.25%
Violet	plus/minus	0.1%

Thus, the nominal ohmic value, tolerance range, and plus and minus deviation of a resistor with red, red, green, gold, and brown bands is a resistance of 22.5 ohms, a tolerance of plus and minus 1%, and a range of 22.72 ohms to 22.27 ohms.

E.I.A. Five-Band Resistor Color Code

The E.I.A. five-band resistor color code reads the same as the four-band resistor color code for the first four bands. The fifth band represents a failure rate (MIL. 39008). Here is how the fifth band reads:

No Band	None	
Brown	M	1.0%/1000 hrs
Red	P	0.1%/1000 hrs
Orange	R	0.01%/1000 hrs
Yellow	S	0.001%/1000 hrs

Thus, for E.I.A. five-band resistors with an orange fifth band, 0.01% will fail in the first 1000 hours.

- ☐ Advanced Microcomputer Systems, Inc.
  1460 SW 3rd Street
  Pompano Beach, FL 33069
  (800) 972-3733
  http://www.advancedmsinc.com
- ☐ Beige Bag Software
  279 E. Libertt
  Ann Arbor, MI 48104
  (734) 332-0487
  http://www.beigebag.com
- ☐ BSoft Software, Inc.
  444 Coltran Road
  Columbus, OH 43207
  (614) 491-0832
  http://www.bsof.com
- ☐ CAD Soft Computer, Inc.
  801 S. Federal Highway
  Delray Beach, FL 33483
  (516) 274-8355
  http://www.cadSoftusa.com
- ☐ CircuitMaker
  5252 N. Edgewood Drive
  Suite 175
  Provo, UT 84604
  (800) 419-4242
  http://www.circuitmaker.com
- ☐ IMSI
  1938 Fourth Street
  San Rafael, CA 94901
  (415) 454-7101
  http://www.imsisoft.com

- ☐ IVEX Design International
  15232 N.W. Green Brier
  Beaverton, OR 97006
  (503) 531-3555
  http://www.ivex.com
- ☐ Mental Automation, Inc.
  5415 136th Place SE
  Bellevue, WA 98006
  (425) 641-2141
  mentala@mentala.com
- ☐ PCBoards
  2110 14th Ave. South
  Birmingham, AL 35205
  (800) 473-7227
  http://www.pcboards.net
- ☐ R4 Systems Inc.
  1100 Gorman Street, Suite 11B-332
  Newmarket, Ontario, Canada L3Y 7V1
  (905) 898-0665
  http://www.R4systems.com
- ☐ Workbench
  111 Peter Street #801
  Toronto, Ontario, Canada M5V 2H1
  (800) 263-5552
  http://www.electronicsworkbench.com

☐ **Cal West Supply, Inc.**
31320 Via Colinas #105
Westlake Village, CA 91362
Phone: (800) 892-8000
Fax: (818) 706-0825
http://www.hallbar.com

☐ **C & S Sales, Inc.**
150 W. Carpenter Avenue
Wheeling, IL 60090
Phone: (708) 541-0710
http://www.elenco.com/cx_sales/

☐ **EKI, Inc.**
140 S. Mountain Way
Orem, UT 85058
Phone: (800) 453-1708

☐ **Electronics Goldmine**
P.O. Box 5408
Scottsdale, AZ 85261
Phone: (800) 445-0697
Fax: (602) 661-8259
http://www.goldmine-elec.com

☐ **Electronic Rainbow, Inc.**
6227 Coffman Road
Indianapolis, IN 46268
Phone: (888) 291-7262
Fax: (317) 291-7269
http://www.rainbowkits.com

☐ **Gibson Tech Ed, Inc.**
1185 South 1480 West
Orem, UT 84058
Phone: (800) 422-1100
http://www.gibsonteched.com

☐ **Graymark International**
P. O. Box 2015
Tustin, CA 92781
Phone: (800) 854-7393
http://www.labvolt.com

☐ **Marlin P. Jones & Associates, Inc.**
P. O. Box 12685
Lake Park, FL 33403-0685
Phone: (800) 652-6733
http://www.mpja.com

☐ **Marcraft International Corp.**
1620 East Hillsbourgh Street
Pasco, WA 99301
Phone: (800) 441-6006

☐ **Nimco, Inc.**
P. O. Box 9-102 Hwy. 81N
Calhoun, KY 42327
Phone: (800) 962-6662
http://www.nimcoinc.com

☐ **Ramsey Electronics, Inc.**
793 Canning Parkway
Victor, NY 14564
Phone: (800) 446-2295
Fax: (716) 924-4555
http://www.ramseyelectronics.com

☐ **Robotkits Direct**
17141 Kingsview Ave., Suite B
Carson, CA 90746
Phone: (310) 515-6800
http://www.robotkitsdirect.com

☐ **Whiterook Products Company**
309 S. Brookshire Avenue
Venture, CA 93003
Phone: (805) 339-0702
http://www.west.net/~wpc

☐ *http://www.CAdSoftUSA.com*
Makers of Eagle 3.5, schematic capture, board layout, and autorouter software.

☐ *http://www.capitaladvanced.com*
Loads of great information on SMT design and fabrication.

☐ *http://www.circuitmaker.com*
Makers of CircuitMaker 2000, a virtual electronics lab.

☐ *http://www.electronicsworkbench.com*
Makers of Multisim and Ultiboard, schematic capture, simulation, and PC board layout software

☐ *http://www.llh-publishing.com*
Publisher of many electronic project books.

☐ *http://www.technologicalarts.com*
Sells modules and accessories.

Answers to Chapter Questions and Safety Test

CHAPTER 1

1. lecture and lab
2. technicians
3. prototype
4. design; drawing; experimenting; prototyping; and testing, troubleshooting, and final documentation
5. Project Report
6. machine
7. input: keyboard, graphics tablet, optical character reader, joystick, mouse; output: video display terminal, printer, plotter
8. light
9. wire
10. gun, pencil

CHAPTER 2

1. danger
2. momentary
3. fumes
4. current
5. death trap
6. hand
7. chemicals
8. dismiss
9. soldering iron
10. static discharge

CHAPTER 3

1. problem, solution
2. input, circuit, output
3. active, passive
4. vacuum tube, transistor, integrated circuit, microprocessor
5. passive
6. digital, analog
7. goals, objectives, responsibilities
8. system, design, packaging
9. design, simulation, analysis
10. graph

CHAPTER 4

1. mechanical, electronic
2. PC board design layout
3. sheet metal
4. working schematic
5. arrow
6. reference
7. template
8. dot
9. bottom, up
10. parts list

CHAPTER 5

1. explain
2. technical
3. problem, solution
4. request for proposal
5. say
6. brainstorming
7. always
8. short

9. people, things
10. three

CHAPTER 6

1. change
2. tin, lead
3. flux
4. (a) apply heat; (b) apply solder; (c) remove solder; (d) remove heat
5. solder, globs, bridges, icicles
6. current
7. modern
8. universal
9. solder misalignment
10. Experiment Results

CHAPTER 7

1. single-sided, double-sided, multilayer
2. any of the following: (a) less room for error; (b) quicker assembly time; (c) greater circuit density; (d) ease in troubleshooting; (e) less highly skilled assemblers are required; (f) automatic assembly equipment can be used; (g) printed circuit board assemblies are less prone to problems caused by vibration; (h) fewer gremlins to worry about
3. design layout drawing
4. not
5. jumpers
6. component, trace
7. freehand, final
8. (1) draw lines;
   (2) draw rectangles;
   (3) draw circles;
   (4) draw donuts;
   (5) draw conductor paths
9. inexpensive
10. reverse

CHAPTER 8

1. exposing, processing
2. developed, stop, fixer, rinsed
3. fabrication
4. hole chart
5. photoresist
6. (a) clean the blank PC board; (b) apply the photoresist; (c) dry the photoresist; (d) expose to ultraviolet light; (e) develop the image (pattern); (f) rinse and dry
7. ferric chloride
8. (a) using fresh acid; (b) agitating the acid; (c) heating the acid

9. shouldn't, copper
10. solder, mask, marking

CHAPTER 9

1. components
2. axial
3. service
4. sheet metal
5. full, template
6. squaring shear, tin snip
7. chassis
8. nibbler
9. brake
10. cleaned (washed)
11. primer
12. rub-on
13. multistrand
14. functional
15. packaging

CHAPTER 10

1. preliminary
2. operational
3. troubleshooting
4. performance
5. construction, component
6. passive, active
7. sight, smell, hearing, touch
8. critic
9. Test Results
10. Summary, Recommendations

CHAPTER 11

1. capture
2. (1) component placement; (2) line drawings; (3) schematic references; (4) design rule checking; (5) support documentation
3. netlist
4. (1) component placement; (2) trace routing; (3) design rule checking; (4) artwork generation; (5) documentation
5. ratsnest
6. library
7. string
8. (1) bill of materials (BOM); (2) netlist (NET); (3) update list (UPD); (4) pinlist (PIN)
9. (1) conductor mask; (2) solder mask; (3) paste mask; (4) silk-screen mask; (5) power/ground plane masks
10. (1) top conductor; (2) top power layer; (3) bottom power layer; (4) bottom conductor

CHAPTER 12

1. surface
2. (1) size reduction; (2) weight reduction; (3) improved performance and reliability; (4) improved manufacturing of circuit boards; (5) cost reduction; (6) new products
3. (1) component availability; (2) lack of complete standards
4. $\frac{1}{4}$, $\frac{1}{10}$
5. terminals
6. DIP
7. (1) PC board inspection; (2) application of solder paste; (3) surface mount component placement; (4) solder reflow; (5) circuit inspection; (6) circuit board cleaning
8. paste
9. (1) PC board inspection; (2) application of adhesive; (3) surface mount component placement; (4) flow (wave) soldering; (5) circuit inspection; (6) circuit board cleaning
10. bonded

CHAPTER 13

1. universal, solderless
2. discrete, integrated
3. The artwork will have to be reduced.
4. trial, final
5. dry, tape
6. 0.015″
7. holes
8. (1) tape; (2) adhesive; (3) tag soldering
9. flux
10. syringe

CHAPTER 14

1. tips
2. convection
3. rework
4. board
5. solder
6. tinned pad, solder paste
7. slotted-spade, tunnel
8. syringe
9. magnifying
10. fillet

APPENDIX A

1. c
2. b
3. a
4. c

5. a
6. b
7. a
8. b
9. b
10. b
11. a
12. a
13. b
14. b
15. c
16. a
17. c
18. a
19. a
20. a
21. c
22. a
23. a
24. b
25. a
26. a
27. a
28. b
29. a
30. a
31. b
32. a
33. c
34. a
35. a
36. b
37. a
38. c
39. b
40. b
41. a
42. b
43. a
44. c
45. b
46. a
47. a
48. a
49. a
50. c

Glossary

acetate   Thin, clear, plasticlike sheet used as a base for the application of printed circuit board artwork materials. Common thicknesses are 0.005″ and 0.007″.

acid   See etchant.

active component   An electronic component capable of controlling voltages or currents to produce gain or switching action in a circuit.

active voice   The form of a verb used to emphasize the agent of action. As its name suggests, the active voice is usually more direct and forceful than the passive voice.

ambient   Surrounding on all sides. Usually used as in: ambient room light.

ammonium persulfate   White powdery acid used to etch printed circuit boards. Highly toxic.

analog circuit   A circuit in which the output varies as a function of the input.

architectural drawing   A set of plans that depicts the construction of a house, office building, or similar structure. The drawings are fairly realistic, especially with regard to building elevations.

artwork   Produced from the printed circuit board design layout, artwork consists of opaque tape, donuts, and component-mounting configuration patterns put down on an acetate sheet, or such patterns generated with CAD software.

artwork materials   Printed circuit board drafting materials used to produce the artwork drawing.

Consist of opaque (usually black) tape, donuts, and component-mounting configuration patterns.

artwork negative   Produced from the artwork, the negative is the reverse of the artwork; that is, the traces and pads are transparent, and the background is opaque (black).

audio frequency (AF)   Any frequency in the audible range. Audio frequencies range roughly from 15 to 20,000 Hz.

automatic insertion equipment   Equipment used to insert electronic components automatically onto a printed circuit board.

automatic placement   The automated assembly of SMT circuits using pick-and-place machinery.

autorouter   A routine within a CAD program that automatically routes the PC board traces. It is rarely more than 90% effective, however.

axial lead component   A component with leads coming out of the ends and along the axes rather than out the side.

ballooned numbers   Numbers used to identify parts or subassemblies on an assembly drawing as part of an indenture identification system.

bending brake   A machine for turning or bending the edges of sheet metal. Has adjustable fingers used to hold the sheet metal during the bending process.

bill of materials   A listing of the components used in the design of a circuit.

black light   Invisible light radiation, either ultraviolet or infrared. The ultraviolet type is used to provide radiation to expose photosensitized printed circuit boards.

block diagram   A system diagram in which the parts or circuits are represented by boxes. The relationship to each box is indicated by appropriate connecting lines.

bond paper   A superior grade of plain, uncoated paper containing all or part rag pulp.

breadboard construction   A system in which electronic components are fastened temporarily to a board for experimental work.

breadboard drawing   Drawing used to depict the component layout for the breadboarded project, usually in sketch form.

breadboarding   The process of constructing a temporary circuit for the purpose of testing and/or modifying a design.

burnishing tool   Tool, similar to a pencil in shape, used to rub down dry-transfer lettering and printed circuit board artwork materials.

buy/make decision   In industry, the choice of whether it is more economical to buy or to make an item.

cabinet   A protective housing for electrical or electronic equipment.

calibration procedure   Method used to ascertain, by measurement or by comparison with a standard, any variation in the reading of another instrument. Often refers to the initial adjustment of a circuit or piece of equipment.

capacitor   A component used to store an electrical charge. Consists of two metal plates separated by a dielectric, or insulating material, such as air, mica, glass, paper, or plastic film.

center punch   A steel punch used for marking a spot where a hole is to be drilled.

central processor   Often referred to as the central processing unit (CPU). The "brain" and therefore the focal point of a computer, where the processing of data takes place.

chassis   A sheet metal box or simple plate on which electronic components and associated circuitry can be mounted.

chassis punch   A tool consisting of three parts: the punch, die, and draw screw. It is used to punch holes in sheet metal.

chip capacitor   A small SMC leadless rectangular capacitor.

chip resistor   A small SMC leadless rectangular resistor.

circuit analysis   The process of determining and identifying circuit parameters.

circuit breaker   A resettable protective device used for interrupting excessive current in an electric circuit.

circuit design sketch   A rough schematic drawing of an electronic circuit. Usually drawn freehand with the aid of graph paper.

circuit ground   A method of grounding whereby the metal chassis that houses the assembly or a large conductive area on a printed circuit board is used as the common or reference point.

circuit simulation   In software, a program that will "run" your virtual circuit.

clean room   A location where dust and dirt are kept to a minimum. Necessary in the fabrication of integrated circuits and other tiny solid-state electronic components.

color monitor   A video screen that displays commands, data, or picture content in color.

color organ   An electronic circuit that takes an audio signal at its input and produces light variations at the output. Usually comes in a three-channel version in which the string of lights attached to each channel responds to a band of audio frequencies: bass, midrange, and treble.

color printer   A device that prints the output from a computer in color.

component identification   Used in the PC assembly drawing where electronic components are shown in outline form and with standard reference designations.

component library   In software, a library of components and component symbols that can be called up at any time.

component-mounting configurations   PC board artwork patterns in a multipad format for transistors, integrated circuits, and similar components. It is much easier to lay down a multipad configuration than to apply individual pads.

component orientation   Refers to the direction in which components are oriented on the printed circuit assembly.

component side   The side of a printed circuit board on which the components are placed.

computer-aided design (CAD)   A computer used to assist in the design of a product. Specifications for the items under design are entered into the computer, and the system then uses graphics capabilities to draw three-dimensional pictures of the object.

computer hardware   The physical components of a computer system.

computer program   A set of instructions telling the computer how to perform a specific task. Usually written in a high-level language and converted into the machine language of the computer.

computer screen   The surface upon which the visible pattern is produced in a cathode ray tube.

computer software   Both the computer program and any data to be manipulated.

computer system   An electronic system consisting of a central processor circuit and at least one input and one output transducer, referred to as peripherals.

computerese   Slang term used to refer to the jargon of computer terminology.

Concepts and Requirements Document   A document that states why a project should be undertaken, the basic design requirements, how those requirements are to be met, and who is to do exactly what and when.

copper-clad board   A phenolic board with a thin sheet of copper (usually 0.0014″ thick) laminated to one or both sides. The basic, unetched printed circuit board.

cut-slash-and-hook (CSH)   Also referred to as point-to-point wiring, an old method in which interconnecting wires are first cut, then stripped (slash), and finally hooked around a terminal strip.

cutter   Also known as a diagonal cutter or dykes; a tool used to cut wires or component leads.

darkroom   A room used for developing photographs. Must be constructed to exclude all unwanted ambient light.

data sheet   A sheet of information, usually supplied by the manufacturer, on a particular electronic component. This document discusses, among other things, component voltage parameters, current-handling capabilities, and minimum and maximum operating temperatures.

datum   A point or line used as a basis for calculations or measurements.

decoder   A circuit, usually digital, that decodes a specific code.

design drawing   A package of three sketches: the system (functional) diagram, circuit design sketch, and packaging plan.

design layout   In printed circuit board layout design, the drawing used to place components and traces prior to the artwork layout.

design methodology   A method used to identify problem and solution categories in the design of a new product.

desktop scanner   A device used to scan documents, artwork, and photographs to be embedded into a computer application.

desoldering pump   Also known as a solder sucker, it uses the action of a spring-loaded plunger to suck up, in one quick action, molten solder.

desoldering wick   A braided copper material that uses capillary action to pull up molten solder. Used in the desoldering process.

developer   A chemical used to bring forth the latent image of a photograph or exposed printed circuit board.

dewetting   A solder defect in which solder lands or leads have lost their adhesive quality.

digital circuit   A device or circuit that manipulates information in discrete steps, such as bits or digits. Usually deals with on or off conditions.

diode   An electronic component that allows current to flow through it in one direction only.

document   A record, in written and graphic forms, used as a reference.

donut pads   Donut-shaped artwork configurations used in traditional printed circuit board artwork masters.

double-sided printed circuit board   A printed circuit board with copper traces and pads on both sides of the board.

drafting machine   A device used to draw straight lines at a determined length and angle. Combines in one instrument the T-square, triangle, protractor, and scale.

drawbridge effect   An SMT term for a solder defect in which a small chip component, such as a chip resistor or capacitor, raises one termination, producing an open. It occurs during flow soldering.

drawing board   A board made of seasoned, straight-grained, soft wood, designed to be used with a T-square and triangles.

drawing notes   Notes placed on a drawing that provide information about materials, fabrication processes, surface finishes, and the like.

Dremel Moto-Tool   A high-speed drill and grinder, particularly adaptable for printed circuit board work.

drill bit   A tool for boring holes in wood, plastic, or metal. Fits into the chuck of a drill.

driver   A tool used to insert or remove a fastener. Usually a screwdriver or nutdriver.

dry-transfer lettering   Also known as rub-on lettering. Consists not only of letters but also numbers and graphic symbols that are rubbed off a clear backing onto a plastic, paper, or metal surface. Used for labeling.

dual wave   An SMT flow-soldering approach that uses two solder waves. The first is a turbulent wave, the second a laminar flow wave.

Dynamark Imaging System   A product of 3M, this labeling system produces high-quality metal or plastic labels using a photosensitive process.

electrical shock   The passage of current through the body.

electricity   The same as current. The movement of free electrons in one direction.

electrocution   Death resulting from electrical shock.

electromechanical   Any device using electrical energy to produce mechanical movement. Also refers to a field of study that involves knowledge of things electrical and mechanical.

electronic circuit   A circuit designed to manipulate electrical current in either an analog or a digital manner.

electronic component   A device that controls electricity. Either active or passive, an active component controls electrons by switching or regulating them; a passive component helps the active component do its job.

electronic design   Creating plans for new electronic circuits and products.

electronic drawing   Shows the "electronics" of a device or product. Uses schematic symbols representing electronic components and their interconnections.

electronic system   An electronic device with one or more electronic circuits and at least one input and one output transducer.

electronics   The control of electricity with control components such as vacuum tubes, transistors, integrated circuits, and microprocessors.

electronics technician   One who installs, repairs, tests, and builds electronic devices.

enclosures   A housing of any electrical or electronic device. As distinguished from the chassis, the outer surface of the box or cabinet.

encoder   A digital circuit that takes an input signal and converts it into a desired code.

engineer   One who is primarily responsible for the design of a product, as opposed to the technician, who builds and repairs the product.

etchant (acid)   A chemical agent that will remove solid material such as copper. Two popular etching materials are ferric chloride and ammonium persulfate.

eutectic solders   Solder alloys that possess an almost negligible plastic stage.

Experiment Results Document   A document that tells what happened when the project was breadboarded.

experimenting stage   Also known as the breadboarding stage; a time to prove out the circuit design by building a temporary version of the electronic circuit.

exposure   The process of reflecting light off the original on the exposure glass through the optics to the photoconductor surface.

ferric chloride   An etching solution used to remove unwanted copper during the printed circuit board fabrication process.

file   A steel tool with a rough, ridged surface for smoothing, grinding down, or cutting through an object.

film   A sheet or roll of a flexible cellulose material covered with a light-sensitive substance and used in taking photographs. Used to produce the PC artwork negative.

film exposure   The process whereby a latent (invisible) image is caused to appear on the film emulsion.

film processing   The process whereby a latent image is developed, or brought forth.

final design layout   The completed PC design layout with all components, connectors and terminals, and traces neatly drawn with pencil, template, and straightedge.

final packaging drawing   Shows how the final project will look from the outside.

fixer solution   The chemical solution used to remove the white areas of the film, leaving transparent areas on the negative.

flux   In soldering, a material used to remove oxide films from metallic surfaces. In rosin core solder, the flux paste is found in the center of the solder wire.

footprint   The land pattern for an SMC.

freehand trial layout sketch   A rough sketch of the proposed printed circuit design layout. Usually drawn freehand without the use of a template.

full clinch   A method whereby the component lead inserted through the printed circuit board is brought flush against the board. It is difficult to remove components that have been installed with a full clinch.

functional assembly drawing   An assembly drawing that is easy to draw but requires special training in order to read. Contrasted with pictorial assembly drawing.

functional wiring diagram   A wiring diagram that is easy to draw but requires special training in order to read. Contrasted with pictorial wiring diagram.

graphics tablet   A device used for inputting graphics with a stylus. Used as an input peripheral with computer-aided design (CAD) systems.

grounded socket   Line cord with a grounded lug. When a soldering iron is so equipped, it reduces the chances of static buildup that can damage sensitive solid-state components.

gull-wing   An SMC lead form that resembles a gull's wing bent downward in flight.

hacksaw   A saw for cutting metal, consisting of a narrow, fine-toothed blade held in a frame.

hand drill   Usually electric, a tool used for drilling holes in PC boards.

indenture identification system   An assembly identification system that uses indented indicators to tell what items are part of a larger grouping or assembly.

inductor (coil)   An electronic component used to introduce inductance in a circuit.

infant mortality   The failure of an electronic device very early in its expected lifetime.

input transducer   An electrical or electronic device that changes nonelectrical energy into electricity.

insertion mount technology   The mounting of electronic components so that their leads protrude through holes in a printed circuit board.

integrated circuit (IC)   A single microscopic chip that contains all the components and interconnections necessary to perform analog or digital circuit functions.

isometric drawing   A three-dimensional drawing in which the vertical and horizontal planes are drawn receding at true length. The standard pictorial drawing used in sketching and technical illustration.

J-lead   A lead found on SMCs that is shaped like the letter J.

joystick   A lever whose motions control the movement of a cursor on a video display terminal.

jumper   A length of wire used to make an electrical connection between two points or terminals of a circuit. Often required on single-sided printed circuit boards.

keyboard   A computer input peripheral with an array of keys that are manually operated to encode data or instructions.

leadless ceramic chip carrier (LCCC)   A surface-mounted component in which the leads are tucked under the body of the integrated circuit.

light box   A box used to house the lighting for a color organ. Usually has a plastic light-diffusing material in the front of the box.

light-emitting diode (LED)   A diode that when conducting current emits light. Most LEDs are red, but orange, yellow, and green LEDs are also available.

light pen   An electronic stylus containing a light sensor that can be used to specify a position on a video display terminal. Used as an input peripheral for computer-aided design (CAD).

light table   A box or table in which light shines up through a glass covering. Used in the printed circuit board tape-up process.

marking mask   Used to aid in printed circuit board component assembly. Silk-screened onto the component side of the PC board, it provides information such as component outlines, positioning, and sequential reference designations.

mechanical drawing   A drawing that shows how an object is to be constructed. Usually done in orthographic projection, in which two or three views of the object are shown at right angles to each other.

MELF   An acronym for Metal Electrode Face (Bonded). A cylindrical package for two-terminal SMCs.

metal-oxide semiconductor (MOS)   A metal-insulator semiconductor structure in which the insulating layer is an oxide of the substrate material. Integrated circuits made in this manner have very low power requirements but are susceptible to damage from static electricity.

microprocessor   A processor small enough to be manufactured on a tiny silicon chip. A microprocessor is the central processing unit of a computer.

mouse   A device that an operator can move over the surface of a graphics tablet. Used as an input peripheral for a computer-aided design (CAD) system.

multilayer printed circuit board   A printed circuit construction method whereby two or more boards having circuit traces on one or both sides are laminated, or sandwiched, together.

multistrand wire   A group of solid cylindrical conductors used to carry current. Multistrand wire, as opposed to single-strand wire, is easier to bend and form.

negative-acting resist   A photoresist that will work with a photographic negative, as opposed to a photographic positive.

negative silk screen   Taken from a positive marking mask, the negative silk screen is used to produce the screened image to aid in PC component assembly.

netlist   In a schematic capture program, a file containing a group of related parameters that are extracted from the schematic.

nibbler   A handheld minishear used to make irregular slots and holes in sheet metal.

noncomponent parts   Usually refers to hardware items such as screws, nuts, and other fasteners.

nutdriver   A handheld driver tool used to turn nuts.

Ohm's law   A law stating that current is directly proportional to voltage and inversely proportional to resistance.

operational testing   The stage during which power is applied for the first time and basic project functioning is determined.

output transducer   An electrical or electronic device that changes electrical energy into another form of energy.

oxidation   The process by which oxide films form when oxygen combines with a metal and forms an oxide layer on the surface of the metal.

packaging plan   This plan is a sketch of what the final project might look like. To be contrasted with the final **packaging drawing**, which is a mechanical drawing of what the final project actually looks like.

pad center   The center of a donut used in producing the PC artwork.

parts carrier   A header designed to hold tiny components such as small resistors, capacitors, and transistors. The header is pressed into a wire-wrap IC socket.

parts list   A list of parts shown on a drawing. Shows the item number, component reference designation, parts description, and quantity required.

passive component   A component that does not create or amplify energy. Resistors, capacitors, and inductors (coils) are passive components.

passive voice   The form of a verb used to emphasize the action instead of the agent. Less direct and forceful than the active voice.

perforated board   Used with the wire-wrapping breadboarding method, perforated board is made of phenolic material, usually $\frac{1}{16}''$ thick, and with holes spaced 0.10″ apart.

performance testing   Project testing under actual user conditions. The stage during which the project is exposed to the real world.

photoconductor   A substance that can accept and hold an applied charge that can be dissipated by exposure to light.

photographic negative   The reverse of the positive PC artwork. In the negative, the traces, donut pads, and so on are transparent, and the remaining surface is opaque (black).

photoresist   A resist material that is sensitive to light. When a PC board that has been sensitized with a photoresist is exposed to ultraviolet energy, the areas exposed will (after development) resist the action of an etchant, or acid.

pictorial assembly drawing   An assembly drawing that is easy to read but difficult and time-consuming to draw. Used where assemblers are untrained in reading functional assembly drawings.

pictorial wiring diagram   A wiring diagram that is easy to read but difficult and time-consuming to draw. Used where assemblers are untrained in reading functional wiring diagrams.

pinholes   Tiny pin-size transparent spots in a photographic negative. They should be covered with an opaquing material.

plastic-leaded chip carrier (PLCC)   Also known as J-leads, a surface-mounted component whose pins, or leads, are curled under the component body in the form of the letter J.

pliers   Small pincers in any of various forms used for gripping small objects or bending wire.

plotter   A fancy printer that makes the hard-copy drawing that is developed using computer-aided design (CAD).

point to point   A method of circuit interconnection whereby wires are strung from one tie point to another. See cut-slash-and-hook.

positive-acting resist   A photoresist material used with a photographic positive, not a photographic negative.

power supply   An electronic circuit that changes alternating current into direct current.

preliminary testing   Tests performed before power is applied to the prototype project.

press-n-peel   A PC board artwork transfer method in which the artwork is literally ironed onto the black PC board.

primary colors   A set of colors from which all other colors are derived. White light comprises the wavelengths of the additive primaries red, green, and blue.

primer   A preliminary coat of paint. Should be applied to all sheet metal before final painting.

printed circuit assembly drawing   Shows where components and parts are to be placed on a PC board.

printed circuit (PC) board   A circuit in which the interconnecting wires have been replaced by conductive strips (traces) of copper left on an etched board.

printed circuit board artwork   A drawing consisting of the actual tape-up of the PC board design layout.

printed circuit board design layout drawing   A drawing that shows the component layout and trace patterns for a printed circuit board.

printed circuit board design template   A drawing template used to lay out the printed circuit board design. Body size and pad locations for all standard components, with both axial and radial leads, are provided.

printed circuit board fabrication drawing   A drawing that shows how the PC board is to be fabricated. All board dimensions, along with hole sizes and quantities, are given.

product   Something that is reliable, manufacturable, profitable, and salable.

project   Something that is self-made for enjoyment, education, or to prove a concept.

Project Report   The report containing all the written and graphic documentation necessary for the design and fabrication of a project.

proposal   Any communication that attempts to sell an idea, a concept, a service, a piece of equipment, a complex system, or the like.

prototype project   A project that is hand-assembled to resemble the final, mass-produced equipment as much as possible.

push-in terminal   A soldering terminal pressed into perforated phenolic board.

radial lead component   A component with leads extending out the side, rather than from the end.

ratsnest   An interconnectivity pattern provided by printed circuit board design software to aid in trace development of PC board layouts.

reamer   A sharp-edged tool for enlarging or tapering holes.

reference designations   Letter–number designations used to identify electronic components on a drawing.

reflow soldering   A solder method that uses solder paste, which is reheated.

relay   An electromagnetically controlled mechanical device in which electrical contacts are open or closed by a magnetizing current.

resist   A material that resists the action of an etchant. Most resists used in the fabrication of printed circuit boards are photoresists.

resistor   A passive electronic component that opposes current flow and drops, or uses up, voltage.

rework   Work done to a circuit to bring it into complete conformance with its original specifications.

rocker tray   Also called a developing tray, used to hold photographic chemicals during the developing process.

safelight   A light, usually red, used during the photographic developing process. Allows one to see in the dark.

schematic   A symbolized diagram of an electrical or electronic circuit.

schematic capture   A CAD schematic drawing program that has the ability to "capture" data from the created schematic. Such data can be transferred seamlessly to a PC design layout program to aid in trace routing.

schematic symbols template   A drawing template with component schematic symbols cut into it.

screwdriver   A hand tool used for turning screws.

self-adhesive vinyl   Contact paper that comes in a variety of printed patterns and is used to cover the surface of a project enclosure.

sensitized PC board   A blank PC board that has been coated with photoresist but not yet exposed to ultraviolet light.

service bend   The bend in a component lead after it passes through the printed circuit board such that it is easy to service, that is, remove. The lead is bent at approximately 30°.

sheet metal drawing   A two-dimensional layout of the pattern required to produce a metal chassis or enclosure.

sheet metal template   A paper template, made from the sheet metal drawing, used as a guide for cutting the sheet metal pattern.

silicon-controlled rectifier (SCR)   An electronic component that acts as a solid-state latching switch.

single-sided printed circuit board   A printed circuit board with traces and pads etched onto only one side.

single-strand wire   Wire with only one strand of metal. Easier in some respects to work with but more difficult to bend than multistrand wire.

small outline   The miniature small-outline IC packages found in surface mount technology.

small-outline integrated circuit (SOIC)   A surface-mounted component with "gull-wing"–shaped leads.

software library   A library of graphic symbols used in creating a drawing with a computer-aided design system.

solder   An alloy of lead and tin used to fuse wires and electronic component leads.

solder coat   A thin (0.00002″ to 0.0005″) coating of lead or a tin–lead alloy placed as a protective coating over the conductive layer on the trace side of a printed circuit board.

solder mask   A polymer coating (usually epoxy resin) screened onto the entire surface of a PC board, except for the pads that require soldering. The use of solder masks reduces the possibility of solder bridging of closely spaced traces.

solder paste   A solder formulation of tiny solder spheres, a liquid flux, a thixotropic vehicle, and a flux activator.

soldering   The joining of metals, such as the leads of electronic components, with a material known as solder.

soldering iron   A tool consisting of a heating element to heat the tip, which in turn melts the solder.

soldering iron tip   The tip of a soldering iron that transfers the heat to the joint to be soldered. Iron-clad tips are the most efficient type.

solderless circuit board   A breadboard consisting of an array of tiny holes spaced 0.10″ apart

that will accept the leads of electronic components and solid-strand wire. With this type of board, no soldering is required.

solderless wiring terminal   Spring-action terminals inserted into perforated board that allow wire and component leads to be inserted without the use of solder.

solenoid   An electromagnet with a core rod capable of moving back and forth within the coil. Used to convert electrical energy into lateral mechanical motion.

solid-state component   A component that can control current without moving parts. All semiconductors are solid-state devices.

spaghetti   Plastic or rubber tubing sometimes used to provide insulation for circuit wiring.

spray paint   Paint in a can under gas pressure. When the opening is depressed, the paint is released in a fine mist.

squaring shear   A large tool used to cut sheet metal.

static electricity   Stationary electricity. Can be uncomfortable to humans and fatal to certain types of solid-state electronic components.

stop bath   A chemical used to stop the action of the developer in the photographic developing process.

straightedge   A piece of wood, metal, or plastic having a perfectly straight edge used in drawing straight lines.

suction bulb   A desoldering tool where the suction of a rubber bulb draws up molten solder that has been reheated with a soldering iron.

Summary and Recommendations Document A concluding document that summarizes the important project developments, from the initial design phase through testing and troubleshooting. It also offers recommendations as to what future direction the project should take.

surface mount component (SMC)   Electronic components one-third the size of standard components that are soldered to both sides of a holeless printed circuit board.

surface mount device (SMD)   An electronic device designed for planar (surface) mounting as opposed to through-hole mounting.

surface mount technology (SMT)   A revolutionary electronic technology that uses surface-mounted components mounted on top of a printed circuit board. Tremendous saving in space is realized with this technology.

switch An electrical or electronic device for opening and closing a current path.

system diagram A block diagram illustrating the functional units that make up the electronic system.

systems approach A design approach that takes into consideration the entire electronic system; with circuits and input and output transducers.

taping up Laying down of tape, donuts, and component-mounting configurations to produce the traditional printed circuit board artwork.

technical jargon The term for the specialized words or phrases used in a particular profession, trade, or scientific field.

technical writing A specialized field of communication whose purpose is to convey technical and scientific information and ideas accurately and efficiently.

template Plastic pattern guides designed to save time and effort in the drafting process. Two common types are needed: the schematic symbols template and the printed circuit board design layout template.

terminal post Posts designed specifically for wire wrapping.

terminal strip A terminal strip consisting of a varying number of lugs spaced $\frac{3}{8}''$ apart on a phenolic or Bakelite strip $\frac{3}{8}''$ wide.

test instruments Equipment used to test electronic circuits. Includes instruments such as the multimeter, signal generator, and oscilloscope.

Test Results Document A document that records the results of the testing and troubleshooting procedures.

thermal coefficient of expansion Also called coefficient of thermal expansion. A measure of the coefficient of expansion of materials under temperature increases.

tin snip A handheld tool used for cutting sheet metal.

tombstone effect See drawbridge effect.

toxic chemicals Chemicals that are dangerous to your health and may cause death if improperly used.

trace routing With the use of a pencil, connecting the various pads with lines on the PC board design layout.

trace side The side of the single-sided PC board that contains the copper traces. The other side is known as the component side.

transfer film A blue-colored sheet with emulsion applied to one side. Used with the press-n-peel PC board artwork transfer method.

transformer An electrical component for transferring alternating current from one circuit to another by magnetic induction.

transistor A semiconductor component that can both switch and regulate current flow. Has replaced the vacuum tube in most applications.

triac An electronic component equivalent to two back-to-back silicon-controlled rectifiers (SCRs).

triangles Drafting instruments used to draw lines at 30, 45, 60, and 90 degrees to the horizontal. Two triangles are usually required: a 45-degree triangle and a 30–60 degree triangle.

troubleshooting A systematic process of isolating, identifying, and correcting a fault in a circuit or system; the application of logical thinking combined with a thorough knowledge of circuit or system operation to find and correct a malfunction.

T-square A drafting instrument used for drawing horizontal lines. Used with a drawing board.

turret punch A machine that contains many different-sized sheet metal die and punch combinations that can easily be swung into place for instant cutting.

ultraviolet light Light rays of wavelengths just beyond the violet end of the visible spectrum. Used in exposing photosensitized printed circuit boards.

universal printed circuit board Wire-wrap perforated boards with copper traces spread throughout.

vacuum tube The first control component that could both switch and amplify an electrical signal. Consists of a filament, cathode, grid, and plate enclosed in a glass or metal tube from which all the air has been removed.

video display terminal (VDT) A computer terminal, usually a cathode ray tube (CRT), used for displaying text and graphics information.

virtual circuit A circuit in name only. A circuit created with computer software.

wave soldering An immersion soldering approach. The circuit board assembly is moved across a bath of molten solder so that the bottom of the board is momentarily immersed.

wetting agent   A material that reduces the surface tension of liquids.

white light   Light from a source such as sunlight that contains all colors of light.

wire   A cylindrical strand of metal having a low resistance to current flow.

wire stripper   Tool used to strip the insulation from wire.

wire wrapping   A technique for terminating conductors that requires the use of special tools, terminals, and wire. No soldering is required.

wiring diagram   Shows how off-board components are connected to one another and to the printed circuit board.

working schematic drawing   Illustrates in a precise and readable manner what the circuit consists of and how it is connected. Drawn with drafting instruments.

wrench   Tool used for turning nuts, bolts, pipes, and the like.

X-Acto knife   Sharp-bladed handheld instrument used for laying down PC artwork materials.

Index